T0171942

Lecture Notes in Mathematics

Volume 2280

This series reports on new developments in all areas of mathematics and their applications - quickly, informally and at a high level. Mathematical texts analysing new developments in modelling and numerical simulation are welcome. The type of material considered for publication includes:

1. Research monographs
2. Lectures on a new field or presentations of a new angle in a classical field
3. Summer schools and intensive courses on topics of current research.

Texts which are out of print but still in demand may also be considered if they fall within these categories. The timeliness of a manuscript is sometimes more important than its form, which may be preliminary or tentative.

More information about this series at http://www.springer.com/series/304

Walter Neumann • Anne Pichon

Editors

Introduction to Lipschitz Geometry of Singularities

Lecture Notes of the International School on Singularity Theory and Lipschitz Geometry, Cuernavaca, June 2018

 Springer

Editors

Walter Neumann
Department of Mathematics
Barnard College, Columbia University
New York, NY, USA

Anne Pichon
Aix Marseille Univ
CNRS, Centrale Marseille, I2M
Marseille, France

ISSN 0075-8434   ISSN 1617-9692   (electronic)
Lecture Notes in Mathematics
ISBN 978-3-030-61806-3   ISBN 978-3-030-61807-0   (eBook)
https://doi.org/10.1007/978-3-030-61807-0

Mathematics Subject Classification: Primary: 32S25, 57M27; Secondary: 32S55, 14B05, 13A18

This Springer imprint is published by the registered company Springer Nature Switzerland AG.
The registered company address is: Gewerbestrasse 11, 6330 Cham, Switzerland

Preface

This book collects the lecture notes of the International School on Singularity Theory and Lipschitz Geometry held in Cuernavaca (Mexico) on June 11–22, 2018. The school consisted in 9 series of lectures, delivered by

Lev Birbrair (Universidade Federal do Ceará, Fortaleza),
José Luis Cisneros Molina (Universidad Nacional Autónoma de México),
Walter Neumann (Columbia University, New York),
Anne Pichon (Université d'Aix Marseille),
Patrick Popescu-Pampu (Université de Lille),
Maria Aparecida Soares Ruas (ICMC, São Carlos),
Jawad Snoussi (Universidad Nacional Autónoma de México),
Bernard Teissier (Institut de Mathématiques de Jussieu, Paris),
David Trotman (Université d'Aix Marseille).

The last part of the volume contains the historic pioneering 1969 work of Frédéric Pham and Bernard Teissier "Fractions Lipschitziennes d'une algèbre analytique complexe et saturation de Zariski" which was still unpublished until now, in an English translation performed by Naoufal Bouchareb (Université d'Aix Marseille) under the supervision of Teissier himself.

Lipschitz geometry of singular sets is an intensively developing subject which started in 1969 with the work of Pham and Teissier on the Lipschitz classification of germs of plane complex algebraic curves. Its essence is the following natural problem. It has been known since the 1972 work of Burghelea and Verona that a real or complex algebraic variety is topologically locally conical. On the other hand it is in general not metrically conical: there are parts of its link with non-trivial topology which shrink faster than linearly when approaching the special point. A natural problem is then to build classifications of the germs up to local bi-Lipschitz homeomorphism, and what we call Lipschitz geometry of a singular space germ is its equivalence class in this category. There are different approaches for this problem depending on the metric one considers on the germ. A real analytic space germ (V, p) has actually two natural metrics induced from any embedding in $\mathbb{R}^N$ with a standard Euclidean metric: the outer metric is defined by the restriction of the

Euclidean distance, while the inner metric is defined by the infimum of lengths of paths in V.

What makes Lipschitz geometry of singular sets attractive is that it gives tame classifications, as conjectured by Siebenmann and Sullivan in 1977: the set of equivalence classes of complex algebraic sets in $\mathbb{C}^N$ defined by polynomial equations of bounded degree is finite. One of the most important results in Lipschitz geometry is the proof of this statement by Mostowski using his theory of Lipschitz stratifications, and then its extension by Parusiński to the subanalytic setting and very recently by Nguyen and Valette to the category of definable sets in polynomially bounded o-minimal structures. In contrast with the Lipschitz geometry of singular varieties, the Lipschitz geometry of germs of maps has continuous moduli, as shown by Henry and Parusiński.

This series of courses was designed to present a broad overview of the important recent progress in this area, which led to the emergence of new ideas and started to build bridges between Lipschitz geometry and several other major areas of singularity theory. Among others, let us mention the surprising discovery by Birbrair and Fernandes that complex singularities of dimension at least two are in general not metrically conical for the inner metric. This started a series of works leading to the complete classification of the inner Lipschitz geometry of germs of normal complex surfaces by Birbrair, Neumann and Pichon, and building on it, to major progress in the study of the outer metric. The ideas developed there inspired for example the complete classification by Birbrair and Gabrielov of the germs of functions from $(\mathbb{R}^2, 0)$ to $(\mathbb{R}, 0)$ up to contact Lipschitz equivalence.

These works pioneered a deep renewal of the field. During the last decade, many researchers in singularity theory started working in Lipschitz geometry and the field is now at its golden age. However, a lot remains to be done; for example, building classifications of Lipschitz geometry in larger settings such as non-isolated and higher dimensional real and complex singularities and in the global, semi-algebraic and o-minimal settings, or continuing the exploration of bridges between Lipschitz geometry and other important aspects of singularity theory such as equisingularity, resolution theory, arc spaces theory, non-Archimedean geometry, toric and tropical geometries, etc. Beyond the interest of the topic in itself, many very recent works already suggest that Lipschitz geometry will give new points of view on these various aspects of singularity theory and will help to solve open problems which are a priori not of metric nature.

The aim of the lecture notes is to introduce to Lipschitz geometry of singularities an audience of graduate students and researchers from other fields of geometry who may want to study in this area the multiple open questions offered by the most recent developments. All the courses are illustrated by many examples and some of them contain exercises.

The first three courses contain introductive lectures to basic tools in singularity theory which are intensively used in the next six advanced courses.

In the first basic course, **José Luis Cisneros Molina** gives a **Geometric Viewpoint of Milnor's Fibration Theorem** in which he presents the local conical structure theorem of singular spaces and gives the proofs of the Milnor Fibration Theorem, the Milnor-Lê Fibration Theorem and the equivalence of these two fibrations, emphasizing the ideas of differential topology involved. He then describes the monodromy of the Milnor fibration of a complex analytic function in two variables with isolated singularity as a quasi-periodic diffeomorphism using a resolution of the singularity. The notes of the course were written with the help of Haydée Aguilar-Cabrera.

The second basic course, by **Jawad Snoussi**, brings the reader in **A Quick Trip into Local Singularities of Complex Curves and Surfaces** which presents the essential tools which are used in the next advanced courses on Lipschitz geometry of complex low dimensional spaces. In particular, it gives the key notions of Puiseux parametrizations and characteristic Puiseux exponents of a germ of plane complex curve, which enable to understand the Lipschitz classification of complex curves presented later. It also introduces the reader to different approaches of resolution of complex surfaces by normalized points blow-ups and by normalized Nash modification and to the related notions of generic hyperplane sections and generic polar varieties, which play a key role in the Lipschitz classifications of complex surface singularities.

In the third basic course, **Walter Neumann** gives a panorama on **3-Manifolds and Links of Singularities** and an overview of general 3-manifold topology, and its implications for links of isolated complex surface singularities. In particular, he worked out the relations between the JSJ decomposition of links of surface singularities and their resolution graphs through plumbing calculus and provided many examples of isolated complex surface singularities whose Lipschitz geometries are explicitly described in some of the advanced courses.

The second part of the volume consists of five advanced courses on Lipschitz geometry of singularities. Each gives its own approach and different doors to enter in the field, but also refers to the other courses. The whole gives a rich picture of the theory and brings the reader to the border of some of the areas which remain unexplored.

The course of **David Trotman** presents the techniques of **Stratifications, Equi-singularity and Triangulations** and gives an introductory overview of Whitney stratifications, Kuo-Verdier stratifications, Mostowski's Lipschitz stratifications and of equisingularity along strata of a regular stratification for the different regularity conditions. In particular, it discusses equisingularity for complex analytic sets including Zariski's problem about topological invariance of the multiplicity of complex hypersurfaces and its bilipschitz counterparts. The last part provides further evidence of the tameness of Whitney stratified sets and of Thom maps, by describing triangulation theorems in the different categories, and including definable and Lipschitz versions.

The course of **Maria Aparecida Soares Ruas** presents some **Basics on Lipschitz Geometry** and starts with an introduction to the main tools used to study the Lipschitz geometry of real and complex singular sets and mappings: the notions of semi-algebraic sets and mappings, and basic notions of Lipschitz geometry. The course then focusses on the real setting, presenting the outer Lipschitz classification of semialgebraic curves, the inner classification of semialgebraic surfaces, the bi-Lipschitz invariance of the tangent cone, ending with a presentation of several results on Lipschitz geometry of.function germs.

Lev Birbrair, in his course entitled **Surface Singularities in $\mathbb{R}^4$: First Steps Towards Lipschitz Knot Theory** presents a new approach to the classification of real surfaces in $\mathbb{R}^4$ based on the following idea. A link of such an isolated singularity is a knot (or a link) in the 3-sphere. Then, he shows that the ambient Lipschitz classification of surface singularities in $\mathbb{R}^4$ can be interpreted as a metric refinement of the topological classification of knots in S^3. In particular, a given knot K gives rise to infinitely many distinct ambient Lipschitz equivalence classes of outer Lipschitz equivalent singularities in $\mathbb{R}^4$ whose links are topologically ambient equivalent to K. The lecture notes of this course were written in collaboration with Andrei Gabrielov.

Anne Pichon gives a course on **Lipschitz Geometry of Complex Singularities** in which she presents the complete classification of outer Lipschitz geometry of complex curves from a geometric point of view. She then describes the thick-thin decomposition of a normal complex surface singularity using resolution theory as a key tool, and based on it, a geometric decomposition of the germ into standard pieces, which is invariant by inner bilipschitz homeomorphisms and which leads to the complete classification of Lipschitz geometry for the inner metric. The course also presents some advanced results towards the classification for the outer metric.

The course of **Bernard Teissier** entitled **The biLipschitz Geometry of Complex Curves, An Algebraic Approach** explores the concept of "generic plane linear projection" of a complex analytic germ of curve in $\mathbb{C}^n$. The notes of the course were written by Arturo Giles Flores and Otoniel Nogueira da Silva under the supervision of Bernard Teissier. The main objective is to prove that all equisingular (topologically equivalent) germs of reduced plane curves are generic projections of a single space curve and that the restriction of a generic projection to a curve is a bi-Lipschitz map with the outer metric. It uses the theory of saturation introduced in his historic joint paper with F. Pham, "Fractions Lipschitziennes d'une algèbre analytique complexe et saturation de Zariski", published at the end of the present book.

Finally, the ninth and last course, by **Patrick Popescu-Pampu**, gives an introduction to the recent theory of **Ultrametrics and Surface Singularities** which he developed in a series of works with Evelia García Barroso, Pedro González Pérez and Matteo Ruggiero. Starting with a theorem of Płoski, it shows how to construct ultrametrics on certain sets of branches drawn on any normal surface singularity from their mutual intersection numbers and how to interpret the associated rooted trees in terms of the dual graphs of adapted embedded resolutions. The text begins by recalling basic properties of intersection numbers and multiplicities on smooth surface singularities and the relation between ultrametrics on finite sets and rooted trees. This theory of ultrametric aspects of intersection theory on normal surface singularities is intimately linked with the valuative and non-Archimedean points of view on singularity theory. It has many promising applications in Lipschitz geometry, as already sketched in several recent works.

We would like to emphasize that this book, which gives a first panorama of the Lipschitz geometry of singularites, is not exhaustive. For example, the recent Lipschitz stratification in power-bounded o-minimal fields developed by Nguyen and Valette and independently by Immanuel Halupczok and Yimu Yin is not presented here, nor the very recent Moderately Discontinuous Homology, a promising Lipschitz invariant constructed by Javier Fernández de Bobadilla, Sonja Heinze, María Pe Pereira and José Edson Sampaio.

The International School on Singularity Theory and Lipschitz Geometry attracted more than 50 participants, largely undergraduate students, PhD students and post-docs, but also several senior researchers; we believe that this international meeting has been rich of useful suggestions and ideas for inspiring new researches and developments in the near future. We wish to thank all the lecturers for their active participation and their valuable contribution. We are also deeply indebted to the Universidad Nacional Autónoma de México and to the Instituto de Matemáticas, Unidad Cuernavaca and in particular, the director Professor Jawad Snoussi and his administrative assistant Elisabeth Dominguez, for their helpful support and for the organization of such a remarkable event in Cuernavaca. We thank the following institutions for their financial support: Consejo Nacional de Ciencia y Tecnología (CONACYT, Mexico), Laboratoire International Solomon Lefschetz (UMI LASOL) of the Centre National de la Recherche Scientifique (CNRS, France), National Scientific Foundation (NSF, USA), Lipschitz geometry of singularities (LISA) of the Agence Nationale de la Recherche (project ANR-17-CE40-0023).

Princeton, NJ, USA
October 30th 2019

Walter Neumann
Anne Pichon

Contents

Contributors

Haydée Aguilar-Cabrera São Carlos, Brasil

Lev Birbrair Dept Matemática, Universidade Federal do Ceará (UFC), Fortaleza-Ce, Brasil

Naoufal Bouchareb Aix Marseille Université, CNRS, Marseille, France

José Luis Cisneros-Molina Instituto de Matemáticas, Universidad Nacional Autónoma de México, Cuernavaca, Morelos, Mexico

Otoniel Nogueira da Silva Instituto de Matemáticas, Universidad Nacional Autónoma de México, Cuernavaca, Morelos, Mexico

Arturo Giles Flores Departamento de Matemáticas y Física, Universidad Autónoma de Aguascalientes, Aguascalientes, Mexico

Andrei Gabrielov Dept Mathematics, Purdue University, West Lafayette, IN, USA

Walter Neumann Barnard College, Columbia University, New York, NY, USA

Nhan Nguyen Basque Center for Applied Mathematics, Bilbao, Basque Country, Spain

Frédéric Pham Université de Nice Sophia Antipolis, CNRS, Nice, France

Anne Pichon Aix Marseille Univ, CNRS, Centrale Marseille, I2M, Marseille, France

Patrick Popescu-Pampu Université de Lille, CNRS Lille, France

Maria Aparecida Soares Ruas Instituto de Ciências Matemáticas e de Computação, Universidade de São Paulo, São Carlos, SP, Brasil

Jawad Snoussi Instituto de Matemáticas, Universidad Nacional Autónoma de México, Cuernavaca, Morelos, Mexico

Bernard Teissier Institut mathématique de Jussieu-Paris Rive Gauche, Paris, France

David Trotman Aix-Marseille Univ, CNRS, Centrale Marseille, I2M, Marseille, France

Chapter 1
Geometric Viewpoint of Milnor's Fibration Theorem

Haydée Aguilar-Cabrera and José Luis Cisneros-Molina

Abstract The main goal of these notes is to give the proofs of Milnor Fibration Theorem, Milnor-Lê Fibration Theorem and the equivalence of these two fibrations, emphasizing the ideas of differential topology involved. We also describe the monodromy of the Milnor fibration of a complex analytic function of two variables with isolated singularity, as a quasi-periodic diffeomorphism using a resolution of the singularity.

1.1   Introduction

These are the lecture notes of the course *Milnor Fibration* given at the *International School on Singularities and Lipschitz Geometry* held in Cuernavaca, Mexico, from 11th to 22nd June, 2018. The main goal is to give the proofs of the existence of the Milnor Fibration, the Milnor-Lê Fibration and their equivalence, emphasizing the ideas of differential topology involved. We also describe the monodromy of the Milnor fibration of a complex analytic function of two variables with isolated singularity as a quasi-periodic diffeomorphism, using a resolution of the singularity.

The Milnor Fibration Theorem is an important result in singularity theory. It is about the topology of the fibres of analytic functions near their critical points. To each singular point of a complex hypersurface $V = f^{-1}(0)$ defined by a holomorphic function $f : \mathbb{C}^n \to \mathbb{C}$, it associates a fibre bundle

$$\phi := \frac{f}{|f|} : \mathbb{S}_\varepsilon \setminus V \longrightarrow \mathbb{S}^1,$$

H. Aguilar-Cabrera
São Carlos, Brazil

J. L. Cisneros-Molina (✉)
Instituto de Matemáticas, Universidad Nacional Autónoma de México, Cuernavaca, Morelos, Mexico
e-mail: jlcisneros@im.unam.mx

© The Author(s), under exclusive license to Springer Nature Switzerland AG 2020
W. Neumann, A. Pichon (eds.), *Introduction to Lipschitz Geometry of Singularities*,
Lecture Notes in Mathematics 2280, https://doi.org/10.1007/978-3-030-61807-0_1

where $\mathbb{S}_\varepsilon$ is a sphere centred at the origin of a sufficiently small radius $\varepsilon > 0$. This fibration is known as the Milnor Fibration. For an overview of the origin, generalizations and connections with other branches of mathematics of the Milnor Fibration Theorem, we recommend to read the recent survey article by Seade [Sea18].

These lecture notes are organized as follows. Section 1.2 gives the definitions and main results of transversality, fibre bundles, vector fields and complex and real gradients. In Sect. 1.3 we present the Conical Structure Theorem and define the link of a singularity. In Sect. 1.4 we state Milnor Fibration Theorem (fibration on the sphere) and give the original proof in [Mil68, §4] emphasizing the ideas of differential topology involved. Then, we define the monodromy of the Milnor Fibration. Afterwards, we prove the existence of the Milnor-Lê Fibration (fibration on the tube) and we prove that the two fibrations are equivalent following [Mil68, §5]. This section appeared in condensed form in [CMSS12, §2]. Sections 1.2, 1.3 and 1.4 present the content of the course given at the *International School on Singularities and Lipschitz Geometry*. The organizers asked to also include the description of the Milnor fibre in the resolution, given by Du Bois and Michel in [DM92]. This is well explained (with nice figures) in the first author's PhD thesis [AC11, §1.5–1.6], so we reproduce it here in Sect. 1.5.

1.2 Preliminaries

In this section we give the definitions and results of differential topology and fibre bundles needed for the proof of the Fibration Theorems. Instead of giving proofs of these basic results we give references to the literature.

1.2.1 Transversality

We assume that the reader knows the basics of differential topology: the definitions of differentiable manifolds (with and without boundary), differentiable maps and of the differential of a differentiable map between tangent spaces. It is enough to know these concepts for manifolds embedded in some Euclidean space $\mathbb{R}^m$, see for instance [GP74] or [Mil65].

Definition 1.2.1 Let $f: M \to N$ be a differentiable map between differentiable manifolds. If the differential $D_x f: T_x M \to T_{f(x)} N$ is surjective for some point $x \in M$, we say that f is a **submersion at** x; a map which is a submersion at every point of M, is simply called a **submersion**. We say that a point $x \in M$ is a **regular point** of f if f is a submersion at x, and a point $y \in N$ is a **regular value** of f if every point of $f^{-1}(y)$ is regular. When a point or value is non-regular, we say it is a **critical** point or value. A critical point $x \in M$ is **isolated** if there exists a neighbourhood U of x where x is the only critical point of f.

Theorem 1.2.2 (Preimage Theorem, [GP74, page 21]) *Let* $f : M \to N$ *be a differentiable map between differentiable manifolds of dimension k and l respectively with $k \geq l$. If $y \in N$ is a regular value, then the set $f^{-1}(y) \subset M$ is a differentiable manifold of dimension $k - l$.*

Definition 1.2.3 Let M and N be differentiable manifolds and let $L \subset N$ be a submanifold of dimension k. A differentiable map $f : M \to N$ is **transverse** to L (denoted by $f \pitchfork L$) if the transversality condition

$$D_x f (T_x M) + T_{f(x)} L = T_{f(x)} N, \tag{1.1}$$

is satisfied for every $x \in f^{-1}(L) \subset M$.

Remark 1.2.4 Let $f : M \to N$ be a differentiable map and let $y \in N$ be a regular value. Consider y as a 0-dimensional submanifold of N. Then f is transverse to y. Therefore, a map transverse to a submanifold generalizes the concept of a regular value of a map.

Remark 1.2.5 Let M and L be submanifolds of N and let $i : M \to N$ be the inclusion. A particular case of Definition 1.2.3 is when $i \pitchfork L$: the submanifolds M and L are **transverse** (denoted by $M \pitchfork L$) if

$$T_x M + T_x L = T_x N, \tag{1.2}$$

is satisfied for every $x \in M \cap L$.

Definition 1.2.6 Let L be a submanifold of N. The **codimension** of L in N, denoted by $\mathrm{codim}_N L$, is defined as $\mathrm{codim}_N L = \dim N - \dim L$.

The next result is a generalization of the Preimage Theorem using transversality.

Theorem 1.2.7 (Transversality Theorem, [GP74, page 28]) *If* $f : M \to N$ *is transverse to a submanifold $L \subset N$ of codimension k and $f^{-1}(L) \neq \emptyset$, then $f^{-1}(L)$ is a submanifold of M of codimension k.*

Corollary 1.2.8 ([GP74, page 30]) *If M and L are transverse submanifolds of N, then $M \cap L$ is also a submanifold of N with $\mathrm{codim}_N (M \cap L) = \mathrm{codim}_N M + \mathrm{codim}_N L$.*

There is also a version of the Transversality Theorem for manifolds with boundary. Given a manifold with boundary M, we denote its boundary by ∂M and its interior $M \setminus \partial M$ by $\mathring{M}$.

Theorem 1.2.9 ([GP74, page 60]) *Let* $f : M \to N$ *be a differentiable map of a manifold with boundary M to a manifold without boundary N. Suppose that both $f : M \to N$ and the restriction $\partial f : \partial M \to N$ are transverse to a submanifold L*

(without boundary) of N. Then $f^{-1}(L)$ *is a manifold with boundary* $\partial(f^{-1}(L)) =$ $f^{-1}(L) \cap \partial M$ *and* $\mathrm{codim}_M f^{-1}(L) = \mathrm{codim}_N L$.

1.2.2   Vector Fields and Integral Curves

Let $M \subset \mathbb{R}^m$ be a differentiable manifold. A **smooth tangent vector field** on M is a smooth map $v \colon M \to \mathbb{R}^m$ such that $v(x) \in T_x M$ for each $x \in M$.

Definition 1.2.10 Let v be a smooth vector field on M and $x \in M$. An **integral curve** of v is a smooth curve $p \colon (a, b) \to M$ such that $p'(t) = v(p(t))$ for all $t \in (a, b)$. Usually we assume that the open interval (a, b) contains 0. In this case, if $p(0) = x$, we say that p is an integral curve of v **starting at** x, and call x the **starting point** of p. To show the dependence of such an integral curve on the initial point x we write $p(t) = H_t(x)$. An integral curve is **maximal** if its domain cannot be extended to a larger interval.

Theorem 1.2.11 ([BJ82, (8.10)]) *Let v be a smooth vector field on M. Let U be an open set in M. For each $x \in U$ there exists a neighbourhood W of x in M, a number $\epsilon > 0$ and a smooth map*

$$H \colon (-\epsilon, \epsilon) \times W \to U$$

$$(t, x) \mapsto H_t(x)$$

such that for each $x \in W$ the function $t \mapsto H_t(x)$ is an integral curve of v starting at x, that is, we have $H_0(x) = x$ and $\frac{d}{dt} H_t(x) = v(H_t(x))$. More over, it satisfies $H_t(H_s(x)) = H_{t+s}(x)$ whenever both sides of the equality are defined.

Definition 1.2.12 The map $H \colon (-\epsilon, \epsilon) \times W \to U$ given in Theorem 1.2.11 is called a **local flow generated by** v about the point x in the neighbourhood U. If a local flow is defined on $\mathbb{R} \times M$ then it is called a **global flow**. A vector field having a global flow is called a **complete vector field**.

If v is a complete vector field, then for every $t \in \mathbb{R}$

$$H_t \circ H_{-t} = H_{-t} \circ H_t = \mathrm{Id}_M,$$

so $H_t \colon M \to M$ is a diffeomorphism. Thus, a complete vector field gives rise to a **one-parameter group of diffeomorphisms of** M that is, to a group homomorphism

$$(\mathbb{R}, +) \to \mathrm{Diff}(M)$$

$$t \mapsto H_t$$

where $(\mathbb{R}, +)$ is the abelian group of the real numbers under addition, and $\mathrm{Diff}(M)$ is the group of diffeomorphism of M onto itself.

Theorem 1.2.13 ([BJ82, (8.10)]) *Let v be smooth vector field on a compact differentiable manifold M. Then v is a complete vector field.*

1.2.3 Fibre Bundles

Now we recall the definition of fibre bundle and state the Ehresmann Fibration Theorem. See for instance [AMR88, §3.4 and §5.5].

Definition 1.2.14 A (smooth) **fibre bundle** (E, π, B, F) consists of the following:

1. A differentiable manifold E called **total space**;
2. A differentiable manifold B called **base space**;
3. A differentiable surjective map $\pi : E \to B$ which is a submersion, called the **fibre bundle projection**;
4. A differentiable manifold F called the **fibre**;
5. There exists a collection $\{(U_\alpha, \varphi_\alpha)\}_{\alpha \in A}$, called **trivializing cover**, where $\{U_\alpha\}_{\alpha \in A}$ is an open cover of B and for every $\alpha \in A$,

$$\varphi_\alpha : \pi^{-1}(U_\alpha) \to U_\alpha \times F$$

is a diffeomorphism such that $\pi \circ \varphi^{-1}(x, f) = x$ for every $(x, f) \in U_\alpha \times F$; that is, the following diagram commutes:

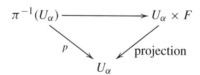

Because of this property, we say that the projection π is **locally trivial**

Example 1.2.15 Examples of fibre bundles are the following:

- Let B and F be differentiable manifolds and consider the projection onto the first factor $\pi : B \times F \to B$. It is clearly a fibre bundle which is called a **product bundle** (see Fig. 1.1a).[1]

---

[1]Figure taken from tex.stackexchange:generate simple cylindrical shape with text in latex (tikz). Answer by hpesoj626.

Fig. 1.1 Fibre bundles. (**a**) Product bundle. (**b**) Möbius band

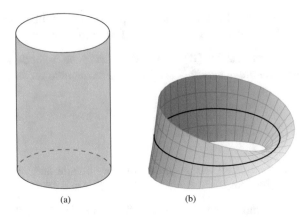

(a)   (b)

- The Möbius band together with the projection onto the central circle is a fibre bundle (see Fig. 1.1b).[2]
- Vector bundles are fibre bundles. For instance, the tangent bundle of a differentiable manifold M in $\mathbb{R}^m$. Its total space is given by

$$TM = \{(x, v) \in M \times \mathbb{R}^m \mid x \in M, v \in T_x M\},$$

and the projection $\pi : TM \to M$ is given by the restriction of the projection onto the first factor.

Theorem 1.2.16 (Ehresmann Fibration for Manifolds with Boundary) *Let M be a manifold with boundary, let N be a closed manifold and $f : M \longrightarrow N$ be a proper surjection. If $f|_{\mathring{M}} : \mathring{M} \longrightarrow N$ and $f|_{\partial M} : \partial M \longrightarrow N$ are submersions, then f is a locally trivial fibre bundle.*

In Theorem 1.2.16 the boundary of M may be empty, in this case we obtain the classical Ehresmann Fibration Theorem. Here we give the idea of the proof of the classical case. See [AMR88, Theorem 5.5.14] or [Wol64, §2] for details of this proof that can be generalized for manifolds with boundary.

Proof (Sketch) We need to prove that f is locally trivial. Since the statement is local, we can replace M and N by chart domains and, in particular, we can assume that $N = \mathbb{R}^l$. In this case we have the basic vector fields $\partial/\partial x_1, \ldots, \partial/\partial x_l$, and we can lift them to obtain vector fields $v_1, \ldots, v_l$ on M, so that, for all $x \in M$ we have $D_x f(v_i(x)) = \partial/\partial x_i$. Locally, the vector fields $v_1, \ldots, v_l$ are easy to find, since f is a submersion there exists coordinate charts such that f has the form of a projection $f : U \times V \to U$ (see [GP74, page 20]), and one obtains the vector fields v_i on all M by glueing together the locally chosen vector fields with a partition of unity (see [GP74, page 52]). Using the hypothesis that f is proper one can prove

---

[2]Figure taken from pgfplots.net: Example: Moebius strip. Example posted by Jake.

that the vector fields $v_1, \ldots, v_l$ are complete. Let $H^1, \ldots, H^l$ be, respectively, the flows generated by the vector fields $v_1, \ldots, v_l$. Set $F = f^{-1}(0)$, let $x \in F$ and $y = (y_1, \ldots, y_l) \in \mathbb{R}^l$. The map $\varphi \colon \mathbb{R}^l \times F \to f^{-1}(\mathbb{R}^l)$ defined by

$$\varphi(y, x) = H^1_{y_1} \circ \cdots \circ H^l_{y_l}(x)$$

gives a local trivialization of f.   □

Theorem 1.2.16 will be used to prove the Milnor-Lê Fibration Theorem (Theorem 1.4.19) in Sect. 1.4.

Definition 1.2.17 Two fibre bundles (E, π, B, F) and (E', π', B, F') with the same base space B are **equivalent** if there exists a diffeomorphism $\Theta \colon E \to E'$ such that the following diagram commutes

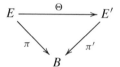

Note that if two fibre bundles are equivalent they have diffeomorphic fibres.

1.2.4 Complex and Real Gradients

Let $v = (v_1, \ldots, v_n)$, $w = (w_1, \ldots, w_n) \in \mathbb{C}^n$, and consider the standard Hermitian inner product $\mathbb{C}^n$ given by

$$\langle v, w \rangle = \sum_{j=1}^{n} v_j \overline{w}_j.$$

The Hermitian vector space can also be thought of as an Euclidean vector space of dimension $2n$ over the real numbers, defining the Euclidean inner product $\langle \, , \, \rangle_{\mathbb{R}}$ to be the real part of the Hermitian inner product. That is,

$$\langle v, w \rangle_{\mathbb{R}} = \mathrm{Re}\langle v, w \rangle.$$

Let $f \colon \mathbb{C}^n \longrightarrow \mathbb{C}$ be a complex analytic function and define its **complex gradient** by

$$\mathrm{grad}\, f = \left(\overline{\frac{\partial f}{\partial z_1}}, \ldots, \overline{\frac{\partial f}{\partial z_n}} \right).$$

With this definition the chain rule for the derivative of f along a path $z = p(t)$ has the form

$$\frac{df}{dt}(p(t)) = \left\langle \frac{dp}{dt}, \operatorname{grad} f \right\rangle. \tag{1.3}$$

In other words, the directional derivative of f along a vector v at the point z is equal to the inner product $\langle v, \operatorname{grad} f(z) \rangle$.

Consider the function $\log f$, we have from the definition that

$$\operatorname{grad} \log f(z) = \frac{\operatorname{grad} f(z)}{\overline{f(z)}}, \tag{1.4}$$

so it is well defined where $f(z) \neq 0$, even though $\log f$ is only locally defined as a single valued function.

Let $f: (\mathbb{C}^n, 0) \longrightarrow (\mathbb{C}, 0)$ be a complex analytic function with $0 \in \mathbb{C}$ a critical value and set $V = f^{-1}(0)$. By the Bertini-Sard Theorem [Ver76, Thm. (3.3)] there exists an open neighbourhood U of $0 \in \mathbb{C}$ such that f restricted to U has $0 \in \mathbb{C}$ as the only critical value. We still denoted $f|_U$ just by f.

Write f as

$$f(z) = |f(z)|e^{i\theta(z)}, \tag{1.5}$$

we can associate to f the real analytic functions

$$\theta: U \setminus V \longrightarrow \mathbb{R},$$

and

$$\log |f|: U \setminus V \longrightarrow \mathbb{R},$$

where θ is locally well-defined as a single valued function.

From (1.5) we have that these functions are related to the function $\log f$ as follows:

$$\log f = \log |f| + i\theta, \tag{1.6}$$

so

$$\theta = \operatorname{Im}(\log f) = \operatorname{Re}(-i \log f) \quad \text{and} \tag{1.7}$$

$$\log |f| = \operatorname{Re}(\log f). \tag{1.8}$$

Proposition 1.2.18 *The real gradients of θ and $\log|f|$ are the vectors*

$$\mathrm{grad}_{\mathbb{R}}\,\theta = i\,\mathrm{grad}\log f$$

and

$$\mathrm{grad}_{\mathbb{R}}\log|f| = \mathrm{grad}\log f.$$

Hence they are normal to the respective level hypersurfaces.

Proof Differentiating (1.6) along a curve $z = p(t)$ we have

$$\frac{d\log f(p(t))}{dt} = \frac{d\log|f(p(t))|}{dt} + i\frac{d\theta(p(t))}{dt}.$$

Using (1.7) and (1.8) we can obtain expressions for these directional derivatives involving $\mathrm{grad}\log f$:

$$
\begin{aligned}
\frac{d\theta(p(t))}{dt} &= \mathrm{Re}\left(\frac{d(-i\log f(p(t)))}{dt}\right)\\[2mm]
&= \mathrm{Re}\left\langle\frac{dp}{dt},\, \mathrm{grad}(-i\log f(p(t)))\right\rangle\\[2mm]
&= \mathrm{Re}\left\langle\frac{dp}{dt},\, i\,\mathrm{grad}\log f(p(t))\right\rangle\\[2mm]
&= \left\langle\frac{dp}{dt},\, i\,\mathrm{grad}\log f(p(t))\right\rangle_{\mathbb{R}}
\end{aligned}
\tag{1.9}
$$

and

$$
\begin{aligned}
\frac{d\log|f(p(t))|}{dt} &= \mathrm{Re}\left(\frac{d(\log f(p(t)))}{dt}\right)\\[2mm]
&= \mathrm{Re}\left\langle\frac{dp}{dt},\, \mathrm{grad}\log f(p(t))\right\rangle\\[2mm]
&= \left\langle\frac{dp}{dt},\, \mathrm{grad}\log f(p(t))\right\rangle_{\mathbb{R}}
\end{aligned}
\tag{1.10}
$$

Remark 1.2.19 Notice that for all $a \in U \setminus V$, the vectors $i\,\mathrm{grad}\log f(z)$ and $\mathrm{grad}\log f(z)$ are orthogonal with respect to the Euclidean inner product.

Let us see how are the level hypersurfaces of θ and $\log|f|$.

1.2.4.1 Level Hypersurfaces of θ

Let $\theta_0 \in \mathbb{R}$ and let $\mathcal{L}_{\theta_0}^+$ be the open real ray in $\mathbb{C}$ with angle θ_0 with respect to the positive real axis (see Fig. 1.2).

The level hypersurface of θ corresponding to the value θ_0 is (see Fig. 1.3):

$$\theta^{-1}(\theta_0) = \{z \in U \setminus V \mid \theta(z) = \theta_0\} = \{z \in U \setminus V \mid f(z) \in \mathcal{L}_{\theta_0}^+\} = f^{-1}(\mathcal{L}_{\theta_0}^+).$$

We denote

$$E_{\theta_0} = f^{-1}(\mathcal{L}_{\theta_0}^+).$$

Remark 1.2.20 Since f is a submersion outside V, E_{θ_0} is a real submanifold of U of real codimension 1.

Corollary 1.2.21 *The vector $i\,\mathrm{grad}\,\log f(z)$ is normal to $E_{\theta(z)}$ at the point $z \in U \setminus V$.*

Proof It is consequence of Proposition 1.2.18. □

Fig. 1.2 The open ray in $\mathbb{C}$ with angle θ_0

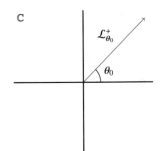

Fig. 1.3 Level hypersurface of θ and a normal vector

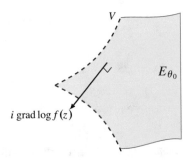

1.2.4.2 Level Hypersurfaces of log $|f(z)|$

Consider $\delta > 0$, let $\mathbb{D}_\delta$ be the disc in $\mathbb{C}$ of radius δ centred at 0 and let $\partial\mathbb{D}_\delta$ be its boundary circle. The level hypersurface of $\log|f|$ corresponding to the value $\log\delta$ is (see Fig. 1.4):

$$
\begin{aligned}
\left(\log|f|\right)^{-1}(\log\delta) &= \{z \in U \setminus V \mid \log|f(z)| = \log\delta\} \\
&= \{z \in U \setminus V \mid |f(z)| = \delta\} \\
&= f^{-1}(\partial\mathbb{D}_\delta).
\end{aligned}
$$

We denote

$$
\mathcal{N}(\delta) = f^{-1}(\partial\mathbb{D}_\delta) \tag{1.11}
$$

Remark 1.2.22 Since f is a submersion outside V, $\mathcal{N}(\delta)$ is a real submanifold of U of real codimension 1.

Corollary 1.2.23 *The vector* $\operatorname{grad}\log f(z)$ *is normal to the tube* $\mathcal{N}(|f(z)|)$ *at the point* $z \in U \setminus V$.

In Sect. 1.4 we will use the intersection of the tube $\mathcal{N}(\delta)$ defined in (1.11) with a ball $\mathbb{B}_\varepsilon$ centred at the origin of radius ε, so we define

$$
\mathcal{N}(\varepsilon, \delta) = \mathbb{B}_\varepsilon \cap \mathcal{N}(\delta) = \mathbb{B}_\varepsilon \cap f^{-1}(\partial\mathbb{D}_\delta). \tag{1.12}
$$

We call $\mathcal{N}(\varepsilon, \delta)$ a **Milnor tube**.

Fig. 1.4 Level hypersurfaces of $\log|f(z)|$ and a normal vector

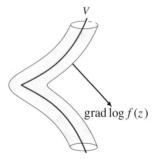

1.3   Conical Structure

Definition 1.3.1 Let $f: (\mathbb{C}^n, 0) \longrightarrow (\mathbb{C}, 0)$ be a holomorphic map germ. The **zero set of** f, given by $V = f^{-1}(0)$, is the **hypersurface defined by** f. The **singularities** of V are the critical points of f contained in V. A point $x \in V$ is an **isolated singularity** if it is an isolated critical point of f.

Let $f: (\mathbb{C}^n, 0) \longrightarrow (\mathbb{C}, 0)$ be a holomorphic map germ with $0 \in \mathbb{C}^n$ an *isolated critical point*, i.e., there exists a neighbourhood U of $0 \in \mathbb{C}$, where 0 is the only critical point of f. As before, set $V = f^{-1}(0)$. Applying Theorem 1.2.2 to the restriction of f to $U \backslash \{0\}$ we get that $V \backslash \{0\}$ is a smooth manifold of real codimension 2.

Let $\mathbb{B}_\varepsilon$ be a closed ball in $\mathbb{C}^n$ centred at 0 of radius $\varepsilon > 0$ and let $\mathbb{S}_\varepsilon = \partial \mathbb{B}_\varepsilon$. Milnor proved [Mil68, Corollary 2.9] that for every sufficiently small ε the sphere $\mathbb{S}_\varepsilon$ intersects $V \backslash \{0\}$ transversely, and by Theorem 1.2.7 the intersection $K_\varepsilon := V \cap \mathbb{S}_\varepsilon$ is a smooth manifold of real dimension $2n - 3$ embedded in the sphere $\mathbb{S}_\varepsilon$. Milnor also proved that in such a ball of sufficiently small radius ε, V has a conical structure given by Theorem 1.3.2 below. Remember that the **cone** of a topological space X is obtained from $X \times [0, 1]$ by collapsing $X \times \{0\}$ to a point. We denote the cone of X by Cone X.

Theorem 1.3.2 ([Mil68, Theorem 2.10])   *For small $\varepsilon > 0$ the intersection of V with $\mathbb{B}_\varepsilon$ is homeomorphic to Cone K_ε. In fact, the pair $(\mathbb{B}_\varepsilon, V \cap \mathbb{B}_\varepsilon)$ is homeomorphic to the pair* (Cone $\mathbb{S}_\varepsilon$, Cone K_ε).

Proof (Sketch)   The flow of the outward radial vector field in $\mathbb{B}_\varepsilon$ gives a homeomorphism between Cone $\mathbb{S}_\varepsilon$ and $\mathbb{B}_\varepsilon$ as follows. Let w be the outward radial vector field on $\mathbb{C}^n$ given by $w(z) = z$ for every $z \in \mathbb{C}^n$. It is easy to see that the integral curves of w are given by $p(t) = tz$, with t a *real variable* (see Definition 1.2.10). For every $z \in \mathbb{S}_\varepsilon$ the integral curve passing throughout z is determined by the initial condition $p(1) = z$. One can see that the flow generated by w is given by $H_t(z) = tz$ and that it is defined on $\mathbb{R} \times \mathbb{C}^n$, so w is complete. Thus, the restriction of the flow

$$H: (0, 1] \times \mathbb{S}_\varepsilon \to \mathbb{C}^n$$

$$(t, z) \mapsto H_t(z),$$

maps diffeomorphically the product $(0, 1] \times \mathbb{S}_\varepsilon$ to the punctured ball $\mathbb{B}_\varepsilon \backslash \{0\}$. Since $H_t(z)$ tends uniformly to $0 \in \mathbb{C}^n$ as t tends to 0, this diffeomorphism extends to a homeomorphism from Cone $\mathbb{S}_\varepsilon$ to $\mathbb{B}_\varepsilon$. The idea is to modify this outward radial vector field such that on points in $V \backslash \{0\}$ it is tangent to $V \backslash \{0\}$. In this way, the integral curve through a point in $V \backslash \{0\}$ is contained in $V \backslash \{0\}$, hence, the homeomorphism given by its flow restricts to a homeomorphism between Cone K_ε and $V \cap \mathbb{B}_\varepsilon$.   □

Definition 1.3.3 From Theorem 1.3.2 follows that for any ε' with $0 < \varepsilon' < \varepsilon$ the pairs of manifolds $(\mathbb{S}_{\varepsilon'}, K_{\varepsilon'})$ and $(\mathbb{S}_\varepsilon, K_\varepsilon)$ are homeomorphic. Hence the

Fig. 1.5 Knots. (**a**) Trefoil knot. (**b**) Figure 8 knot

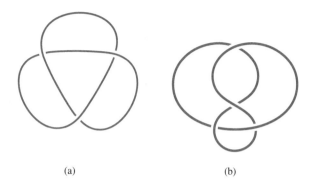

(a)   (b)

homeomorphism class of the pair $(\mathbb{S}_\varepsilon, K_\varepsilon)$ is independent of the radius of the sphere, and we call it the **embedded link of** f. We also call the homeomorphism type of K_ε the **abstract link of** f and we denote it simply by K.

Example 1.3.4 For $n = 2$, the links are topological knots or links embedded in the 3-sphere $\mathbb{S}_\varepsilon$. They are torus or iterated torus knots or links also called cable knots [BK86, §8.3 Theorem 12], for instance, the trefoil knot (see Fig. 1.5a) is the link of the singularity given by $x^2 + y^3 = 0$. They cannot be hyperbolic knots, for instance the figure-eight knot (see Fig. 1.5b) [Thu82, Corollary 2.5].

Example 1.3.5 For $n = 3$, the links are graph manifolds embedded in the 5-sphere $\mathbb{S}_\varepsilon$ [Neu81]. For instance, the link of the singularity $x^2 + y^3 + z^5 = 0$ is the Poincaré homology 3-sphere (see [KS77, Description 3, p. 116]). They cannot be hyperbolic manifolds, for instance, the complement of the figure-eight knot (see [Neu19] in the present volume).

Definition 1.3.6 Let $f\colon (\mathbb{C}^n, 0) \longrightarrow (\mathbb{C}, 0)$ and $g\colon (\mathbb{C}^n, 0) \longrightarrow (\mathbb{C}, 0)$ be holomorphic map germs with an isolated critical point at 0. Let $V_f = f^{-1}(0)$ and $V_g = g^{-1}(0)$ be, respectively, the hypersurfaces defined by f and g. We say that f and g are **R-equivalent** if there is a germ of self-homeomorphism $\Psi\colon (\mathbb{C}^n, 0) \to (\mathbb{C}^n, 0)$ such that $f = g \circ \Psi$; if we only have that $\Psi(V_f) = V_g$ then we say that f and g **have the same topological type**.

By Theorem 1.3.2 if two germs f and g have the same embedded link, then they have the same topological type. Therefore, the embedded link $(\mathbb{S}_\varepsilon, K_\varepsilon)$ and the abstract link K are invariants of the topological type of a singularity.

1.3.1 Whitney Stratifications

For the case when $0 \in \mathbb{C}^n$ is not an isolated critical point of f, it is possible to endow $\mathbb{C}^n$ with a Whitney stratification, such that V is union of strata. For more details on stratifications see [Tro20] on the present volume.

A **stratification** of a subset X of $\mathbb{C}^n$ is a *locally finite* partition $\{S_\alpha\}$ of X into smooth, connected submanifolds of $\mathbb{C}^n$ called **strata** which satisfy the **frontier condition**, that if S_α and S_β are strata with $S_\alpha \cap \overline{S}_\beta \neq \emptyset$, then $S_\alpha \subset \overline{S}_\beta$.

We say that a stratification $\{S_\alpha\}$ of X is **complex analytic** if all the strata are smooth complex analytic varieties.

Now consider a triple (y, S_α, S_β), where S_α and S_β are strata of X with $y \in S_\alpha \subset \overline{S}_\beta$. We say that the triple (y, S_α, S_β) is **Whitney regular** if it satisfies the **Whitney** (b) **condition**: given

1. a sequence $\{x_n\} \subset S_\beta$ converging in $\mathbb{C}^n$ to $y \in S_\alpha$ such that the sequence of tangent spaces $T_{x_n} S_\beta$ converges to a subspace $T \subset \mathbb{C}^n$; and
2. a sequence $\{y_n\} \subset S_\alpha$ converging to $y \in S_\alpha$ such that the sequence of lines (secants) $l_{x_i y_i}$ passing through x_i and y_i converges to a line l;

then one has $l \subset T$.

By convergence of tangent spaces or secants we mean convergence of the translates to the origin of these spaces, so these are points in the corresponding Grassmannian.

There is also a **Whitney** (a) **condition**. It will not be used in the sequel, we give it here for completeness.

Given a sequence $\{x_n\} \subset S_\beta$ converging in $\mathbb{C}^n$ to $y \in S_\alpha$ such that the sequence of tangent spaces $T_{x_n} S_\beta$ converges to a subspace $T \subset \mathbb{C}^n$; then T contains the space tangent to S_α at y.

It is an exercise to show that condition (b) implies condition (a).

Definition 1.3.7 The stratification $\{S_\alpha\}$ of X is **Whitney regular** (also called a **Whitney stratification**) if every triple (y, S_α, S_β) as above, is Whitney regular.

The existence of Whitney stratifications for every analytic space X was proved by Whitney in [Whi65, Thm. 19.2] for complex varieties, and by Hironaka [Hir73] in the general setting.

Let $f : (\mathbb{C}^n, 0) \longrightarrow (\mathbb{C}, 0)$ be a holomorphic map germ with $0 \in \mathbb{C}^n$ a non-isolated critical point. Endow $\mathbb{C}^n$ with a Whitney stratification, such that V is union of strata. Then for every sufficiently small ε the sphere $\mathbb{S}_\varepsilon$ intersects every stratum transversely, in this case, the link is no longer a smooth manifold but a stratified set. However Theorem 1.3.2 is still true in this more general case (see [BV72, Lemma 3.2]).

Example 1.3.8 Figure 1.6 shows the Whitney umbrella, given by $x^2 - y^2 z = 0$. The set of singular points are the z-axis. A Whitney stratification is given as follows: the origin is one 0-dimensional stratum, one 1-dimensional stratum is the positive z-axis, another 1-dimensional stratum is the negative z-axis, and the rest is one 2-dimensional stratum.

Fig. 1.6 Whitney stratification of the Whitney umbrella $x^2 - y^2 z = 0$

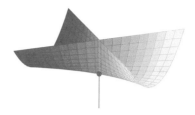

1.4 Fibration Theorems

Milnor proved his Fibration Theorem for polynomials, but his proof also works for complex analytic functions with some minor modifications (see for instance [BV72, Section 2]).

1.4.1 Milnor Fibration

Let $f : (\mathbb{C}^n, 0) \longrightarrow (\mathbb{C}, 0)$ be a holomorphic map-germ with $0 \in \mathbb{C}^n$ a critical point (not necessarily isolated). As in Sect. 1.2.4, let U be a neighbourhood of $0 \in \mathbb{C}^n$ such that f restricted to U has $0 \in \mathbb{C}$ as the only critical value and set $V = f^{-1}(0) \cap U$. Let $\mathbb{B}_\varepsilon$ be a closed ball in U centred at 0 of a sufficiently small radius ε, let $\mathbb{S}_\varepsilon = \partial \mathbb{B}_\varepsilon$ and let $K = V \cap \mathbb{S}_\varepsilon$ be the link of $0 \in \mathbb{C}^n$. The aim of this subsection is to prove Milnor Fibration Theorem.

Theorem 1.4.1 (Milnor Fibration) *The map*

$$\phi := \frac{f}{|f|} : \mathbb{S}_\varepsilon \setminus K \longrightarrow \mathbb{S}^1 \tag{1.13}$$

is the projection of a smooth fibre bundle.

The proof consists of two main steps:

1. To prove that ϕ is a submersion.
2. To prove that ϕ is locally trivial.

Step 1
To show that ϕ is a submersion, Milnor characterizes its possible critical points. Consider the map

$$\Phi := \frac{f}{|f|} : U \setminus V \longrightarrow \mathbb{S}^1. \tag{1.14}$$

Lemma 1.4.2 *The map Φ is a submersion.*

Proof Φ can be seen as the composition of the restriction of f to $U \setminus V$ and the projection $\pi : \mathbb{C} \setminus \{0\} \longrightarrow \mathbb{S}^1$ given by $\pi(x) = \frac{x}{|x|}$. The lemma follows since both maps are submersions. $\qquad\square$

Remark 1.4.3 The fibres of Φ are precisely the E_θ defined in Sect. 1.2.4.1. Given $e^{i\theta} \in \mathbb{S}^1$

$$\Phi^{-1}(e^{i\theta}) = f^{-1}(\pi^{-1}(e^{i\theta})) = f^{-1}(\mathcal{L}_\theta^+) = E_\theta$$

Therefore, given $z \in E_\theta$ we have that $\ker D_z \Phi = T_z E_\theta$.

Now we can state Milnor's characterization of the critical points of ϕ.

Lemma 1.4.4 ([Mil68, Lemma 4.1]) *The critical points of the map ϕ are precisely those points $z \in \mathbb{S}_\varepsilon \setminus K$ for which the vector $i \operatorname{grad} \log f(z)$ is a real multiple of the vector z.*

Proof The map ϕ is the restriction of Φ to $\mathbb{S}_\varepsilon \setminus K$. Thus, a point $z \in \mathbb{S}_\varepsilon \setminus K$ is a critical point of ϕ if and only if

$$T_z(\mathbb{S}_\varepsilon \setminus K) = \ker D_z \Phi = T_z E_\theta$$

since $\mathbb{S}_\varepsilon \setminus K$ and E_θ have both real codimension 1 in $\mathbb{C}^n$. But $T_z(\mathbb{S}_\varepsilon \setminus K) = T_z E_\theta$ if and only if the normal vector to E_θ at z is a real multiple of the normal vector to $\mathbb{S}_\varepsilon \setminus K$ at z. By Proposition 1.2.18 the normal vector to E_θ at z is $i \operatorname{grad} \log f(z)$ and the normal vector to $\mathbb{S}_\varepsilon \setminus K$ at z is z itself. $\qquad\square$

Remark 1.4.5 For each line $\mathcal{L}_\theta$ through the origin in $\mathbb{C}$ we can consider the set

$$X_\theta := \{z \in \mathbb{C}^n \mid f(z) \in \mathcal{L}_\theta\}.$$

Then each X_θ is a real analytic hypersurface with singular set equal to the singular set of V. We have

$$X_\theta \setminus V = E_\theta \cup E_{\theta+\pi}.$$

The family $\{X_\theta\}$ is called the **canonical pencil** of f. We can reinterpret Lemma 1.4.4 as

Proposition 1.4.6 *The critical points of ϕ are the points in $\mathbb{S}_\varepsilon \setminus K$ where the elements $X_\theta \setminus V$ of the canonical pencil are tangent to the sphere $\mathbb{S}_\varepsilon$.*

Using Lemma 1.4.4, to prove that ϕ is a submersion for every sufficiently small ε, we only need to prove.

Lemma 1.4.7 ([Mil68, Lemma 4.2]) *There exists $\varepsilon_0 > 0$ such that for every $z \in U \setminus V$ with $\|z\| < \varepsilon_0$, the two vectors z and $i \operatorname{grad} \log f(z)$ are linearly independent over $\mathbb{R}$.*

Fig. 1.7 The sphere $\mathbb{S}_{\|z\|} \setminus V$
and $E_{\theta(z)}$ are transverse at z

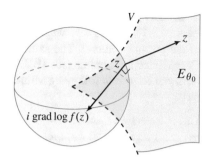

Remark 1.4.8 Lemma 1.4.7 is equivalent to say that for every $z \in U \setminus V$ with $\|z\| < \varepsilon_0$, the manifolds $\mathbb{S}_{\|z\|} \setminus V$ and $E_{\theta(z)}$ are transverse at z, where $\mathbb{S}_{\|z\|}$ is the sphere of radius $\|z\|$. In other words in $\mathbb{B}_{\varepsilon_0}$ all the spheres $\mathbb{S}_\varepsilon \setminus V$ with $\varepsilon \leq \varepsilon_0$ and all the E_θ are transverse (see Fig. 1.7).

To prove Lemma 1.4.7 Milnor proves a stronger statement:

Lemma 1.4.9 ([Mil68, Lemma 4.3]) *There exists $\varepsilon_0 > 0$ so that, for all $z \in U \setminus V$ with $\|z\| \leq \varepsilon_0$, the two vectors z and $i \operatorname{grad} \log f(z)$ are either linearly independent over the complex numbers or else*

$$\operatorname{grad} \log f(z) = \lambda z$$

where λ is a non-zero complex number such that $|\arg \lambda| < \dfrac{\pi}{4}$.

It is easy to see that Lemma 1.4.9 implies Lemma 1.4.7:

- If z and $i \operatorname{grad} \log f(z)$ are linearly independent over $\mathbb{C}$ then they are linearly independent over $\mathbb{R}$.
- If $\operatorname{grad} \log f(z) = \lambda z$ and $|\arg \lambda| < \dfrac{\pi}{4}$, then $\operatorname{Re} \lambda > 0$ so λ cannot be a pure imaginary number (see Fig. 1.8), then z and $i \operatorname{grad} \log f(z)$ cannot be linearly dependent over $\mathbb{R}$.

Fig. 1.8 $|\arg \lambda| < \dfrac{\pi}{4}$

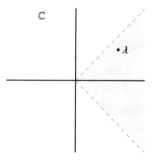

Milnor proved Lemma 1.4.9 using the Curve Selection Lemma ([Mil68, Lemma 3.1]) for the case when f is a polynomial (see also [Rua20, Lemma 1.19] in the present volume). The same proof follows using the analytic version of the Curve Selection Lemma (see [BV72, Proposition 2.2] or [Loo84, (2.1)]), which in turn, can be proved following Milnor's proof with minor modifications (see [BV72, §2]).

Curve Selection Lemma *Let* $\Omega \subset \mathbb{R}^n$ *be an open neighbourhood of* 0, *let* $g_1, \ldots, g_p: \Omega \to \mathbb{R}$ *be analytic maps and let* $A \subset \Omega$ *be an analytic set. Also define de open set* $U = \{x \in \Omega \mid g_i(x) > 0, i = 1, \ldots, p\}$. *If* $0 \in \overline{A \cap U}$, *then there exists a real-analytic curve* $p: [0, \delta) \to \Omega$ *such that* $p(0) = 0$ *and* $p(t) \in A \cap U$ *for* $t \in (0, \delta)$.

Proof (Sketch of the Proof of Lemma 1.4.9) The proof is a nice application of the Curve Selection Lemma and it is done by contradiction: suppose that there were points $z \in \mathbb{C}^n \setminus V$ arbitrarily close to the origin with

$$\operatorname{grad} \log f(z) = \lambda z \neq 0, \tag{1.15}$$

and with $|\arg \lambda| > \frac{\pi}{4}$. In other words, assume that λ lies in the open half-plane $\operatorname{Re}\big((1 + i)\lambda\big) < 0$ or the open half-plane $\operatorname{Re}\big((1 - i)\lambda\big) < 0$.

Then consider the set A of points $z \in \mathbb{C}^n$ for which the vectors $\operatorname{grad} f$ and z are linearly dependent. We have that A is an analytic set. Using (1.4) is easy to see that $z \in \mathbb{C}^n \setminus V$ is in A if and only if Eq. (1.15) holds for some complex number $\lambda(z)$. Let U_+ (respectively U_-) be the open set consisting of all z satisfying the inequality $\operatorname{Re}\big((1 + i)\lambda'(z)\big) < 0$ (respectively $\operatorname{Re}\big((1 - i)\lambda'(z)\big) < 0$), where $\lambda'(z)$ is some real positive multiple of $\lambda(z)$ defined by a real analytic function, and thus $\lambda(z)$ and $\lambda'(z)$ have the same argument. The original supposition implies that there exists points z arbitrarily close to the origin with $z \in A \cap (U_+ \cup U_-)$. By the Curve Selection Lemma there exists a real analytic curve $p: [0, \delta) \to \mathbb{C}^n$ with $p(0) = 0$ and $p(t) \in A \cap (U_+ \cup U_-)$ for all $t > 0$. This proves Lemma 1.4.9 since the existence of such a curve contradicts Lemma 1.4.10 below. $\qquad\square$

Lemma 1.4.10 ([Mil68, Lemma 4.4]) *Let* $p: [0, \delta) \to \mathbb{C}^n$ *be a real analytic path with* $p(0) = 0$ *such that, for each* $t > 0$, *the number* $f\big(p(t)\big)$ *is non-zero and the vector* $\operatorname{grad} \log f\big(p(t)\big)$ *is a complex multiple* $\lambda(t)p(t)$. *Then the argument of the complex number* $\lambda(t)$ *tends to zero as* $t \to 0$.

Proof By (1.4) we have that

$$\operatorname{grad} f(p(t)) = \lambda(t)p(t)\overline{f(p(t))}. \tag{1.16}$$

Consider the Taylor expansions of $p(t)$, $f\big(p(t)\big)$ and $\operatorname{grad} f\big(p(t)\big)$ denoting their corresponding non-zero leading coefficients by a, b and c, and their corresponding leading exponents by α, β and γ, which are integers with $\alpha \geq 1$, $\beta \geq 1$ and $\gamma \geq 0$. Substituting these Taylor expansions in (1.16) on can prove that $\lambda(t)$ has a Taylor expansion of the form

$$\lambda(t) = \lambda_0 t^{\gamma - \alpha - \beta} (1 + k_1 t + k_2 t^2 + \dots),\tag{1.17}$$

and the leading coefficients satisfy the equation

$$c = \lambda_0 a \bar{b}.\tag{1.18}$$

Substituting (1.18) in the power series expansion of the identity

$$\frac{df}{dt} = \left(\frac{dp}{dt}, \text{grad } f \right),$$

and comparing leading coefficients we obtain

$$\beta = \alpha \|a\|^2 \bar{\lambda}_0,$$

which proves that λ_0 is a positive real number. Therefore

$$\lim_{t \to 0} \frac{\lambda(t)}{|\lambda(t)|} = \frac{\lambda_0 t^{\gamma - \alpha - \beta} (1 + k_1 t + k_2 t^2 + \dots)}{\lambda_0 t^{\gamma - \alpha - \beta} \|1 + k_1 t + k_2 t^2 + \dots\|}$$

$$= \lim_{t \to 0} \frac{1 + k_1 t + k_2 t^2 + \dots}{\|1 + k_1 t + k_2 t^2 + \dots\|} = 1.$$

Hence $\arg \lambda(t) \to 0$ as $t \to 0$.   □

Corollary 1.4.11 *If $\varepsilon \le \varepsilon_0$ then the map*

$$\phi := \frac{f}{|f|} : \mathbb{S}_\varepsilon \setminus K \longrightarrow \mathbb{S}^1$$

is a submersion.

Remark 1.4.12 By Remark 1.4.3 the fibres of Φ are the E_θ. Since ϕ is the restriction of Φ to $\mathbb{S}_\varepsilon \setminus K$, we have that the fibres of ϕ are given by

$$F_\theta := E_\theta \cap \mathbb{S}_\varepsilon.$$

By Remark 1.4.8 this intersection is transverse and by Corollary 1.2.8 F_θ is then a smooth $(2n - 2)$-dimensional manifold.

Step 2
Since we are removing the link K from the sphere $\mathbb{S}_\varepsilon$ in the domain of the map ϕ, its fibres are not compact and therefore ϕ is not proper. Thus, we cannot use the Ehresmann Fibration Theorem (Theorem 1.2.16) to prove that ϕ is locally trivial.

Instead, we will construct a *complete* vector field and use the flow generated by it, as we did in the proof of Ehresmann Fibration Theorem.

Proposition 1.4.13 *If $\varepsilon \leq \varepsilon_0$ then the map*

$$\phi := \frac{f}{|f|} : \mathbb{S}_\varepsilon \setminus K \longrightarrow \mathbb{S}^1$$

is locally trivial.

Proof To prove that ϕ is locally trivial it is enough to give a complete vector field w on $\mathbb{S}_\varepsilon \setminus K$ which projects under $D\phi$ to the unit vector field u tangent to $\mathbb{S}^1$ given by

$$u(e^{i\theta}) = i e^{i\theta}$$

so w is transverse to the fibres of ϕ (see Fig. 1.9).

Given such a w, let $p(t)$ be the integral curve of w. The function $p(t)$ depends smoothly on t and on the initial value $z_0 = p(0)$. We denote this dependency by using the flow generated by w as in Sect. 1.2.2

$$p(t) = H_t(z_0).$$

Fig. 1.9 Fibres of ϕ and vector field

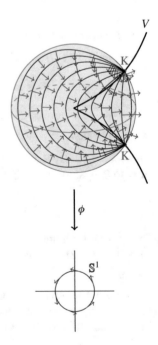

Then each H_t is a diffeomorphism

$$H_t : \mathbb{S}_\varepsilon \setminus K \to \mathbb{S}_\varepsilon \setminus K \tag{1.19}$$

which sends the fibre F_θ onto the fibre $F_{\theta+t}$, that is

$$H_t(F_\theta) = F_{\theta+t}. \tag{1.20}$$

Let $e^{i\theta} \in \mathbb{S}^1$ and let W be a small neighbourhood of $e^{i\theta}$. Then the correspondence

$$W \times F_\theta \longrightarrow \phi^{-1}(W)$$

$$(e^{i(\theta+t)}, z) \longmapsto H_t(z)$$

for $|t| <$ constant and $z \in F_\theta$ is a diffeomorphism, proving the local triviality of ϕ.   □

Since ϕ is a submersion we can take w to be a lifting of the vector field u on $\mathbb{S}^1$ to $\mathbb{S}_\varepsilon \setminus K$. The main difficulty is to guarantee that w is *complete*, since $\mathbb{S}_\varepsilon \setminus K$ is non-compact we need to insure that $p(t)$ cannot tend to K as t tends to some finite limit t_0. This is equivalent to guarantee that $f(p(t))$ cannot tend to zero or that $\log |f(p(t))|$ cannot tend to $-\infty$ as t tends to a finite value t_0.

One way to do this is to make $\log |f(p(t))|$ to increase or decrease "slowly" by keeping its derivative small in absolute value (compare with [Mil68, Lemma 4.7]). Suppose that

$$\left| \frac{d \log |f(p(t))|}{dt} \right| < 1.$$

Then

$$\left| \log |f(p(t_0))| - \log |f(p(0))| \right| = \left| \int_0^{t_0} \frac{d \log |f(p(t))|}{dt} dt \right|$$

$$\leq \int_0^{t_0} \left| \frac{d \log |f(p(t))|}{dt} \right| dt$$

$$< \int_0^{t_0} dt = t_0.$$

Then $\log |f(p(t))|$ cannot tend to $-\infty$ as t tends to any finite limit t_0. Hence we need w so that

$$\left| \frac{d \log |f(p(t))|}{dt} \right| = |\mathrm{Re} \langle w(p(t)), \mathrm{grad} \log f(p(t)) \rangle| < 1$$

Also w has to project under $D\phi$ to the unit tangent vector field u on $\mathbb{S}^1$. That is, the integral curves $p(t)$ of w need to project under ϕ to the path on $\mathbb{S}^1$ which winds around the unit circle in the positive direction with unit velocity, i.e.,

$$\theta(p(t)) = t + \text{constant}.$$

This is equivalent to

$$\frac{d\theta(p(t))}{dt} = \text{Re}\left\langle \frac{dp(t)}{dt}, i \operatorname{grad}\log f(p(t)) \right\rangle = 1, \quad i.e.,$$

$$\frac{d\theta(p(t))}{dt} = \text{Re}\langle w(p(t)), i \operatorname{grad}\log f(p(t)) \rangle = 1.$$

In summary, we need a vector field w on $\mathbb{S}_\varepsilon \setminus K$ with the following properties:

1. $\text{Re}\langle w(z), z \rangle = 0$ (tangent to $\mathbb{S}_\varepsilon$).
2. $|\text{Re}\langle w(z), \operatorname{grad}\log f(z) \rangle| < 1$ (complete).
3. $\text{Re}\langle w(z), i \operatorname{grad}\log f(z) \rangle = 1$ (projects onto u).

Milnor solves the problem using Lemma 1.4.9 to construct w as follows:

Lemma 1.4.14 *There exists a smooth vector field w on $\mathbb{B}_{\varepsilon_0} \setminus V$ such that for every $\varepsilon \leq \varepsilon_0$ and every $z \in \mathbb{S}_\varepsilon \setminus V$, the vector $w(z)$ is tangent to $\mathbb{S}_\varepsilon \setminus V$ at z and the complex inner product*

$$\langle w(z), i \operatorname{grad}\log f(z) \rangle$$

is non-zero and the absolute value of its argument is less than $\dfrac{\pi}{4}$.

Remark 1.4.15 The condition $\langle w(z), i \operatorname{grad}\log f(z) \rangle \neq 0$ guarantees that $w(z)$ is transverse to the fibres of ϕ, because if $\langle w(z), i \operatorname{grad}\log f(z) \rangle = 0$ then we have that $\text{Re}\langle w(z), i \operatorname{grad}\log f(z) \rangle = 0$ so $w(z) \in T_z E_{\theta(z)}$, since $w(z) \in T_z \mathbb{S}_\varepsilon$ then $w(z) \in T_z F_{\theta(z)} = T_z \mathbb{S}_\varepsilon \cap T_z E_{\theta(z)}$.

Proof It suffices to construct such a vector field locally, in a neighbourhood of some given point $z \in \mathbb{B}_{\varepsilon_0} \setminus V$. We have two cases:

Case 1
The vectors z and $\operatorname{grad}\log f(z)$ are linearly independent over $\mathbb{C}$. Then z does not lie in the complex line generated by $\operatorname{grad}\log f(z)$ (where it also lies $i \operatorname{grad}\log f(z)$). So z and $\operatorname{grad}\log f(z)$ are also linearly independent over $\mathbb{R}$. Then the sphere $\mathbb{S}_{\|z\|}$ is transverse to the Milnor tube $N(\varepsilon_0, |f(z)|)$ at z (see Sect. 1.2.4). The intersection $T_z \mathbb{S}_{\|z\|} \cap T_z N(\varepsilon_0, |f(z)|)$ is the vector space orthogonal (with respect to $\langle \, , \, \rangle_{\mathbb{R}}$) to the real plane generated by z and $\operatorname{grad}\log f(z)$. It has real dimension $2n - 2$ and it does not coincide with

$$T_z F_{\theta(z)} = T_z \mathbb{S}_{\|z\|} \cap T_z E_{\theta(z)}$$

because in that case it would be orthogonal to $i \operatorname{grad} \log f(z)$ contradicting that z does not lie in the complex line generated by $\operatorname{grad} \log f(z)$. Take $w(z)$ in $T_z \mathbb{S}_{\|z\|} \cap T_z \mathcal{N}(\varepsilon_0, |f(z)|)$ such that it satisfies property 3. It satisfies property 1 by construction. It satisfies property 2 because the directional derivative of $\log |f|$ along a vector tangent to $\mathcal{N}(\varepsilon_0, |f(z)|)$ is zero because $\mathcal{N}(\varepsilon_0, |f(z)|)$ is the level hypersurface of $\log |f(z)|$.

Milnor constructs w in this case taking the simultaneous solution to the linear equations

$$\langle w(z), z \rangle = 0 \quad \text{and}$$

$$\langle w(z), i \operatorname{grad} \log f(z) \rangle = 1.$$

The first one implies property 1 and the second one implies properties 2 and 3 since

$$\operatorname{Re}\langle w(z), \operatorname{grad} \log f(z) \rangle = \operatorname{Im}\langle w(z), i \operatorname{grad} \log f(z) \rangle = 0 \quad \text{and}$$

$$\operatorname{Re}\langle w(z), i \operatorname{grad} \log f(z) \rangle = 1.$$

Hence the vector field w constructed in **Case 1** satisfies the conditions in the statement of Lemma 1.4.14.

Case 2

There exists λ such that $\operatorname{grad} \log f(z) = \lambda z$ with $|\arg \lambda| < \dfrac{\pi}{4}$. The point z is in the complex line generated by $\operatorname{grad} \log f(z)$ (see Fig. 1.10).

Take $w(z) = \dfrac{iz}{\operatorname{Re}(\bar{\lambda}) \|z\|^2}$ which is tangent to $\mathbb{S}_{\|z\|}$, so it satisfies property 1. On the other hand, we have that

$$\langle w(z), i \operatorname{grad} \log f(z) \rangle = \langle \frac{iz}{\operatorname{Re}(\bar{\lambda}) \|z\|^2}, i \operatorname{grad} \log f(z) \rangle$$

$$= \frac{1}{\operatorname{Re}(\bar{\lambda}) \|z\|^2} \langle z, \operatorname{grad} \log f(z) \rangle$$

Fig. 1.10 $\operatorname{grad} \log f(z) = \lambda z$ with $|\arg \lambda| < \dfrac{\pi}{4}$

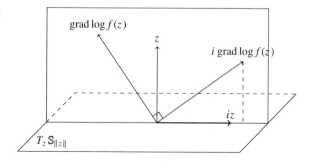

Fig. 1.11 Case 2: w satisfies
property 2

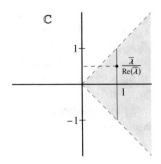

$$= \frac{1}{\mathrm{Re}(\bar{\lambda})\|z\|^2} \langle z, \lambda z \rangle$$

$$= \frac{\bar{\lambda}\|z\|^2}{\mathrm{Re}(\bar{\lambda})\|z\|^2} = \frac{\bar{\lambda}}{\mathrm{Re}(\bar{\lambda})}.$$

Hence $\left| \arg \dfrac{\bar{\lambda}}{\mathrm{Re}(\bar{\lambda})} \right| \leq \dfrac{\pi}{4}$ and $\dfrac{\bar{\lambda}}{\mathrm{Re}(\bar{\lambda})}$ has real part equal to 1, so w satisfies
property 3. It also satisfies property 2 because (see Fig. 1.11)

$$|\mathrm{Re}\langle w(z), \operatorname{grad} \log f(z) \rangle| = |\mathrm{Im}\langle w(z), i \operatorname{grad} \log f(z) \rangle| < 1.$$

In either case one can choose a local tangential vector field $v(z)$ which takes the
constructed value v at z. The condition

$$| \arg \langle w(z), i \operatorname{grad} \log f(z) \rangle | < \frac{\pi}{4}$$

will hold throughout a neighbourhood of z. Using a partition of unity we obtain a
global vector field $w(z)$ having the same property.   □

1.4.2 Monodromy

The fibre F of the Milnor Fibration (1.13) is a differentiable manifold of real
dimension $2(n-1)$. Consider the one-parameter group of diffeomorphisms H_t
of $\mathbb{S}_\varepsilon \setminus K$ given in (1.19), defined by the vector field w of Lemma 1.4.14. The
geometric monodromy on the total space $\tilde{h} \colon \mathbb{S}_\varepsilon \setminus K \to \mathbb{S}_\varepsilon \setminus K$ is defined by
$\tilde{h} = H_{2\pi}$. By (1.20) the fibre $F := F_0 = F_{2\pi}$ is invariant under $\tilde{h}$, so the **geometric
monodromy on the fibre** $h \colon F \to F$ is defined by the restriction $h = \tilde{h}|_F$.

The geometric monodromy $h \colon F \to F$ depends on the choice of lifting w of the
unit tangent vector field on $\mathbb{S}^1$, but its isotopy class does not depend on the choice

of w. Hence, the homomorphism induced by h on homology

$$h_{*,i} : H_i(F; \mathbb{A}) \to H_i(F; \mathbb{A}) \quad \text{for } 0 \leq i \leq 2(n-1),$$

is independent of w. The coefficients $\mathbb{A}$ are usually $\mathbb{Z}$, $\mathbb{Q}$, $\mathbb{R}$ or $\mathbb{C}$ depending on the situation. The isomorphism $h_{*,i}$ is called the i-th **monodromy isomorphism**. This also defines a well-defined homomorphism

$$\rho_i : \pi_1(\mathbb{S}^1) \to Aut(H_i(F; \mathbb{A})) \quad \text{for } 0 \leq i \leq 2(n-1),$$

where the image of the canonical generator of $\pi_1(\mathbb{S}^1)$, represented by a counter-clockwise loop, is the monodromy isomorphism $h_{*,i}$.

1.4.3   Open books

The Milnor Fibration $\phi : \mathbb{S}_\varepsilon \setminus K \longrightarrow \mathbb{S}^1$ together with the link K give an open book structure to the sphere $\mathbb{S}_\varepsilon$, where K is *the binding* and the fibres of the Milnor Fibration are *the pages*.

The formal definition of open book was introduced by Winkelnkemper in [Win73] and has become an important concept in topology (see for example [Ran98, Appendix]). Open books allow to describe an arbitrary closed manifold in terms of lower dimensional ones.

Definition 1.4.16 ([Sea06, Def. 5.1]) Let M be a smooth closed n-manifold and let N be a codimension 2 submanifold of M with trivial normal bundle. Let

$$\pi : M \setminus N \to \mathbb{S}^1$$

be a map such that

- π is a locally trivial fibration and
- there exists a tubular neighbourhood of N diffeomorphic to $N \times \mathbb{D}^2$ such that the restriction of π to $N \times (\mathbb{D}^2 \setminus \{0\})$ is the map $(x, y) \mapsto y/||y||$.

The map π is called an **open-book fibration** of M, N is called the **binding** and the fibres of π are called the **pages**.

It follows that the pages are all diffeomorphic and each page $\mathcal{F}$ can be compactified by attaching the binding N as its boundary, thus getting a compact manifold with boundary.

1.4.4 Thom a_f-condition

Let $f: (\mathbb{C}^n, 0) \longrightarrow (\mathbb{C}, 0)$ be a holomorphic map-germ with $0 \in \mathbb{C}^n$ a critical point (not necessarily isolated), as before, set $V = f^{-1}(0)$. Take a Whitney stratification of U such that V is union of strata and $U \setminus V = S_\beta$ is a stratum. Let S_α be a stratum contained in V such that $S_\alpha \subset \overline{S_\beta}$, $z \in S_\alpha$, $\{z_n\} \subset S_\beta$ such that $z_n \to z$. Let F_n be the fibre of f which contains z_n. Let $T = \lim_{n \to \infty} T_{z_n} F_n$ (taking a subsequence if necessary). The map f satisfies the **Thom a_f-condition** if (see Fig. 1.12)

$$T_z S_\alpha \subset T.$$

We have the following result by Hironaka [Hir77, §5, Corollary 1].

Proposition 1.4.17 *All complex analytic maps $f: \mathbb{C}^n \to \mathbb{C}$ satisfy Thom's a_f-condition.*

Let $\mathbb{D}_\delta$ be a closed disc in $\mathbb{C}$ of radius δ centred at 0.

Corollary 1.4.18 *Let $\mathbb{B}_\varepsilon$ be a ball such that $\mathbb{S}_\varepsilon = \partial \mathbb{B}_\varepsilon$ is transverse to all strata of V. Then all the nearby fibres of f are also transverse to $\mathbb{S}_\varepsilon$, so there exist δ with $0 < \delta \ll \varepsilon$ such that all the fibres over $\mathbb{D}_\delta \subset \mathbb{C}$ are transverse to $\mathbb{S}_\varepsilon$.*

Proof In the case that f has 0 as an isolated critical point, since V is transverse to the sphere $\mathbb{S}_\varepsilon$, the lemma follows by the Implicit Function Theorem. In the general case, by Proposition 1.4.17 f satisfies Thom's a_f-condition, which implies that all the fibres near-by V are also transverse to $\mathbb{S}_\varepsilon$.   □

Fig. 1.12 Thom a_f-property

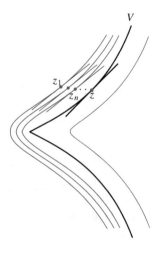

1.4.5   Milnor-Lê Fibration

There is a second fibre bundle associated to a holomorphic map-germ called the
Milnor-Lê Fibration. In this subsection we will prove its existence using Ehresmann
Fibration Theorem (Theorem 1.2.16).

Let $f : (\mathbb{C}^n, 0) \longrightarrow (\mathbb{C}, 0)$ be a holomorphic map-germ with $0 \in \mathbb{C}^n$ a critical
point. Let U be a neighbourhood of $0 \in \mathbb{C}^n$ such that f restricted to U has $0 \in \mathbb{C}$
as the only critical value and set $V = f^{-1}(0) \subset U$. Let $\mathbb{B}_\varepsilon$ be a closed ball in U
centred at 0 of a sufficiently small radius such that $\mathbb{S}_\varepsilon = \partial \mathbb{B}_\varepsilon$ is transverse to (all the
strata of) V.

Let $\varepsilon_0 > 0$ as in Lemma 1.4.9. Let δ be such that $0 < \delta << \varepsilon$ and $\mathbb{D}_\delta$ be as
in Corollary 1.4.18 and let $\partial \mathbb{D}_\delta$ be the boundary circle. Consider the Milnor tube
$\mathcal{N}(\varepsilon, \delta) = \mathbb{B}_\varepsilon \cap f^{-1}(\partial \mathbb{D}_\delta)$ defined in (1.12), and the restriction of f to it (see
Fig. 1.13).

Theorem 1.4.19 (Milnor-Lê Fibration)   *The restriction*

$$f : \mathcal{N}(\varepsilon, \delta) \longrightarrow \partial \mathbb{D}_\delta \qquad\qquad (1.21)$$

is the projection of a smooth fibre bundle.

Milnor proved this theorem for the case when f has an isolated critical point at
0 (see proofs of [Mil68, Theorem 11.2] or [Mil66, Theorem 2]). The general case
was proved by Lê in [Lê77, Theorem (1.1)]. This is the reason why this fibration is
called the Milnor-Lê Fibration.

Fig. 1.13 Milnor-Lê
Fibration

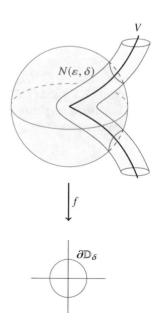

Proof By Corollary 1.4.18 the points in $\partial \mathcal{N}(\varepsilon, \delta)$ are regular points of the restriction $f|_{\mathbb{S}_\varepsilon}$. Therefore $f|_{\mathbb{S}_\varepsilon} \pitchfork \partial \mathbb{D}_\delta$, and since $f|_{\mathbb{B}_\varepsilon}$ is a submersion $f|_{\mathbb{B}_\varepsilon} \pitchfork \partial \mathbb{D}_\delta$. Then, by Theorem 1.2.9, $\mathcal{N}(\varepsilon, \delta)$ is a compact submanifold with boundary of $\mathbb{B}_\varepsilon$. Hence $f : \mathcal{N}(\varepsilon, \delta) \longrightarrow \partial \mathbb{D}_\delta$ is proper and by Theorem 1.2.16 it is a fibre bundle.   □

1.4.6   Equivalence of the Fibrations

Here we will prove that the Milnor fibration (1.13) and the Milnor-Lê fibration (1.21) are equivalent. Let $\pi : \mathbb{C} \setminus \{0\} \to \mathbb{S}^1$ be the projection given by $\pi(x) = \frac{x}{|x|}$. Recall that the composition $\pi \circ f$ is the map Φ defined in (1.14), thus, taking the composition of the Milnor-Lê fibration (1.21) with π we get the map

$$\Phi|_{\mathcal{N}(\varepsilon, \delta)} = \pi \circ f : \mathcal{N}(\varepsilon, \delta) \longrightarrow \mathbb{S}^1. \tag{1.22}$$

Since π sends $\partial \mathbb{D}_\delta$ diffeomorphically onto $\mathbb{S}^1$, the map (1.22) is the projection of a fibre bundle equivalent to (1.21). Thus, now we can compare the fibre bundles (1.13) and (1.22) since now they have the same base space.

By Definition 1.2.17 we need to find a diffeomorphism between the total space $\mathbb{S}_\varepsilon \setminus K$ of (1.13) to the total space $\mathcal{N}(\varepsilon, \delta)$ of (1.22) which commutes with the projections. This is not possible since the latter is compact while the former is not. So consider the restriction of the map ϕ defined in (1.13) to the subspace $\mathbb{S}_\varepsilon \setminus f^{-1}(\mathring{\mathbb{D}}_\delta)$, where $\mathring{\mathbb{D}}_\delta$ is the interior of the disc $\mathbb{D}_\delta$. Since the map ϕ is also a restriction of the map Φ defined in (1.14) we get

$$\Phi|_{\mathbb{S}_\varepsilon \setminus f^{-1}(\mathring{\mathbb{D}}_\delta)} : \mathbb{S}_\varepsilon \setminus f^{-1}(\mathring{\mathbb{D}}_\delta) \longrightarrow \mathbb{S}^1, \tag{1.23}$$

We have that $\mathbb{S}_\varepsilon \setminus f^{-1}(\mathring{\mathbb{D}}_\delta)$ is a compact manifold with boundary. Now, we want to prove that (1.22) and (1.23) are equivalent fibre bundles, so we need to prove that there is a diffeomorphism $\Theta : \mathcal{N}(\varepsilon, \delta) \to \mathbb{S}_\varepsilon \setminus f^{-1}(\mathring{\mathbb{D}}_\delta)$ which makes the following diagram commute

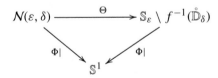

To prove this, suppose there is a vector field v on $U \setminus V$ such that

1. It is transverse to all the spheres $\mathbb{S}_\varepsilon$ with $\varepsilon \leq \varepsilon_0$.
2. It is transverse to all Milnor tubes.
3. If $t \mapsto p(t)$ is an integral curve of v then $f(p(t))$ has constant argument for all t.

Once we have such a vector field, given an integral curve $t \mapsto p(t)$ of v we have that $f(p(t))$ lies on the ray $\mathcal{L}_\theta^+$ with $\theta = \arg f(z)$. Let z be a point in $\mathcal{N}(\varepsilon, \delta)$ and let $p(t)$ be the integral curve of v through z. Following $p(t)$ we travel "away" form the origin, transversely to the tubes and to the spheres until we reach a point z' on $\mathbb{S}_\varepsilon \setminus K$. Since $f(p(t))$ has constant argument along $p(t)$ we have that

$$\Phi(z) = \frac{f(z)}{|f(z)|} = \frac{f(z')}{|f(z')|} = \Phi(z').$$

Thus, the correspondence $z \mapsto z'$ is a diffeomorphism which gives the equivalence between the fibre bundle (1.22) and the fibre bundle (1.23).

Milnor constructed such a vector field in the following lemma. Originally Milnor used this vector field just to prove that the interior of the fibre in the tube is diffeomorphic to the fibre in the sphere, since he did not have the fibration on the tube in the general case.

Lemma 1.4.20 ([Mil68, Lemma 5.9]) *Let $\varepsilon \leq \varepsilon_0$. There exists a smooth vector field v on $\mathbb{B}_\varepsilon \setminus V$ so that*

$$\langle v(z), \operatorname{grad} \log f(z) \rangle$$

is real and positive for all $z \in \mathbb{B}_\varepsilon \setminus V$ and so that the inner product $\langle v(z), z \rangle$ has positive real part.

Proof The proof is analogous to the proof of Lemma 1.4.14. It suffices to construct such a vector field locally, in the neighbourhood of some given point $z \in \mathbb{B}_\varepsilon \setminus V$.

Case 1
z and $\operatorname{grad} \log f(z)$ linearly independent over $\mathbb{C}$. Take the simultaneous solution to

$$\langle v, \operatorname{grad} \log f(z) \rangle = 1 \quad \text{and}$$

$$\langle v, z \rangle = k, \quad k \in \mathbb{C} \text{ with } \operatorname{Re}(k) > 0.$$

Case 2
$\operatorname{grad} \log f(z) = \lambda z$ with $|\arg \lambda| < \dfrac{\pi}{4}$. Take $v = \operatorname{grad} \log f(z)$

$$\langle \operatorname{grad} \log f(z), \operatorname{grad} \log f(z) \rangle = \langle \lambda z, \lambda z \rangle = \|\lambda z\|^2 \in \mathbb{R}^+,$$

$\langle \operatorname{grad} \log f(z), z \rangle = \langle \lambda z, z \rangle = \lambda \|z\|^2$ so $|\arg \lambda \|z\|^2| < \dfrac{\pi}{4}$ so its real part is positive (see Fig. 1.8). $\qquad \square$

Corollary 1.4.21 *The vector field v of Lemma 1.4.20 satisfies properties 1, 2 and 3.*

Proof Let $t \mapsto p(t)$ be an integral curve of v, that is $\frac{dp}{dt} = v(p(t))$. By equality (1.3) we have that

$$\frac{d \log f(p(t))}{dt} = \langle \frac{dp}{dt}, \operatorname{grad} \log f(p(t)) \rangle.$$

The condition that $\langle \frac{dp}{dt}, \operatorname{grad} \log f(p(t)) \rangle = c$ with c real and positive, by equalities (1.9) and (1.10), implies that

$$\frac{d \log |f(p(t))|}{dt} = \operatorname{Re} \langle \frac{dp}{dt}, \operatorname{grad} \log f(p(t)) \rangle = c > 0, \quad \text{and} \tag{1.24}$$

$$\frac{d\theta(p(t))}{dt} = \operatorname{Im} \langle \frac{dp}{dt}, \operatorname{grad} \log f(p(t)) \rangle = 0. \tag{1.25}$$

From (1.24) we have that $\log |f(p(t))| = ct + k$ with k constant, hence

$$|f(p(t))| = e^{ct+k},$$

whose derivative with respect to t does not vanish. Therefore the integral lines $p(t)$ of v are transverse to the Milnor tubes and w satisfies 2. From (1.25) we have that

$$\theta(p(t)) = \theta_0, \qquad \text{for a constant value } \theta_0,$$

in other words, $f(p(t))$ has constant argument as required by 3. On the other hand, the condition that $\operatorname{Re}\langle v(z), z \rangle > 0$ implies that

$$\frac{d \|p(t)\|^2}{dt} = 2 \operatorname{Re} \langle \frac{dp}{dt}, p(t) \rangle > 0,$$

therefore the integral lines $p(t)$ are transverse to all the spheres and w satisfies 1.

$\square$

Remark 1.4.22 Since the fibre bundles (1.22) and (1.23) are equivalent, their respective fibres, which are manifolds with boundary, are diffeomorphic. Thus, also their respective interiors, which are manifolds without boundary, are also diffeomorphic. Milnor proved, that the fibre of (1.23), which is part of the fibre of (1.13), is diffeomorphic to the whole fibre of (1.13) (see [Mil68, Theorem 5.11]). Hence, the Milnor-Lê Fibration (1.21) restricted to the interior of the Milnor tube $N(\varepsilon, \delta)$ is equivalent to the Milnor Fibration (1.13).

1.5 Resolution and Monodromy

In this section we will see the monodromy of the Milnor fibration of a complex analytic function $f \colon (\mathbb{C}^2, 0) \to (\mathbb{C}, 0)$ with isolated singularity as a quasi-periodic diffeomorphism using a resolution of the singularity as is made in [DM92, 1.2 to 1.11]. For this, we will describe the sphere $\mathbb{S}^3$ as the boundary of a plumbing.

1.5.1 Plumbing

Here we present a method to construct manifolds "gluing" disc bundles; in particular we are interested in the construction of 4-manifolds gluing 2-disc bundles over 2-manifolds. We use this construction in Sect. 1.5.3 to describe the monodromy of the Milnor fibration of a curve singularity.

As Bredon in [Bre93, Ch. VI, Sect. 18] and Hirzebruch and Neumann in [HNK71, § 8], we will first describe plumbing in arbitrary dimensions before going into more detail in the case of our interest, namely plumbing of 2-disc bundles over 2-manifolds. See [OPP16] for a more general notion of plumbing and a historical evolution of this concept.

Let $\xi = (E, p, M)$ and $\kappa = (E', p', N)$ be two smooth n-disc bundles over smooth connected n-manifolds M and N. Let $A \cong \mathbb{D}^n$ be a neighbourhood of a point in M and take a trivialization

$$\zeta \colon E_A \to \mathbb{D}^n \times \mathbb{D}^n$$

where E_A is the total space of the bundle ξ restricted to A, such that the following diagram commutes:

$$
\begin{array}{ccc}
E_A & \overset{\cong}{\longrightarrow} & \mathbb{D}^n \times \mathbb{D}^n \\
{\scriptstyle p}\big\downarrow & & \big\downarrow{\scriptstyle p_2} \\
A & \overset{\cong}{\longrightarrow} & \mathbb{D}^n
\end{array}
$$

where $p_2(x, y) = x$ is the projection on the second coordinate.

Similarly, let $B \cong \mathbb{D}^n$ be a neighbourhood of a point in N and take a trivialization

$$\eta \colon E_B' \to \mathbb{D}^n \times \mathbb{D}^n.$$

Let $\chi \colon \mathbb{D}^n \times \mathbb{D}^n \to \mathbb{D}^n \times \mathbb{D}^n$ be defined by the change of factors $\chi(x, y) = (y, x)$ and let $\vartheta \colon E_B' \to E_A$ be the composition given by

$$\vartheta \colon E_B' \overset{\eta}{\longrightarrow} \mathbb{D}^n \times \mathbb{D}^n \overset{\chi}{\longrightarrow} \mathbb{D}^n \times \mathbb{D}^n \overset{\zeta^{-1}}{\longrightarrow} E_A$$

Fig. 1.14 Plumbing P^{2n} of E and E'

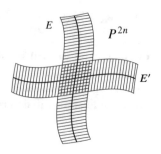

Definition 1.5.1 The **plumbing** of E and E' is defined as the identification

$$P^{2n} = E \bigcup_{\vartheta} E' .$$

The (common) images of E_A and E'_B on the quotient P^{2n} is called the **plumbing polydisc** Note that the identification ϑ matches the base of one bundle with the fibre of the other (see Fig. 1.14).

The space P^{2n} is a topological $2n$-manifold with boundary and is close to being a smooth manifold, but it has "corners". There is a canonical way to smooth these corners and so to produce P^{2n} as a smooth manifold (see [Mil07a, pp. 86–87] and [HNK71, § 8]).

We will describe how to plumb several bundles together according to a finite tree (more generally a connected graph).

Let T be a tree. For each vertex in T one takes a n-disc bundle over $\mathbb{S}^n$, a plumbing of two bundles is made if and only if there is an edge joining the corresponding vertices.

If several edges of T meet in one vertex v, one chooses the corresponding neighbourhoods in $\mathbb{S}^n$ to be disjoint. A theorem of Thom (see [Mil07b, Th. 1.1]) assures that the plumbing is independent of the choice of these neighbourhoods.

Example 1.5.2 Given the tree T in Fig. 1.15, let us take a trivial 1-disc bundle over $\mathbb{S}^1$ for each vertex in T and make the plumbing of two of them when there is an edge in T joining the corresponding vertices.

Fig. 1.15 Graph T, where each vertex represents a trivial 1-disc bundle of $\mathbb{S}^1$

Fig. 1.16 Plumbing $P^2(T)$
according to the tree T

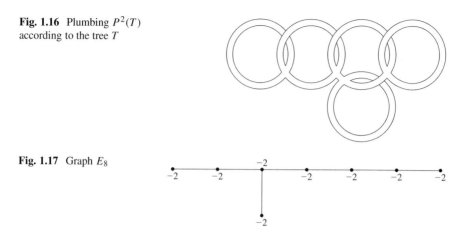

Fig. 1.17 Graph E_8

Let $P^2(T)$ be the result of the plumbing (see Fig. 1.16).

We will now restrict to the case in which the bundles are 2-disc bundles over 2-manifolds.

Let T be a tree weighted in each vertex v by two integers: e_v and $g_v \geq 0$. Let ξ_v be a 2-disc bundle with Euler number e_v over the surface of genus g_v and let $P^4(T)$ be the 4-manifold with boundary obtained by the plumbing according to T; i.e., to plumb two bundles ξ_v and $\xi_{v'}$ when there is an edge in T joining the corresponding vertices v and v'.

Example 1.5.3 Let E_8 be the tree in Fig. 1.17 weighted at each vertex with -2 (see [HNK71, pp. 61,62]). The other weight is not written because it is equal to zero, then we consider -2 as the Euler number of the 2-disc bundles taken for each vertex in T; i.e., we take a 2-disc bundle over the sphere $\mathbb{S}^2$ for each vertex in T.

Figure 1.18 presents schematically the plumbing according to the tree E_8.

In fact, the boundary of the manifold $P^4(E_8)$ is the Poincaré sphere mentioned in Example 1.3.5 (see [KS77, Description 1, p. 114]).

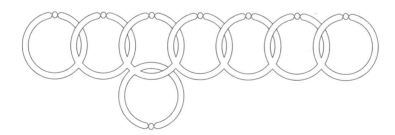

Fig. 1.18 Plumbing $P^4(E_8)$ according to the tree E_8

In [PP07] Popescu-Pampu describes the relation between 3-manifolds which are the boundary of the plumbing of 2-disc bundles over 2-manifolds, graph manifolds and the JSJ decomposition. In the lecture notes by Walter Neumann [Neu19, §4.3] in the present volume, plumbing will be used to describe Seifert manifolds.

1.5.2 Resolution and Blowup

Here we recall that a curve singularity can be resolved by successive blowups.

Definition 1.5.4 Let V be an analytic space and let R be the regular points of V. A **resolution of the singularities of** V consists of a complex manifold M and a proper analytic map $\pi: M \to V$ such that π is biholomorphic on the inverse image of R and such that $\pi^{-1}(R)$ is dense in M.

Let S be a complex surface and let $p \in S$, the **blowup of** p **in** S consists on constructing a new surface B and a map $\phi: B \to S$ such that $\phi^{-1}(p)$ is a curve E, ϕ gives an isomorphism between $B \setminus E$ and $S \setminus \{p\}$, and the points in E correspond to different directions in S at p.

Firstly we do it for the case $S = \mathbb{C}^2$ and $p = (0, 0)$.

Consider the projective space $\mathbb{CP}^1$ of complex lines through the origin in $\mathbb{C}^2$ and define the map

$$\ell: \mathbb{C}^2 \setminus \{(0, 0)\} \to \mathbb{CP}^1$$

$$x \mapsto \ell_x$$

where ℓ_x is the complex line which passes through x and the origin. Let B be the closure of the graph of ℓ in $\mathbb{C}^2 \times \mathbb{CP}^1$ and $\phi: B \to \mathbb{C}^2$ the restriction to B of the projection onto the first factor. Notice that for any point $x \in \mathbb{C}^2 \setminus \{(0, 0)\}$ the preimage $\phi^{-1}(x)$ only consists of the point (x, ℓ_x), while $\phi^{-1}(0, 0) = \{(0, 0)\} \times \mathbb{CP}^1$, so $\phi^{-1}(0, 0)$ is a curve E isomorphic to $\mathbb{CP}^1$ called the **exceptional curve** of the blowup.

For a general surface S we can take a holomorphic chart around a smooth point p, this gives a biholomorphic equivalence between a neighbourhood U of p in S and a neighbourhood V of $(0, 0) \in \mathbb{C}^2$. The blowup of S is obtained by gluing together $S \setminus \{p\}$ and $\phi^{-1}(V)$ using the equivalence of $U \setminus \{p\}$ and $\phi^{-1}(V \setminus \{(0, 0)\})$ via the equivalence of each with $V \setminus \{(0, 0)\}$. We need this more general case because after blowing up the origin of $\mathbb{C}^2$ to get B we will need to blow up a point p_1 in B to get a surface B_1, and then, we will need to blow up a point p_2 in B_1, and so on. All the computations can be made taking local coordinates in these surfaces, consult [Wal04, §3.2] for details.

Theorem 1.5.5 ([Lau71, Theorem 1.1]) *Let V be a 1-dimensional subvariety in a complex surface S. There exists a complex surface M obtained from S by successive blowups, a map $\pi: M \to S$, such that if R is the set of regular points on V,*

$\pi : \overline{\pi^{-1}(R)} \to V$ *is a resolution of the singularities of V. Locally, M is obtained from S by only a finite number of blowups.*

Definition 1.5.6 The map $\pi : M \to S$ is an **embedded resolution** of V.

In the lecture notes by Anne Pichon [Pic19] in the present volume, resolutions will be use to describe the Lipschitz geometry of a complex singularity.

1.5.3   Description of the Milnor Fibre in the Resolution

Let $f : (\mathbb{C}^2, 0) \to (\mathbb{C}, 0)$ be a holomorphic reduced germ with isolated singularity at the origin and let $V = f^{-1}(0)$. Let $\mathcal{U}$ be a neighbourhood of the origin in $\mathbb{C}^2$ and let $\pi : \mathcal{W} \to \mathcal{U}$ be a embedded resolution of V at the origin, given by a finite number of blowups in points.

Definition 1.5.7 Let $\widehat{E} = \pi^{-1}(0) \subset \pi^{-1}(\mathcal{U})$ be the **exceptional divisor** of π. Let $\widetilde{E_0} \subset \pi^{-1}(\mathcal{U})$ be the closure of the complement of $\widehat{E}$ in $\pi^{-1}(V \cap \mathcal{U})$; i.e.,

$$\widetilde{E_0} = \overline{\pi^{-1}(V \cap \mathcal{U}) \setminus \widehat{E}},$$

$\widetilde{E_0}$ is called the **strict transform** of $f^{-1}(0)$. Let us denote by E_i the irreducible components of $\widehat{E}$, with $i = 1, \ldots, k$.

Each irreducible component E_i of $\widehat{E}$ is non-singular and $\widehat{E}$ has **normal crossings**; i.e., if $i \neq j$, E_i intersects E_j in at most one point where they meet transversely and no three components of $\widehat{E}$ intersect. Also $\widehat{E} \cup \widetilde{E_0}$ has normal crossings (see Fig. 1.19).

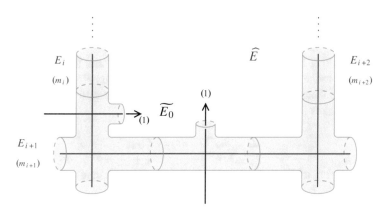

Fig. 1.19 The preimage $\pi^{-1}(\mathcal{U})$

Definition 1.5.8 Let $p_{i,j}$ be the intersection $E_i \cap E_j$ when this intersection is not empty. A point $p \in E_i$ is called **smooth** if $p \notin E_j$ for any $j \neq i$.

Let i with $0 < i \leq k$, at each smooth point $p \in E_i$ of $\widehat{E}$ there exist some local coordinates (u, v) centred at p such that $u = 0$ is a local equation of E_i and, locally,

$$(f \circ \pi)(u, v) = u^{m_i} \iota(u, v) \tag{1.26}$$

where ι is an unity in the ring of convergent power series $\mathbb{C}\{\{u, v\}\}$. After performing a change of coordinates, one can assume that $\iota = 1$.

Definition 1.5.9 Let i with $0 < i \leq k$, the order m_i of $f \circ \pi$ in a small neighbourhood of a smooth point of E_i is called the **multiplicity** at E_i.

Let b be the number of branches of f. If the neighbourhood $\mathcal{U}$ is small enough, $\widetilde{E}_0$ consists of b curves transverse to $\widehat{E}$. Let b_i be the number of branches of f intersecting E_i. As f is reduced, the multiplicity m_0 at each component of $\widetilde{E}_0$ is equal to one.

We choose an open neighbourhood of E_i such that it is a fibration of discs with base E_i The fibres of this fibration are called the **fibres** of E_i.

By Theorem 1.4.19, there exists $\varepsilon > 0$ small enough and δ with $0 < \delta \ll \varepsilon$ also small such that f restricted to the Minor tube $N(\varepsilon, \delta)$ defines a locally trivial fibration over $\partial \mathbb{D}_\delta$ (see Fig. 1.13).

Let us choose δ small enough such that if $z \in \mathbb{C}$ with $|z| = \delta$, then the manifold $\mathcal{F}_z = \pi^{-1}\big(f^{-1}(z) \cap \mathbb{B}^4_\varepsilon\big)$ is transverse to the fibres of the irreducible components of $\widehat{E}$ around a small neighbourhood of $\widetilde{E}_0$.

Let $z \in \mathbb{C}$ be fixed and let

$$\mathcal{F} = \mathcal{F}_z, \qquad\qquad\qquad X = \pi^{-1}\big(f^{-1}(\mathbb{B}^2_\delta) \cap \mathbb{B}^4_\varepsilon\big),$$

$$\mathcal{F}_0 = \pi^{-1}\bigg(\mathbb{S}^3_\varepsilon \cap \bigcup_{0 \leq t \leq 1} f^{-1}(tz)\bigg), \qquad E_0 = \widetilde{E}_0 \cap X.$$

The manifold X is a closed neighbourhood of $\widehat{E}$ in $\mathcal{W}$ and the boundary ∂X is diffeomorphic to the sphere $\mathbb{S}^3_\varepsilon$ (see Fig. 1.20). The boundary $\partial \mathcal{F}$ is equal to $\mathcal{F} \cap \mathcal{F}_0$ and there is an isotopy between the identity in $\mathbb{S}^3_\varepsilon$ and the diffeomorphism on the sphere which takes $\partial \mathcal{F}$ to the boundary

$$L = \partial(\mathcal{F} \cup \mathcal{F}_0) = \partial \widetilde{E}_0.$$

Let $\mathcal{U}_0$ be a union of fibres of E_i (with $E_i \cap E_0 \neq \emptyset$) such that

$$\pi\big(\mathcal{U}_0 \cap \partial X\big) \subset \mathbb{S}^3_\varepsilon \cap f^{-1}(\mathbb{B}^2_\delta).$$

Fig. 1.20 Milnor fibration in the embedded resolution of f

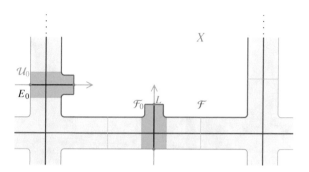

Now we construct a manifold $\overline{X}$ in the following way: For all i with $0 < i \leq k$, let $\overline{X}_i$ be the total space of a fibration of real discs with base E_i, $\overline{X}_i$ is isomorphic to a closed neighbourhood of E_i in X.

Then $\overline{X}$ will be the manifold obtained after doing plumbing in a neighbourhood of the intersection points $p_{i,j}$, let $\mathbf{B}_{i,j}$ be the corresponding plumbing polydisc and let $\mathbf{T}_{i,j}$ be the **plumbing torus** defined as the intersection $\mathbf{B}_{i,j} \cap \partial \overline{X}$.

Proposition 1.5.10 ([DM92, Prop 1.4]) *There exists a diffeomorphism with corners* $\rho : \overline{X} \to X$ *such that:*

- *ρ is the identity on $\widehat{E}$,*
- *$\rho^{-1}(E_0)$ is a union of fibres of the $\overline{X}_i$; these fibres are outside of the plumbing polydiscs $\mathbf{B}_{i,j}$ for any i, j,*
- *If Δ is a fibre of E_i outside of $\mathcal{U}_0$, $\rho^{-1}(\Delta)$ is a fibre of $\overline{X}_i$.*

Let $X_i = \rho(\overline{X}_i)$, $B_{i,j} = \rho(\mathbf{B}_{i,j})$ and $T_{i,j} = \rho(\mathbf{T}_{i,j})$ (see Fig. 1.21), where $T_{i,j}$ is the image under ρ of the plumbing torus $\mathbf{T}_{i,j}$ with $0 < j < i \leq k$ and $T_{i,0}$ will denote the union of tori in the boundary ∂X_i such that

$$\bigcup_{i>0} T_{i,0} = \pi^{-1}\left(\mathbb{S}_\varepsilon^3 \cap f^{-1}(\mathbb{S}_\delta^1)\right).$$

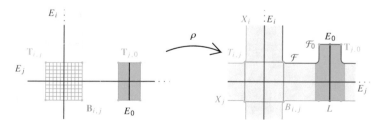

Fig. 1.21 The plumbing of the fibred X_i's is diffeomorphic to X

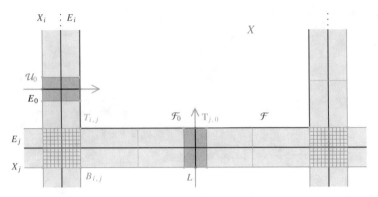

Fig. 1.22 The manifold X as a plumbing

Then we obtain Fig. 1.22.

Definition 1.5.11 An orientation preserving diffeomorphism $h: \mathcal{F} \to \mathcal{F}$ is called **quasi-periodic** if there is a family C of disjoint simple closed curves in $\mathcal{F}$ and a small neighbourhood $\mathcal{U}(C) \subset \mathcal{F}$ of C such that

- for each curve $c \in C$, $\mathcal{U}(c)$ is a small annulus neighbourhood of c in $\mathcal{F}$,
- for any pair of curves $c_i, c_j \in C$, we have that $\mathcal{U}(c_i) \cap \mathcal{U}(c_j) = \emptyset$,
- $h(C) = C$
- $h(\mathcal{U}(c)) = \mathcal{U}(c)$,
- the restriction of h to the complement of

$$\mathcal{U}(C) = \bigcup_{c \in C} \mathring{\mathcal{U}}(c)$$

is periodic, where $\mathring{\mathcal{U}}(c)$ is the interior of $\mathcal{U}(c)$.

The family C is called a **reduction system** of curves for the diffeomorphism h.

Let $h: \mathcal{F} \to \mathcal{F}$ be a quasi-periodic diffeomorphism and let C be a reduction system for h. Let $c \in C$ be a simple closed curve in $\mathcal{F}$ an let $\mathcal{U}(c) \subset \mathcal{F}$ be as before, then there exists an orientation preserving diffeomorphism $\mu: [-1, 1] \times \mathbb{S}^1 \to \mathcal{U}(c)$ such that $\mu(\{0\} \times \mathbb{S}^1) = c$.

Let N be the smallest integer such that

$$h^N|_{\mathcal{F} \setminus \mathcal{U}(C)} = id_{\mathcal{F} \setminus \mathcal{U}(C)},$$

then the restriction h^N to $\mathcal{U}(c)$ is a **Dehn twist** and the restriction of h is characterised by a rational number t in the following way:

Consider the path γ in $\mathcal{U}(c)$ defined by $\gamma(s) = \mu(s, e^{i\theta})$ where θ is fixed and $s \in [-1, 1]$. We orient γ by $[-1, 1]$ and then, we orient c in such a way that

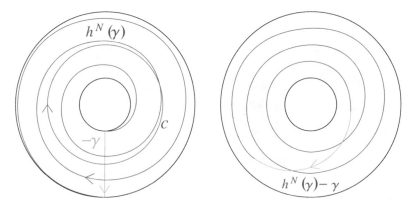

Fig. 1.23 The cycles Kc and $h^N(\gamma) - \gamma$ are homologous

$\gamma \cdot c = +1$ in $H_1(\mathcal{U}(c), \mathbb{Z})$. Then there exists $K \in \mathbb{Z}$ such that the cycles Kc and $h^N(\gamma) - \gamma$ are homologous in $\mathcal{U}(c)$ (see Fig. 1.23).

Definition 1.5.12 The rational number $t = \frac{K}{N}$ is called the **twist number** of h along c.

Let $\mathcal{F}_i = \mathcal{F} \cap \partial X_i$; we then have $\mathcal{F} = \cup_{i>0} \mathcal{F}_i$.

The intersection of the fibres of E_i with the boundary ∂X_i endows ∂X_i with a fibration in circles. Let $h_i : \mathcal{F}_i \to \mathcal{F}_i$ be the first return diffeomorphism on $\mathcal{F}_i$ along the fibres of ∂X_i.

By Eq. (1.26), in the local coordinates (u, v), a fibre of E_i over a smooth point is given by the equation $v = c$, where c is a constant. Then the intersection of this fibre with F_i consists of m_i points (u, c), where c is solution of

$$u^{m_i} = c .$$

The diffeomorphism h_i permutes cyclically these m_i points and $h_i^{m_i}$ is the identity, then the order of h_i is m_i (see Fig. 1.24).

Let $h_0 \colon \mathcal{F}_0 \to \mathcal{F}_0$ be the identity on $\mathcal{F}_0$. Now we take the following family of curves:

$$C = \cup_{0 \le j < i} \mathcal{F}_i \cap \mathcal{F}_j .$$

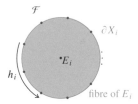

Fig. 1.24 The diffeomorphism h_i in a fibre of E_i

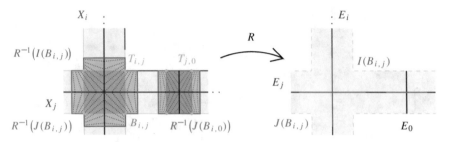

Fig. 1.25 The deformation retract R from X to E

Let i, j be such that $E_i \cap E_j \neq \emptyset$, then let $\widehat{m}_{i,j} = \gcd(m_i, m_j)$. If $0 < j < i$, $C_{i,j} = \mathcal{F}_i \cap \mathcal{F}_j$ is a collection of $\widehat{m}_{i,j}$ simple closed curves of $\mathcal{F}$.

Now we will construct the neighbourhood $\mathcal{U}(C)$: For each point $p_{i,j}$ (with $0 \leq j < i \leq k$), we choose closed discs $I(B_{i,j})$ and $J(B_{i,j})$, neighbourhoods of $B_{i,j} \cap E_i$ in E_i and $(B_{i,j} \cap E_j)$ in E_j respectively.

Let $E = \widehat{E} \cup E_0$. There exists a deformation retract $R \colon X \to E$ (see Fig. 1.25) such that

- the plumbing torus $T_{i,j}$ goes to the point $p_{i,j}$,
- if $x \in (E_i \setminus \bigcup I(B_{i,j}))$, $R^{-1}(x)$ is the fibre of E_i in the point x,
- if $x \in I(B_{i,j}) \setminus \{p_{i,j}\}$, $R^{-1}(x)$ is a curve transverse to E_i at the point x.

Let

$$V = \bigcup_{0 \leq j < i} \left(R^{-1}(I(B_{i,j})) \cup R^{-1}(J(B_{i,j})) \right) \quad \text{and} \quad \mathcal{U}(C) = V \cap \mathcal{F}.$$

The next step is "to glue" the diffeomorphisms h_i. On the $\widehat{m}_{i,j}$ curves in $C_{i,j}$, h_i is a permutation of these curves and $h_i^{\widehat{m}_{i,j}}$ is the identity. On the boundary of the annuli $\mathcal{U}(C_{i,j}) \cap \mathcal{F}_i$, the diffeomorphism $h_i^{\widehat{m}_{i,j}}$ is a rotation and $h_i = h$.

Proposition 1.5.13 ([MMA11, Theorem 7.3 (iv)], [BFP19, Proposition 4.12]) *Let $n_i = m_i/\widehat{m}_{i,j}$ and $n_j = m_j/\widehat{m}_{i,j}$. The twist number of $h^{\widehat{m}_{i,j}}$ restricted to the anulus $\mathcal{U}(c)$, with c a curve in $C_{i,j}$ is given by $t_{i,j} = -\frac{1}{n_i n_j}$.*

Then, we extend h from the boundary $\partial(\mathcal{U}(C_{i,j}) \cap \mathcal{F}_i)$ to the boundary $\partial \mathcal{F}_i$ by an isotopy; this is possible because $\mathcal{U}(C_{i,j}) \cap \mathcal{F}_i$ is a disjoint union of annuli.

Then we have the following result.

Proposition 1.5.14 ([DM94, Prop. 1.5]) *Let h be the monodromy of the Milnor fibration of f. Its restriction $h|_{\mathcal{F}_i}$ is the diffeomorphism h_i.*

Remark 1.5.15 The representative of the monodromy found in this way depends on the resolution; in order to obtain a canonical quasi-periodic monodromy, we take the minimal embedded resolution and we proceed in the same way.

Let us finish this section with two results about the topology of $\mathcal{F}$.

For all i with $0 < i \leq k$, let $r_i = \gcd(\widehat{m}_{i,j})_{E_i \cap E_j \neq \emptyset}$, let $\overset{\circ}{\mathcal{F}}_i$ be the interior of $\mathcal{F}_i$ and let $\overset{\circ}{E}_i$ be the set of smooth points of E_i. Let v be the restriction to $\mathcal{F}$ of the deformation retract R and let $v_i : \overset{\circ}{\mathcal{F}}_i \to \overset{\circ}{E}_i$ be the restriction to $\overset{\circ}{\mathcal{F}}_i$.

Proposition 1.5.16 ([DM94, Prop. 1.6]) *The restriction v_i is the finite cyclic covering of order m_i, defined by the homomorphism*

$$\rho_i : H_1(\overset{\circ}{E}_i; \mathbb{Z}) \longrightarrow \mathbb{Z}_{m_i}$$

$$([C_{i,j}]) \longmapsto \omega_j$$

where $\omega_j \equiv -m_j \pmod{m_i}$.

Proposition 1.5.17 ([DM92, Prop 1.11])

1. *The number of connected components of $\mathcal{F}_i$ is r_i,*
2. *$\mathcal{F}_0$ is the disjoint union of b annuli, where b is the number of branches of f.*
3. *if $s_i = 1$, $\mathcal{F}_i$ is a union of discs and if $s_i = 2$, $\mathcal{F}_i$ is a union of annuli,*

where s_i is the number of irreducible components E_j of $\widehat{E}$ which intersect E_i.

Acknowledgments The second author is Regular Associate of the Abdus Salam International Centre for Theoretical Physics, Trieste, Italy. He was supported by UNAM-DGAPA-PASPA sabbatical scholarship and by CONACYT 253506. The second author would like to thank the organizers of the *International School on Singularities and Lipschitz Geometry* for the invitation to give the course *Milnor Fibration* and to write these lecture notes. Both authors thank the referee for her/his comments which improved the presentation of this chapter.

References

[AC11] H. Aguilar-Cabrera, Topology of singularities of real analytic functions. PhD thesis, Universidad Nacional Autónoma de México, April 2011 2

[AMR88] R. Abraham, J.E. Marsden, T. Ratiu, *Manifolds, Tensor Analysis, and Applications*. Applied Mathematical Sciences, vol. 75, 2nd edn. (Springer, Berlin, 1988) 5, 6

[BFP19] A. Belotto da Silva, L. Fantini, A. Pichon, Inner geometry of complex surfaces: a valuative approach, May 2019. ArXiv:1905.01677 [math.AG] 40

[BJ82] T. Bröcker, K. Jänich, *Introduction to Differential Topology* (Cambridge University Press, Cambridge, 1982) 4, 5

[BK86] E. Brieskorn, H. Knörrer, *Plane Algebraic Curves*. Modern Birkhäuser Classics (Birkhäuser/Springer Basel AG, Basel, 1986). Translated from the German original by J. Stillwell, 2012 reprint of the 1986 edition 13

[Bre93] G.E. Bredon, *Topology and Geometry, Graduate Texts in Mathematics*, vol. 139 (Springer, New York, 1993) 31

[BV72] D. Burghelea, A. Verona, Local homological properties of analytic sets. Manuscr. Math. **7**, 55–66 (1972) 14, 15, 18

[CMSS12] J.L. Cisneros-Molina, J. Seade, J. Snoussi, Milnor fibrations and the concept of d-regularity for analytic map germs, in *Real and Complex Singularities, Contemp. Math.*, vol. 569 (Amer. Math. Soc., Providence, 2012), pp. 1–28 2

[DM92] P. Du Bois, F. Michel, Filtration par le poids et monodromie entière. Bull. Soc. Math. France **120**(2), 129–167 (1992) 2, 31, 37, 41

[DM94] P. Du Bois, F. Michel, The integral Seifert form does not determine the topology of plane curve germs. J. Algebr. Geom. **3**, 1–38 (1994) 40, 41

[GP74] V. Guillemin, A. Pollack, *Differential Topology* (Prentice-Hall, Englewood Cliffs, 1974) 2, 3, 6

[Hir73] H. Hironaka, Subanalytic sets, in *Number Theory, Algebraic Geometry and Commutative Algebra, in Honor of Yasuo Akizuki*, Kinokuniya, Tokyo (1973), pp. 453–493 14

[Hir77] H. Hironaka, Stratification and flatness, in *Real and Complex Singularities (Proc. Ninth Nordic Summer School/NAVF Sympos. Math., Oslo, 1976)* (Sijthoff and Noordhoff, Alphen aan den Rijn, 1977), pp. 199–265 26

[HNK71] F. Hirzebruch, W.D. Neumann, S.S. Koh, *Differentiable Manifolds and Quadratic Forms* (Marcel Dekker, New York, 1971). Appendix II by W. Scharlau, Lecture Notes in Pure and Applied Mathematics, vol. 4 31, 32, 33

[KS77] R.C. Kirby, M.G. Scharlemann, Eight faces of the Poincaré homology 3-sphere, in *Geometric Topology*, ed. by J.C. Cantrell (Academic Press, New York, 1977), pp. 113–146. Proceedings of the 1977 Georgia Topology Conference 13, 33

[Lau71] H.B. Laufer, *Normal Two-Dimensional Singularities* (Princeton University Press, Princeton; University of Tokyo Press, Tokyo, 1971). Annals of Mathematics Studies, No. 71 34

[Lê77] D.T. Lê, Some remarks on relative monodromy, in *Real and Complex Singularities (Proc. Ninth Nordic Summer School/NAVF Sympos. Math., Oslo, 1976)*, ed. by P. Holm (Sijthoff and Noordhoff, Alphen aan den Rijn, 1977), pp. 397–403 27

[Loo84] E.J. N. Looijenga, *Isolated Singular Points on Complete Intersections*. London Mathematical Society Lecture Note Series, vol. 77 (Cambridge University Press, Cambridge, 1984) 18

[MMA11] Y. Matsumoto, J.M. Montesinos-Amilibia, *Pseudo-Periodic Maps and Degeneration of Riemann Surfaces*. Lecture Notes in Mathematics, vol. 2030 (Springer, Heidelberg, 2011) 40

[Mil65] J. Milnor, *Topology from the Differentiable Point of View* (The University Press of Virginia, Charlottesville, 1965) 2

[Mil66] J. Milnor, On isolated singularities of hypersurfaces. Preprint, June 1966 27

[Mil68] J. Milnor, *Singular Points of Complex Hypersurfaces*. Annals of Mathematics Studies, No. 61 (Princeton University Press, Princeton, 1968) 2, 12, 16, 17, 18, 21, 27, 29, 30

[Mil07a] J. Milnor, Differentiable manifolds which are homotopy spheres, in *Collected papers of John Milnor*, vol. III. Differential Topology (American Mathematical Society, Providence, 2007), pp. 65–88 32

[Mil07b] J. Milnor, Lectures on differentiable structures, in *Collected papers of John Milnor*, vol. III. Differential Topology (American Mathematical Society, Providence, 2007), pp. 177–190 32

[Neu81] W.D. Neumann, A calculus for plumbing applied to the topology of complex surface singularities and degenerating complex curves. Trans. Am. Math. Soc. **268**(2), 299–344 (1981) 13

[Neu19] W.D. Neumann, 3-manifolds and links of singularities, in *Introduction to Lipschitz Geometry of Singularities. Lecture Notes of the International School on Singularity Theory and Lipschitz Geometry, Cuernavaca, June 2018*, vol. 2280 (Springer, Cham, 2020). https://doi.org/10.1007/978-3-030-61807-0 13, 34

[OPP16] B. Ozbagci, P. Popescu-Pampu, Generalized plumbings and Murasugi sums. Arnold Math. J. **2**(1), 69–119 (2016) 31

[Pic19] A. Pichon, An introduction to Lipschitz geometry of complex singularities, in *Introduction to Lipschitz Geometry of Singularities. Lecture Notes of the International School on Singularity Theory and Lipschitz Geometry, Cuernavaca, June 2018*, vol. 2280 (Springer, Cham, 2020). https://doi.org/10.1007/978-3-030-61807-0 35

[PP07] P. Popescu-Pampu, The geometry of continued fractions and the topology of surface singularities, in *Singularities in Geometry and Topology 2004*. Adv. Stud. Pure Math., vol. 46 (Math. Soc. Japan, Tokyo, 2007), pp. 119–195 34

[Ran98] A. Ranicki, *High-Dimensional Knot Theory*. Springer Monographs in Mathematics (Springer, New York, 1998). Algebraic surgery in codimension 2, With an appendix by E. Winkelnkemper 25

[Rua20] M.A.S. Ruas, Basics on Lipschitz geometry, in *Introduction to Lipschitz Geometry of Singularities. Lecture Notes of the International School on Singularity Theory and Lipschitz Geometry, Cuernavaca, June 2018*, vol. 2280 (Springer, Cham, 2020). https://doi.org/10.1007/978-3-030-61807-0 18

[Sea06] J. Seade, *On the Topology of Isolated Singularities in Analytic Spaces. Progress in Mathematics*, vol. 241 (Birkhäuser, Basel, 2006) 25

[Sea18] J. Seade, On Milnor's fibration theorem and its offspring after 50 years. Bull. Am. Math. Soc. (2018). https://doi.org/10.1090/bull/1654 2

[Thu82] W.P. Thurston, Three-dimensional manifolds, Kleinian groups and hyperbolic geometry. Bull. Am. Math. Soc. **6**(3), 357–381 (1982) 13

[Tro20] D. Trotman, Stratifications, equisingularity and triangulation, in *Introduction to Lipschitz Geometry of Singularities. Lecture Notes of the International School on Singularity Theory and Lipschitz Geometry, Cuernavaca, June 2018*, vol. 2280 (Springer, Cham, 2020). https://doi.org/10.1007/978-3-030-61807-0 13

[Ver76] J.-L. Verdier, Stratifications de Whitney et théorème de Bertini-Sard. Invent. Math. **36**, 295–312 (1976) 8

[Wal04] C.T.C. Wall, *Singular Points of Plane Curves*. London Mathematical Society Student Texts, vol. 63 (Cambridge University Press, Cambridge, 2004) 34

[Whi65] H. Whitney, Tangents to an analytic variety. Ann. Math. (2) **81**, 496–549 (1965) 14

[Win73] H.E. Winkelnkemper, Manifolds as open books. Bull. Am. Math. Soc. **79**, 45–51 (1973) 25

[Wol64] J.A. Wolf, Differentiable fibre spaces and mappings compatible with Riemannian metrics. Mich. Math. J. **11**, 65–70 (1964) 6

Chapter 2
A Quick Trip into Local Singularities of Complex Curves and Surfaces

Jawad Snoussi

Abstract In these notes we give a summary of some of the properties of curve and surface singularities needed in the study of Lipschitz geometry of singular varieties. In particular, we describe normalization and resolution processes, and we introduce the concepts of polar curves and exceptional tangents for surfaces.

2.1 Introduction

Low dimensional singularities are an appropriate field for experiencing and understanding new tools or concepts. This is the case of the metric study of singularities and Lipschitz geometry of singular varieties.

The wide understanding of the geometry, topology, analytic and algebraic aspects of low-dimensional singularities is then fundamental. An extensive literature is available for this purpose.

However we took the challenge in these notes to propose a short text with an introduction to curve and surface singularities. We are far from pretending to expose the state of art in this theme. Instead we try to describe some of the aspects of one- and two-dimensional singularities that are most used in the actual study of their metric aspects. We give almost no proof, we prefer instead to refer to some of the appropriate references. We focus on the description of the objects, concepts, tools and procedures that are also most used in the other lectures of this school in these dimensions.

We present some examples, with detailed computations, as we do believe they are the best way to understand the procedures that are followed in many of the other texts in this volume.

The text is divided into three parts: some general aspects of local singularities, such as normality and normalization, tangent cone, secants, blowups. Then we

J. Snoussi (✉)

Instituto de Matemáticas, Universidad Nacional Autónoma de México, Cuernavaca, Morelos, Mexico

e-mail: jsnoussi@im.unam.mx

© The Author(s), under exclusive license to Springer Nature Switzerland AG 2020
W. Neumann, A. Pichon (eds.), *Introduction to Lipschitz Geometry of Singularities*,
Lecture Notes in Mathematics 2280, https://doi.org/10.1007/978-3-030-61807-0_2

describe these concepts in the case of plane curves, with an explicit description of normalization, point blowup and resolution. We also make a brief mention of Newton-Puiseux parametrizations and state the theorem of the equivalence between the topological type, the embedded resolution and the combinatorial data given by the characteristic exponents.

For surfaces, we also describe the normalization and the point blowup. Then we discuss some procedures of resolution of singularities and give explicit examples of resolution by Jung's method. We also establish the relation between base points of hyperplane sections and polar curves with the point blowup and Nash modification and we define and describe the exceptional tangents, since they are fundamental objects in the Lipschitz geometry of normal surface singularities.

These notes are extracted from a series of lectures given by the author at the "International school on singularities and Lipschitz geometry". The author is grateful to the audience for interesting feedback, and also would like to thank Otoniel da Silva for fruitful discussions and strong help with technical aspects.

2.2   General Settings

By an **analytic space** X we mean a space defined in an open set U of $\mathbb{C}^N$ as the zero set of a finite number of functions $f_1, \ldots, f_r$ holomorphic in U. This space comes together with its ring of holomorphic functions defined as the quotient $O(U)/(f_1, \ldots, f_r)$, where $O(U)$ is the ring of holomorphic functions in U and $(f_1, \ldots, f_r)$ is the ideal generated by $f_1, \ldots, f_r$ in $O(U)$.

By a **germ** of analytic set (X, x) we mean the equivalence class of analytic spaces under the relation where two analytic spaces defined in open sets containing x are equivalent if they coincide on a common smaller open set containing that point. The ring associated to a germ (X, x) is the local ring $O_{X,x} := O_{N,0}/(f_1, \ldots, f_r)$, quotient of the local ring of holomorphic functions in $\mathbb{C}^N$ near x by the ideal generated by the equations of a representative of (X, x). The maximal ideal of $O_{X,x}$ will be denoted by $M_{X,x}$.

Consider a germ of analytic space (X, x); its **dimension** can be defined as the Krull dimension of its local ring $O_{X,x}$. The point x is singular if and only if the ring $O_{X,x}$ is not regular, in other words, the ideal $M_{X,x}$ cannot be generated by $\dim(X, x)$ functions. This property can also be characterized by the Jacobian criterion, that is, if $(X, x) \subset (\mathbb{C}^N, 0)$ is defined by $f_1, \ldots, f_r$ then, the rank of the Jacobian matrix of the f_i's at x is strictly smaller than $N - \dim(X, x)$.

We will say that an analytic space X is **singular** at a point x if the corresponding germ (X, x) is singular. A point will be called **regular** or **smooth** when it is non-singular.

The direction of tangent space of an analytic space X at a regular point $x \in X$ is the vector space defined by the kernel of the Jacobian matrix of the functions $f_1, \ldots, f_r$ at x. It has the same dimension as (X, x). When x is singular the dimension of this kernel is strictly bigger than the dimension of (X, x).

2.2.1 Multiplicity and Tangent Cone

The **tangent cone** to X at x is the space defined by the ideal of initial forms of the ideal $(f_1, \ldots, f_r)$ in $O_{X,x}$. More precisely, consider an element $f \in I$ for some ideal $I \subset O_{N,x}$; the initial form of f at x, is the polynomial $In_x(f)$ defined as the homogeneous polynomial of lowest degree in the Taylor expansion of f around x. The initial ideal of I at x is the ideal generated by all the initial forms of elements of I; $In_x(I) :=< \{In_x(f), f \in I\} >$. This ideal is generated by homogeneous polynomials, so the tangent cone is in fact a homogeneous algebraic variety. We will denote it by $T_x X$.

When the point x is non-singular, the tangent cone at x coincides with the tangent space. When X is equidimensional at x the tangent cone has the same dimension as (X, x).

We define the multiplicity of an analytic germ as follows. Let $(X, x) \subset (\mathbb{C}^N, x)$ be analytic of dimension d. Choose a linear space $L \subset \mathbb{C}^N$ of codimension d, in such a way that $L \cap T_x X = \{0\}$. One can prove that a translate L_ϵ of L close to x intersects X transversally in a number of points that does not depend either on ϵ or on the choice of L. This number of points is called the **multiplicity** of (X, x) or the multiplicity of X at x and we denote it by $m(X, x)$.

An analytic germ (X, x) is non-singular if and only if $m(X, x) = 1$. In case (X, x) is a **hypersurface**, i.e., defined by a function $f \in O_{N,x}$, then $m(X, x) = \text{ord}_x(f) := \deg In_x(f)$.

The multiplicity of a germ at a point coincides with the multiplicity of its tangent cone at the same point. For more details and equivalent definitions of the multiplicity, see [dJP00, IV].

2.2.2 Blowups

Consider an ideal $I = (g_1, \ldots, g_k) \subset O_{X,x}$, a representative X of (X, x) and call $V(I)$ the zero locus of I in X. Define the map:

$$\lambda : X \setminus V(I) \to \mathbb{P}^{k-1}$$
$$z \mapsto [g_1(z) : \ldots : g_k(z)]$$

and call X_I the closure of the graph of λ in $X \times \mathbb{P}^{k-1}$. The projection on the first factor induces a map

$$e_I : X_I \to X$$

called the **blowup** of the ideal I in X. It is a proper map which induces an isomorphism outside $V(I)$. The inverse image $e_I^{-1}(V(I))$ is called the **exceptional locus**, or **exceptional divisor** of the blowup e_I. When the support of the blown-up ideal is a point, the inverse image $e_I^{-1}(V(I))$ is also called the **exceptional fiber**. For another description and properties of the blowup see [Har77, p. 163].

The blowup of the maximal ideal $\mathcal{M}_{X,x}$ is also called the blowup of the point x in X. In this case, the exceptional fiber is the projective space associated to the tangent cone, that is $\mathrm{Proj}(T_x X)$. In other words, every line l of the tangent cone passing through the vertex x is a limit of a sequence of secant lines of the form (xz_n), where (z_n) converges to x; see [Whi65, 5.8].

Example 2.2.1 Consider the curve C defined in $\mathbb{C}^2$ by the function $f(x, y) = 0$, where $f(x, y) = y^2 - x^2 - x^3$. The origin is the only singular point of C. The initial form of f is $y^2 - x^2$, which in this case generates the initial ideal associated to the ideal generated by f. So the tangent cone $T_0 C$ is defined by the equation $y^2 - x^2 = 0$; it is the union of two lines.

We will compute the blowup of the origin in C. The maximal ideal is generated by x and y. The map λ is defined from $C \setminus \{0\}$ to $\mathbb{P}^1$ by $\lambda(x, y) = [x : y]$. Its graph in $C \times \mathbb{P}^1$ is defined as:

$$\{((x, y), [u : v]) \in \mathbb{C}^2 \times \mathbb{P}^1, (x, y) \neq (0, 0), (x, y) \in C \text{ and } [x : y] = [u : v]\}$$

$$= \{((x, y), [u : v]) \in \mathbb{C}^2 \times \mathbb{P}^1, (x, y) \neq (0, 0), y^2 - x^2 - x^3 = 0 \text{ and } xv = yu\}.$$

Decomposing the projective space $\mathbb{P}^1$ into affine charts $(u \neq 0) \cup (v \neq 0)$, the closure of the graph of λ can be decomposed as the union of two pieces:

$$U_1 = \text{the closure of } \{((x, y), [u : v]) \in \mathbb{C}^2 \times \mathbb{P}^1, (x, y) \neq (0, 0),$$
$$y = x\tfrac{v}{u} \text{ and } x^2\left(\tfrac{v}{u}\right)^2 - x^2 - x^3 = 0\},$$
$$U_2 = \text{the closure of } \{((x, y), [u : v]) \in \mathbb{C}^2 \times \mathbb{P}^1, (x, y) \neq (0, 0),$$
$$x = y\tfrac{u}{v} \text{ and } y^2 - y^2\left(\tfrac{u}{v}\right)^2 - y^3\left(\tfrac{u}{v}\right)^3 = 0\}.$$

In U_1 the main equation factorizes into $x^2((\tfrac{v}{u})^2 - 1 - x) = 0$ and in U_2 into $y^2(1 - (\tfrac{u}{v})^2 - y(\tfrac{u}{v})^3) = 0$. In both cases the lines defined by $x^2 = 0$ and by $y^2 = 0$ do not belong to the closure space. So they may be removed from the equations and then the charts of the blown-up space U_1 and U_2 are respectively defined by the functions

$$\left(y - x\frac{v}{u}, \left(\frac{v}{u}\right)^2 - 1 - x\right) \text{ and } \left(x - y\frac{u}{v}, 1 - \left(\frac{u}{v}\right)^2 - y\left(\frac{u}{v}\right)^3\right).$$

The blown-up space is a curve in $\mathbb{C}^2 \times \mathbb{P}^1$ and one can check that it is non-singular in both charts.

Let us now compute the exceptional divisor, inverse image of the origin. In U_1, it is defined by $x = y = 0$ and $(\frac{v}{u})^2 = 1$, and in U_2, by $x = y = 0$ and $(\frac{u}{v})^2 = 1$. These are the points $((0, 0), [1 : 1])$ and $((0, 0), [1 : -1])$, each of them appearing in both charts.

The affine lines corresponding to these two projective points in the affine space $\mathbb{C}^2$ with coordinates x and y are given respectively by the equation $x = y$ and $x = -y$. These are precisely the lines of the tangent cone T_0C.

When $Y \subset X$ is an analytic subspace of X the inverse image $e_I^{-1}(Y)$ is called the **total transform** of Y by e_I. We call the **strict transform** of Y the closure of $e_I^{-1}(Y \setminus V(I))$ in X_I.

Note that in the example above, the blowup of the curve C coincides with the strict transform of C under the blowup of the origin in $\mathbb{C}^2$.

2.2.3   Normality and Normalization

An analytic space X is said to be **normal** at x if every meromorphic bounded function on X near x extends to a holomorphic function on X near x; in other words, X satisfies the Riemann extension property near x. This is equivalent to the following algebraic property on integral dependence of rings: the **total ring of fractions** of a ring A is the ring obtained by making invertible all non-zero-divisors of A (see [PT69]); we denote it by $\text{Tot}(A)$. A ring A is said to be **normal** if it is integrally closed in its total ring of fractions; in other words, whenever $f \in \text{Tot}(A)$ satisfies a polynomial monic equation with coefficients in A, then f is actually in A. An analytic space X is normal at x if and only if the local ring $O_{X,x}$ is normal, see [Nar66, VI. 2].

One can prove that a Unique Factorization Domain is normal, so if (X, x) is non-singular then it is normal.

A **normalization** of an analytic space X is a finite (hence proper) map $n : Y \to X$ such that Y is a normal analytic space and n induces an isomorphism outside the non-normal points of X.

We say that an analytic space is **reduced** if its ring of holomorphic functions is **reduced**, i.e., it has no nilpotent element. A normalization of reduced analytic spaces always exists, it is unique up to isomorphism and is characterized by the following **universal property**: every morphism from a normal analytic space to a given analytic space X factors through the normalization. For proofs and more see [dJP00, 4.4].

The **normal closure** of a ring A is the ring of all elements of $\text{Tot}(A)$ that satisfy polynomial monic equations with coefficients in A. One can prove that if X is an analytic space, then the normal closure of the ring O_X is isomorphic to the ring of holomorphic functions on the normalization of X.

A normal analytic space may be singular. For example, the surface given by the equation $z^2 + x^3 + y^5$ is normal and singular at the origin. However the singular locus in a normal space has codimension bigger or equal than 2. For more details on normality and normalization we recommend [GLS07, I.9].

2.2.4   Resolution of Singularities

A **modification** of an analytic space X is a proper map $\mu : Y \rightarrow X$ which induces an isomorphism (biholomorphism) over the complement of a nowhere dense analytic subspace of X. The blowup of an ideal and the normalization are modifications.

A **resolution of singularities** of an analytic space X is a modification $\rho : Z \rightarrow X$ where Z is a non-singular analytic space.

H. Hironaka proved in [Hir64] that every algebraic space defined over a field of characteristic zero admits a resolution of singularities. An analytic version of this result was proved in [AHV77].

Resolution of singularities is an active subject of research. Different tools have been developed in order to produce resolutions of singularities. The problem is still open over fields of positive characteristic. We recommend some texts on the subject, such as [Cut04, Kol07, HLOQ00].

In next sections we will refer to some particular methods, such as Jung's method, Nash modification, and point blowups, for curves and surfaces.

2.3   Complex Curve Singularities

In this section, we will review some properties of plane curve singularities related to normalization and parametrization. We will also describe the way to get an embedded resolution and to construct its dual graph.

A **curve singularity** is a germ of an analytic space $(C, 0)$ of dimension one. When it is a hypersurface in $\mathbb{C}^2$, i.e., defined by one analytic function $f \in O_{2,0}$, we call it a plane curve singularity, otherwise, one can talk about space curve singularity.

Any function $f \in O_{2,0}$ has a decomposition $f = f_1^{\alpha_1} \cdots f_r^{\alpha_r}$ into irreducible holomorphic functions, corresponding to the decomposition $(C, 0) = (C_1, 0) \cup \ldots (C_r, 0)$ into irreducible components, or **branches** , defined by the respective f_i. The component $(C_i, 0)$ is non-reduced if and only if $\alpha_i > 1$.

2.3.1   Normality and Normalization

In dimension one, the normalization produces a non-singular curve. More precisely:

Theorem 2.3.1 *The normalization of a reduced curve singularity* $(C, 0)$ *is a map* $n : (T_1, 0) \sqcup \ldots \sqcup (T_r, 0) \to (C, 0)$, *where each germ* $(T_i, 0)$ *is a non-singular (irreducible) curve germ, and the restriction of* n *to any* $(T_i, 0)$ *is the normalization of a branch* $(C_i, 0)$ *of* $(C, 0)$.

We refer to [dJP00, 4.4] for the proof and details on normalization.

Remark 2.3.2

(a) Since all the T_i's in Theorem 2.3.1 are non-singular, they are isomorphic to $(\mathbb{C}, 0)$. So the normalization provides a one-to-one parametrization on each irreducible component, with parameter in $\mathbb{C}$.
(b) When the curve $(C, 0)$ is non-reduced, the normalization is not well defined as a modification. However when the only non-reduced point of the curve is the origin itself, one can define a normalization as the normalization of the reduced associated curve, i.e., the curve defined by the radical ideal defining $(C, 0)$; see for example [Gre17].

Example 2.3.3 Consider the plane curve singularity $(C, 0)$ defined by the function $(x, y) \mapsto y^2 - x^2 - x^3$ with local ring $R := O_{C,0} = \mathbb{C}\{x, y\}/(y^2 - x^2 - x^3)$.

One can see that $t := \frac{y}{x}$ is in the field of fractions of R and satisfies the polynomial equation with coefficients in R: $t^2 - (1 + x) = 0$. So R is not normal. Its integral closure, $\overline{R}$, in its field of fractions must then contain t. Let us call R_1 the minimal subring of $\overline{R}$ containing R and the element t. So $R_1 = \mathbb{C}\{x, y\}[t]/(y - tx, t^2 - (1 + x))$. It is then isomorphic to $\mathbb{C}\{x\}[t]/(t^2 - (1 + x))$. Since the ideals $(t - \sqrt{1 + x})$ and $(t + \sqrt{1 + x})$ are co-prime in $\mathbb{C}\{x\}[t]$ then the ring R_1 is isomorphic to $\mathbb{C}\{x\}[t]/(t - \sqrt{1 + x}) \oplus \mathbb{C}\{x\}[t]/(t + \sqrt{1 + x})$. Each of the summands is isomorphic to $\mathbb{C}\{x\}$. So the normalization map corresponds to

$$n : (\mathbb{C}, 0) \sqcup (\mathbb{C}, 0) \to (C, 0)$$
$$u \mapsto (u, u\sqrt{1 + u})$$
$$v \mapsto (v, -v\sqrt{1 + v}).$$

2.3.2   Tangent Cone and Blowups

The tangent cone of a curve singularity is a homogeneous algebraic variety of dimension one, so it is a finite union of lines. When it is a plane curve singularity defined by a function $f \in O_{2,0}$ the tangent cone is defined by the initial form of f. This form is a homogeneous polynomial in two variables. So it factors into a product of linear forms; the zero set of each of them is a line.

Proposition 2.3.4 *The lines of the tangent cone of a curve singularity are precisely the limits of tangent lines to the curve.*

Proof Recall from Sect. 2.2.2 that the lines of the tangent cone correspond to the limits of secants. Consider a reduced curve $(C, 0) \subset (\mathbb{C}^N, 0)$ and its normalization

$$n : \bigsqcup_i (\mathbb{C}, 0) \to (C, 0).$$

The restriction n_i of n to each copy of $(\mathbb{C}, 0)$ is given by holomorphic functions $(\varphi_{i,1}, \ldots, \varphi_{i,N})$. A limit of secant lines, respectively tangent lines, can always be taken as a limit over points lying in the same irreducible component, and hence images of a sequence of points (u_k) in the same complex line by a map n_i. So the sequence of secants is given by the sequence of projective points in $\mathbb{P}^{N-1}$ defined by $[\varphi_{i,1}(u_k) : \ldots : \varphi_{i,N}(u_k)]$ and the sequence of tangent lines is given by the sequence of points in the dual projective space $\check{\mathbb{P}}^{N-1}$ defined by $[\frac{d\varphi_{i,1}}{du}(u_k) : \ldots : \frac{d\varphi_{i,N}}{du}(u_k)]$. By l'Hospital's rule, these two sequences have the same limit when $k \to \infty$.   □

From the proof of the proposition above, one can see that for each branch there is one and only one limit of secants or tangents. This limit will be determined by the functions $\varphi_{i,k}$ of minimal order when k varies. So there is no ambiguity in talking about the tangent line to a branch at a singular point. However two different branches may have the same limit of tangents.

Consider a germ of curve $(C, 0)$ and the blowup of the origin $e_0 : X \to C$. The inverse image of the origin corresponds to the projectivization of the tangent cone, so in this case it is a finite number of points; as many as the number of distinct tangents to C at the origin. The blowup is then a finite map and the curve X is a multi-germ. In Example 2.3.3, the normalization map coincides with the blowup of the origin in the curve.

One can also consider the blowup of the ambient space and follow the strict transform of the curve. This procedure is particularly useful in the case of plane curve singularities for its tight relation with the embedded topology of the curve.

Consider an open ball U of $\mathbb{C}^2$ around the origin and a representative C of $(C, 0)$ in U. Call $\pi_0 : \tilde{U} \to U$ the blowup of the origin in U. The tangent cone of U at the origin is $\mathbb{C}^2$, so the exceptional fiber of π_0 is the projective line $\mathbb{P}^1$; we will call it E. The inverse image of C by π_0 is the **total transform** of C by π_0; note that it contains the exceptional fiber E. The closure in $\tilde{U}$ of $\pi_0^{-1}(C \setminus \{0\})$ is a curve in $\tilde{U}$; it is the **strict transform** of C by π_0. It will be denoted by C^* whenever there is no ambiguity. In this case, the restriction of π_0 to C^* coincides with the blowup of the origin in the curve C. So the intersection points of C^* with the exceptional fiber E correspond to the distinct tangent lines to the branches of C. Two branches are separated by π_0 if and only if they have different tangent lines.

Consider a system of local coordinates (x, y) in U and homogeneous coordinates $[u : v]$ in $\mathbb{P}^1$. The surface $\tilde{U}$ is then defined by the equation $xv - yu = 0$ in $U \times \mathbb{P}^1$.

The exceptional fiber E is locally defined by one equation; namely, in the chart $u \neq 0$, by $x = 0$ and in $v \neq 0$, by $y = 0$.

Consider now a representative of a smooth curve S in U defined by an equation $f = 0$ with $f \in O_2(U)$. The Taylor expansion of f will be of the form $f = f_1 + f_2 + \cdots$, where f_i is homogeneous of degree i and $f_1 \neq 0$ since S is non-singular at the origin.

We will describe the total and strict transforms of S in both charts of $\widetilde{U}$.

In the chart $u \neq 0$, denote by $v_1 = \frac{v}{u}$ the local coordinate of $\mathbb{P}^1$. The surface $\widetilde{U}$ is then defined by $y = xv_1$. Note that in this chart it is isomorphic to an open neighborhood of the origin in $\mathbb{C}^2$. The total transform of S is defined by the equations $y = xv_1$ and $f_1(x, xv_1) + f_2(x, xv_1) + \cdots = 0$. Since every f_i is homogeneous of degree i, the second equation can be written as $xf_1(1, v_1) + x^2 f_2(1, v_1) + \cdots = 0$. The total transform of S has two components, the exceptional fiber E, defined by $x = 0$ in this chart, and the strict transform S^* defined in this chart by $y = xv_1$ and $f_1(1, v_1) + xf_2(1, v_1) + \ldots = 0$. We have two possibilities, either $f_1(1, v_1)$ is a non zero constant, in which case the strict transform S^* does not intersect E in this chart, or $f_1(1, v_1) = \alpha v_1$ and in this case, S^* intersects E transversally in one point. In the other chart, the situation is completely symmetric. We need to make precise that if in one chart the initial form of the strict transform is a constant, then in the other chart it is not. In other words, if the initial form is a constant, then the strict transform intersects the exceptional fiber at infinity with respect to that chart.

In conclusion, the strict transform of a representative of a smooth curve intersects the exceptional fiber transversally at one point.

When we consider a singular curve the situation can be very different:

Example 2.3.5 Consider the curve singularity C defined in a ball U around the origin by $x^3 + y^5 = 0$. This is called the E_8 curve singularity. The blown-up surface $\widetilde{U} \subset U \times \mathbb{P}^1$ is defined as above by the equation $xv - yu = 0$. In the chart $u \neq 0$ with local coordinates $(x, y, v_1 := \frac{v}{u})$, the exceptional fiber is defined by $x = 0$ and the total transform of C is defined by $y = xv_1$ and $x^3(1 + x^2v_1^5) = 0$. So the strict transform of C in this chart is given by $y = xv_1$ and $1 + x^2v_1^5 = 0$ which does not intersect the exceptional fiber.

In the other chart, $v \neq 0$ with local coordinates $(x, y, u_1 := \frac{u}{v})$, the exceptional fiber is defined by $y = 0$ and the total transform is given by $x = yu_1$ and $y^3(u_1^3 + y^2) = 0$. So the strict transform is given by $x = yu_1$ and $y^2 + u_1^3 = 0$; it is a singular curve, intersecting the exceptional fiber at the origin of this chart, and being tangent to the exceptional fiber. This blowup is illustrated in the first map of Fig. 2.1.

2.3.3 Resolution and Dual Graph

We have already seen that the normalization of a curve provides a resolution of singularities. However, in the case of plane curve singularities, one is interested

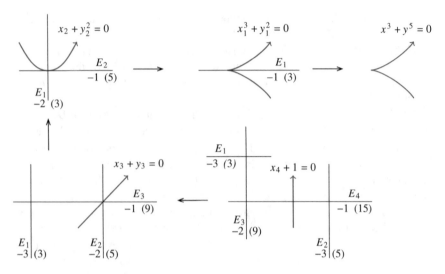

Fig. 2.1 Embedded resolution of the E_8 curve singularity

in a resolution taking into account the ambient space. This is called an **embedded resolution**.

Definition 2.3.6 Consider a representative C of a plane curve singularity in an open ball U around the origin in $\mathbb{C}^2$. An **embedded resolution** of C is a modification $\mu : W \to U$ such that the total transform of C by μ is a union of smooth irreducible components having at most normal crossing intersections, i.e., the components intersect transversally and at most two of them intersect in one point.

Theorem 2.3.7 *An embedded resolution of a plane curve singularity can be obtained after a finite number of blowups of points.*

See for example [Wal04, Chap. 3] for a proof and more information.

Example 2.3.8 In Example 2.3.5, we obtained after a first blowup, a surface $\tilde{U}$ with an exceptional fiber E_1 tangent to the strict transform C_1 of the curve C, and C_1 is given by the local equations $x = yu_1$, $u_1^3 + y^2 = 0$ in coordinates x, y, u_1 of the chart $v \neq 0$. In this chart, the surface $\tilde{U}$ is isomorphic to an open neighborhood U_1 of the origin of $\mathbb{C}^2$, with local coordinates $y_1 = y$ and $x_1 = u_1$. In these coordinates, E_1 and the strict transform are respectively defined by $y_1^3 = 0$ and $x_1^3 + y_1^2 = 0$.

We will then apply a second blowup at the intersection point $C_1 \cap E_1$. We obtain a new surface $\tilde{U}_2 \subset U_1 \times \mathbb{P}^1$ with two charts, both isomorphic to an open neighborhood of the origin of $\mathbb{C}^2$. In one of the charts the strict transform C_2 of C_1 does not intersect the new exceptional fiber E_2. Meanwhile in the other chart U_2, where $y_1 = x_1 v_2$, the exceptional fiber E_2 is defined by $x_1^2 = 0$, the strict transform C_2 is defined by $x_1 + v_2^2 = 0$ and the strict transform of E_1 by $v_2^3 = 0$; for simplicity we will still call it E_1. We name a new coordinate system: $x_2 = x_1$ and $y_2 = v_2$.

The strict transform C_2 of C is non-singular, intersecting both exceptional fibers E_1 and E_2 at the origin, and being tangent to E_2. So we are not yet at a situation of normal crossings.

We blow up the origin, obtaining a surface $\widetilde{U}_3$ with two charts. In the chart $x_2 = y_2 u_3$, the new exceptional fiber E_3 is defined by $y_2 = 0$ and the strict transform C_3 of C is defined by $u_3 + y_2 = 0$. The strict transform of E_2 is defined by $u_3 = 0$ and the strict transform of E_1 does not intersect E_3 in this chart. In the other chart, the strict transform of E_1 intersects transversally the exceptional fiber E_3. So in the first chart, taking new coordinates: $x_3 = u_3$ and $y_3 = y_2$, we still have a non-normal crossing situation, since $C_3 \cap E_2 \cap E_3 = \{0\}$.

We need one more blowup. Now all the considered curves are smooth (in their reduced structure), and they all intersect transversally. So the new blowup produces an exceptional fiber E_4 intersecting transversally E_2, E_3 and the strict transform C^* of C at distinct points. The strict transform of E_1 has not been modified by this blowup. See Fig. 2.1, where the values a and (b) associated to exceptional curves correspond respectively to the self-intersection (negative value) and the valuation.

One associates to an embedded resolution a graph constructed in the following way:

Definition 2.3.9 Let $\rho : W \to U$ be an embedded resolution of a representative of a plane curve germ $(C, 0)$. Consider the decomposition of the exceptional divisor into irreducible components, $\rho^{-1}(0) = \cup_i E_i$ and call $C_1^*, \ldots, C_r^*$ the strict transforms of the branches of $(C, 0)$ by ρ. The graph associated to ρ consists of vertices, edges and arrows as follows:

- a vertex is associated to each component E_i; call it e_i,
- an arrow is associated to each component C_j^*; call it c_j^*,
- an edge will link the vertices e_i and e_j if and only if $E_i \cap E_j \neq \emptyset$,
- and an arrow c_j^* is attached to the vertex e_i if and only if $E_i \cap C_j^* \neq \emptyset$.

Moreover, to each vertex e_i we associate two numbers: the self-intersection w_i of the curve E_i, and the valuation v_i of the function f defining $(C, 0)$; that is, the order of the composition map $f \circ \rho$ at a generic point of the curve E_i. Such a weighted graph is called the **dual graph** associated to an embedded resolution.

The computation of the numbers w_i and v_i associated to the graph can be made in the following way: all the blowups consist in blowing-up a point in a non-singular surface. So at each step, the exceptional fiber produced is a smooth rational curve (isomorphic to $\mathbb{P}^1$) that will have self-intersection -1; see for example [Wal04, Lemma 8.1.2] or [BK86, p. 531]. So the first blowup produces an exceptional curve E_1 with self-intersection -1 and valuation equal to the multiplicity of the curve at the origin of $\mathbb{C}^2$. For the second blowup, the new exceptional curve E_2 will have self-intersection -1 and the strict transform of E_1 will have self intersection -2 since the blown-up point belongs to E_1. The valuation v_1 remains unchanged, and the new valuation v_2 is the order of f at the blown-up point, that is the sum of v_1 with the multiplicities of the strict transform of the irreducible components of C at

Fig. 2.2 Dual graph
associated to the minimal
embedded resolution of the
E_8 curve singularity

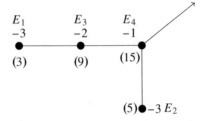

the blown-up point. So at each step, the self-intersection of the created exceptional
fiber is -1 and the strict transform of each irreducible component of the exceptional
fiber containing the blown-up point has its self-intersection decreasing by 1. The
valuation of f along the strict transform of an exceptional fiber does not change,
and the valuation along the created one is the order of f at the blown-up point.

A **minimal embedded resolution** ρ of a plane curve singularity is an embedded
resolution that factors through any other embedded resolution; that is, whenever π
is an embedded resolution of the plane curve singularity, there exists a sequence of
point blowups φ such that $\pi = \rho \circ \varphi$.

For each plane curve singularity, a minimal embedded resolution exists and is
unique up to isomorphism. This minimal embedded resolution can be obtained from
any embedded resolution by blowing-down any irreducible component with self
intersection -1 which does not intersect the strict transform of the curve. This is due
to the fact that contracting a smooth rational projective curve with self-intersection
-1 in a smooth surface creates a non-singular surface; this is precisely the inverse
process of blowing-up a point in a non-singular surface that is locally isomorphic to
$\mathbb{C}^2$.

Example 2.3.10 The resolution of the curve singularity of Example 2.3.8 is the
minimal embedded resolution and its dual graph is shown in Fig. 2.2, where the
integer values a and (b) refer respectively to the self-intersection and the valuation
of the corresponding irreducible component of the exceptional divisor. The vertex
E_i corresponds to the exceptional fiber appearing at the i'th blowup.

2.3.4 Newton-Puiseux Parametrization, Characteristic Exponents and Topological Type

All the information about the dual graph associated to the minimal embedded
resolution of a plane curve singularity can be recovered from another process
associated to the singularity, namely Newton-Puiseux parametrization.

We have already seen that the normalization of a curve singularity provides a
parametrization and a resolution of singularity, however it does not bring all the
information about the embedded resolution.

Definition 2.3.11 Let $(C, 0)$ be a reduced and irreducible germ of a plane curve singularity defined by an irreducible holomorphic function $f \in \mathbb{C}\{x, y\}$. One may proceed to a linear change of coordinates to ensure that the x-axis coincides with the tangent cone to C at 0. A **Newton-Puiseux parametrization** of $(C, 0)$ is a map $\varphi : \mathbb{C} \to \mathbb{C}^2$ with $\varphi(t) = (t^m, \sum_{i>m} a_i t^i)$, with $f(\varphi(t)) = 0$ for $t \in \mathbb{C}$ small enough and m a non-zero integer, being the smallest one for which such a parametrization is possible.

Such a parametrization always exists, and can be computed using the method of Newton Polygon, see for example [Wal04, Chap. 2].

From the Newton-Puiseux parametrization one can extract significant combinatorial data. First, the integer m in Definition 2.3.11 is the multiplicity of the curve at the origin; we set $\beta_0 = m$. The smallest power i for which $a_i \neq 0$ and i is not multiple of β_0 will be denoted by β_1; if there is none the process stops. Call l_1 the greatest common divisor of β_0 and β_1. If $l_1 = 1$ the process stops, otherwise, we call β_2 the smallest power $i > \beta_1$ for which $a_i \neq 0$ and l_1 does not divide i. And we repeat the process. It will end-up after a finite number of steps. The values in the ordered sequence $\{\beta_0, \beta_1, \ldots, \beta_g\}$ are called the **Puiseux characteristic exponents** of the branch $(C, 0)$.

These exponents, or equivalent combinatorial data, characterize the topology of an irreducible branch in $\mathbb{C}^2$. Let us explain it briefly: two plane curve singularities are topologically equivalent if there is a homeomorphism of $\mathbb{C}^2$ near the origin sending a curve surjectively into the other. We know by the local conical structure theorem (see [Mil68, 2.10]), that a curve embedded in a small ball of $\mathbb{C}^2$ is homeomorphic to the cone over its intersection with a small sphere in $\mathbb{R}^4$, this intersection is called the **link** of the singularity. For plane curves the link is a one-dimensional smooth manifold in $\mathbb{S}^3$. Each connected component of this manifold is a knot corresponding to the link of a branch.

So the topology of a plane curve singularity is determined by the topology of each connected component of the link, together with the linking number of each pair of components. This linking number is equal to the **intersection multiplicity** of the corresponding branches. Let us recall that for two plane curve singularities defined by f and g in O_2, the intersection multiplicity is defined as the dimension of $O_2/(f, g)$ as $\mathbb{C}$-vector space.

Theorem 2.3.12 *For a reduced plane curve singularity, the following data are equivalent:*

1. *The characteristic exponents of each branch of the curve given by the Newton-Puiseux parametrization together with the intersection multiplicity of each two branches*
2. *The dual graph of the minimal resolution of the singularity*
3. *The isotopy type of the link of the singularity.*

We refer to the classical book by Brieskorn and Knörrer [BK86, Theorem 21 section 8.5] as well as well as [Wal04, Theorem 5.5.9].

2.3.5   Some Invariants: δ-Number and Milnor Number

Some other numerical invariants are associated to plane curve singularities. We will introduce some of them that are used in other chapters of this volume.

We have already talked about the multiplicity of a plane curve singularity, which is the sum of the multiplicities of all its branches, hence the sum of the first characteristic exponents of the branches.

The δ-invariant is defined as follows. First consider a reduced plane curve singularity $(C, 0)$ and its normalization $n : (\bar{C}, \bar{0}) \to (C, 0)$; where $(\bar{C}, \bar{0})$ is a multi-germ having $\bar{O}_C$ as a multi-local ring of functions. Since the normalization map is finite, the dimension $\dim \bar{O}_C / O_{C,0}$ as vector space over $\mathbb{C}$ is finite. This number is called the δ-**invariant**, and is denoted by $\delta(C, 0)$.

In the case of an irreducible branch the normalization induces an inclusion $n^* : O_{C,0} \hookrightarrow \mathbb{C}\{t\}$. Since the normalization is finite, there is an integer ν such that for any $\tau \geq \nu$ one has $t^\tau \in n^*(O_{C,0})$, so in this case, $\delta(C, 0)$ is the number of powers of t that are not in the image of $O_{C,0}$.

The **Milnor number** of a reduced curve singularity is defined as follows. If $(C, 0)$ is defined by $f \in \mathbb{C}\{x, y\}$, then $\mu(C, 0) := \dim_{\mathbb{C}} \mathbb{C}\{x, y\}/(\frac{\partial f}{\partial x}, \frac{\partial f}{\partial y})$ is the Milnor number of C at 0.

This number is related to the δ-invariant by the formula $\mu(C, 0) = 2\delta(C, 0) - r(C, 0) + 1$; where $r(C, 0)$ is the number of branches of the curve at the origin; see [Mil68, Chap. 10].

These two numbers are analytic invariants of curve singularities, and they play an important role in the study of equisingularity properties of families of curves.

Example 2.3.13 Consider the curve singularity E_8 defined by $y^3 + x^5$. A Newton-Puiseux parametrization is given by $x = t^3$ and $y = t^5$. So there are only two characteristic exponents, namely $(3, 5)$. The ring of holomorphic functions is isomorphic to $\mathbb{C}\{t^3, t^5\}$, the missing powers of t are t, t^2, t^4 and t^7. So $\delta(E_8, 0) = 4$. The Milnor number is the dimension of $\mathbb{C}\{x, y\}/(y^2, x^4)$, which is 8.

Note that these definitions extend to reduced space curves ([BG80]) and also to curves with an embedded component ([BG90]).

2.4   Complex Surface Singularities

In this section we will deal with surfaces and explain the normalization and the point blowup process. We describe some methods of resolution. We introduce the concept of exceptional tangents and its relation with the polar curves and Nash modification.

A **surface singularity** is a germ $(S, 0)$ of an analytic space of Krull dimension 2. When it is defined by one function $f \in O_{3,0}$ we talk about a **hypersurface** in $\mathbb{C}^3$. However, a surface need not be either a hypersurface, or a **complete intersection** (defined by an ideal with as many generators as the codimension).

A reduced surface germ can have either isolated or non-isolated singularities, and its different irreducible components could have different dimensions. When all the irreducible components are surfaces we say that the surface is **equidimensional**, or is **of pure dimension** . A hypersurface or a complete intersection is always equidimensional, but still it could have isolated or non-isolated singularities.

Example 2.4.1 The ideal $(xz, xy) \subset \mathbb{C}\{x, y, z\}$ defines a surface with two components, the two-dimensional plane defined by $x = 0$ and the x-axis. It has an isolated singularity at the origin.

The function $x^2 - y^2 z$ defines a hypersurface having the z-axis as singular locus; it is known as the Whitney umbrella.

The function $z^2 + x^3 + y^5$ defines a hypersurface with an isolated singularity at the origin, this is known as the two-dimensional E_8-singularity.

2.4.1   Normality and Normalization

In general, a normal analytic space has a singular locus of codimension at least two. So a normal surface has isolated singularities. More precisely, a surface germ $(S, 0)$ is normal if and only if it has an isolated singularity and the ring $O_{S,0}$ is **Cohen-Macaulay**, see for example [dJP00, 4.4]. The algebraic property of being Cohen-Macaulay is the same as saying that the ring has a regular sequence of length two; in other words, there exist two elements f and g in the maximal ideal $M_{S,0}$ such that f is not a zero devisor in $O_{S,0}$ and g is not a zero devisor in $O_{S,0}/f$. Moreover, it is proved in [ZS75, Appendix 6, corollary. 1] that $O_{S,0}$ is Cohen-Macaulay if and only if, whenever $f \in M_{S,0}$ is not a zero divisor, the quotient $O_{S,0}/f$ has a non-zero divisor in its maximal ideal. In other words, a surface germ is not Cohen-Macaulay if and only if, any (or some) curve defined on it by one function has an embedded component.

Hypersurfaces and more generally complete intersections have always Cohen-Macaulay local rings of holomorphic functions ([ZS75, Appendix 6, Theorem 2]. So, for a complete intersection surface singularity, normality is equivalent to isolated singularity.

Example 2.4.2 The surface $(S, 0)$ defined in $(\mathbb{C}^4, 0)$ by the equations $y^2 - zx^2 = 0$, $yz - tx = 0$, $z^2 x - ty = 0$ and $t^2 - z^3 = 0$ has an isolated singularity and is not Cohen-Macaulay. One can see that for every $f \in O_{S,0}$, the curve defined by f has an embedded component. For instance, the function x defines a curve with ring of functions $\mathbb{C}\{y, z, t\}/(y^2, yz, yt, t^2 - z^3)$. The zero-ideal of this ring has the following primary decomposition $(0) = (y, z^3 - t^2) \cap (y^2, yz, z^3, t)$, showing that the maximal ideal is an associated prime.

The normalization map of this surface is given by: $(u, v) \mapsto (u, uv, v^2, v^3)$. It is a homeomorphism which is not an isomorphism.

Unlike the case of curves, the normalization does not always provide a resolution of singularities. In particular when a surface is normal and singular, its normalization is the identity. In some cases the normalization could be non-singular. This is the case for the Whitney umbrella where the map $n : (u, v) \mapsto (uv, u, v^2)$ is the normalization.

Since a germ of a normal surface is a domain, the normalization process separates irreducible analytic components. More precisely, let $n : X \to S$ be the normalization of a representative of $(S, 0)$. For every point $s \in S$ the fiber $n^{-1}(s)$ has as many points as irreducible components of S at s. In terms of germs, the normalization of $(S, 0)$ produces a multi-germ depending on the number of irreducible components of the germ $(S, 0)$.

In the case of the Whitney umbrella, one can see that $(S, 0)$ is irreducible and $n^{-1}(0) = \{0\}$, however, for a point in the singular locus other than 0, we have $n^{-1}(0, 0, z) = (0, \pm\sqrt{z})$. In fact a representative of $(S, 0)$ has two irreducible components at a generic point of the singular locus.

2.4.2 Blowups and Resolution

When we blow up the origin, or equivalently the maximal ideal of a surface, we obtain a proper morphism $e_0 : S' \to S$ which is an isomorphism outside the origin, and has an exceptional fiber, which is a divisor in S' equal to the projective curve associated to the tangent cone of S at 0, $e_0^{-1}(0) = \mathrm{Proj}(T_0 S)$.

When one makes the computation using local coordinates, it is useful first to apply the blowup E_0 of the origin in the ambient space $\mathbb{C}^N$ of $(S, 0)$, and consider the strict transform of S by E_0. This strict transform is precisely the surface S' obtained by the blowup e_0. The exceptional divisor $e_0^{-1}(0)$ is the intersection $E_0^{-1}(0) \cap S'$. Note that in this case $E_0^{-1}(0) = \mathbb{P}^{N-1}$.

Example 2.4.3 Consider the surface $(S, 0)$ defined by $x^2 + y^4 + z^4 = 0$. It is a hypersurface with isolated singularity, hence normal. We blow up the origin in $\mathbb{C}^3$ and we obtain a three-dimensional space $C \subset \mathbb{C}^3 \times \mathbb{P}^2$ defined by the functions $xv - yu$, $xw - zu$ and $yw - zv$, where $[u : v : w]$ are homogeneous coordinates in $\mathbb{P}^2$. In the chart $v \neq 0$, for simplicity we will say $v = 1$, the total transform of S is defined by the functions $x - yu$, $z - yv = 0$ and $y^2(u^2 + y^2 + y^3 w^5) = 0$. In this chart, the exceptional divisor is defined by $y = 0$, so the strict transform of S is isomorphic to the surface of $\mathbb{C}^3$ with equation $u^2 + y^2 + y^3 w^5 = 0$; this is the surface obtained by the blowup e_0 of the origin in S. The exceptional divisor of e_0 is the intersection of this strict transform with $\mathbb{P}^2$, in this chart, it is given by $y = u = 0$. Note that every point of the exceptional divisor is singular. This is an example where the blowup of the origin of a normal surface produces a surface with non-isolated singularities.

There exists a class of surface singularities, called **absolutely isolated singularities**, that are normal and their blowup is still normal. This class contains the so-called **rational surface singularities** for which the blowup still produces normal surfaces with rational singularities. They are a useful source of examples and counter-examples; see [Art66] for definitions and properties.

We have already seen that every analytic space admits a resolution of singularities. More specific results are achieved for surfaces.

Theorem 2.4.4 (O. Zariski [Zar39]) *A normal surface singularity can be resolved by the iteration of a finite number of point blowups composed with normalization.*

This result provides a particular resolution of singularities for normal surfaces. In [BL02], the authors give another proof of this result, bringing an interesting description of normal surface singularities.

Consider a resolution $\rho : X \to S$ of singularities of a representative of a germ of surface $(S, 0)$. If Σ denotes the singular locus of S, the exceptional locus of ρ is the inverse image $\rho^{-1}(\Sigma)$. When $(S, 0)$ is normal, $E := \rho^{-1}(0)$ is called the exceptional fiber of the resolution, and is a connected curve; see [Har77, 11.4].

Suppose $(S, 0)$ is normal and consider $E = \cup_i E_i$, the decomposition of E into its irreducible components. They are all projective curves, but they need not be smooth and they may have a non-zero genus. If one of these irreducible components, E_i, happens to be rational, smooth and with self-intersection -1, we know it can be obtained by the blowup of a point in a non-singular surface. So, contracting, or blowing down, the component E_i produces a non-singular surface X_1 and a map $\rho_1 : X_1 \to S$ that is still a resolution of singularities of S, and ρ factors through ρ_1. This leads to the concept of minimal resolution.

Definition 2.4.5 A resolution of singularities is **minimal** if any other resolution factors through it.

Proposition 2.4.6 *Every surface singularity admits a minimal resolution; it is unique up-to isomorphism.*

For a proof, see for example [BPVdV84, III. 6.2].

Such a minimal resolution can be obtained from any resolution of the normalized surface, by blowing down each component of the exceptional fiber which is smooth, rational and of self-intersection -1. Repeating this process one obtains the minimal resolution.

The existence of a minimal resolution is a particular property of surface singularities. It does not extend to higher dimensions.

Note that the resolution of a surface, obtained by normalized blowups as in Theorem 2.4.4, need not be the minimal resolution. For some particular surface singularities, such as the class of rational surface singularities, the minimal resolution is the one obtained by normalized point blowups, but this is not general.

In the minimal resolution of a normal surface, the irreducible components of the exceptional fiber may be singular and may have non-normal crossings. Applying point blowups one can achieve a new resolution of the surface in which the exceptional fiber has at most normal crossing singularities. Such a resolution is called a **good resolution**.

To such a good resolution we can associate a weighted **dual graph** in the following way: to each irreducible component of the exceptional fiber we associate a vertex, edges will correspond, one-to-one, to intersection points of exceptional curves, linking the corresponding vertices. To each vertex we attach two values: the genus and the self-intersection of the corresponding irreducible component. See Example 2.4.19.

The factorization of a modification through the blowup of the singularity is related to hyperplane sections and their base points.

If $(S, 0)$ is a normal surface singularity embedded in $(\mathbb{C}^N, 0)$, then a **hyperplane section** is the curve obtained as the intersection $(H \cap S, 0)$, where H is a hyperplane in $\mathbb{C}^N$. The family of all hyperplane sections is a linear system determined by the maximal ideal $\mathcal{M}_{S,0}$. The elements of this linear system are defined by linear combinations of a local system of coordinates of S near 0. The parameter space of this linear system is $\mathbb{P}^{N-1}$ since a hyperplane section is defined by a unique linear combination up to multiplication by a non-zero constant.

Now let $\mu : Y \to S$ be a modification of a representative of $(S, 0)$ over the origin. We say that a point $\eta \in \mu^{-1}(0)$ is a **base point**, or a fixed point, of the linear system of hyperplane sections if η belongs to almost all strict transforms of hyperplane sections by μ, that is hyperplane sections defined by a parameter in an open dense set of $\mathbb{P}^{N-1}$.

Using the **universal property of the blowup** of an ideal, which states that it is the minimal morphism in which the pull-back of the ideal is locally principal (see for example [Har77, 7.14]), one can prove the following:

Proposition 2.4.7 *Let $\mu : Y \to S$ be a modification of S over 0 with Y normal, then μ factors through the blowup of 0 if and only if the linear system of hyperplane sections has no base point by μ.*

The minimal resolution of a normal surface singularity may have base points of the hyperplane sections; see Example 2.4.18 below.

2.4.3 Nash Modification, Exceptional Tangents and Polar Curves

We will now construct another modification that also leads to a resolution of singularities in dimension two.

Consider a representative S of an equidimensional surface singularity embedded in $\mathbb{C}^N$ and define the map:
$\gamma : S \setminus \text{Sing}(S) \to G(2, N)$, sending every non-singular point $x \in S$ to the direction of the tangent space to S at x, $\gamma(x) := T_x S$, seen as a point in the Grassmannian, $G(2, N)$, of two-dimensional spaces in $\mathbb{C}^N$. Call $\widetilde{S}$ the closure of the graph of γ in

$S \times G(2, N)$. The restriction of the first projection to $\widetilde{S}$ defines a map

$$\nu : \widetilde{S} \to S$$

called the **Nash modification** of S. It is a modification since it is defined as the closure of a graph.

If $p \in S$ is a singular point, the inverse image $\nu^{-1}(p)$ is made of points of the form $(p, T) \in S \times G(2, N)$ where T is a limit of directions of tangent spaces taken over a sequence of non-singular points of S converging to p. The set of all these planes is called the set of **limits of tangent spaces** to S at p.

The normalized Nash modification is the composition $\nu \circ n$ where n is the normalization of the surface $\widetilde{S}$. One has a similar result to the one of Theorem 2.4.4, due to M. Spivakovsky in [Spi90].

Theorem 2.4.8 *A normal surface singularity can be resolved by the iteration of a finite number of normalized Nash modifications.*

The normalization process is fundamental in this theorem, since it is still unknown whether this result holds without normalization.

Nash modification satisfies a **universal property** with respect to a family of curves, called the polar curves as follows:

Consider a linear projection $P : \mathbb{C}^N \to \mathbb{C}^2$ and its restriction to a representative S of $(S, 0)$, $\pi : S \to \mathbb{C}^2$. Let us call $L := \ker P$ the kernel of P; it is an $(N-2)$-plane in $\mathbb{C}^N$. For a generic choice of L in the Grassmannian $G(N-2, N)$, the map π is finite. The closure in S of the critical locus of the restriction of π to the non-singular locus of S will be denoted by $P_L(S, 0)$; recall that the critical locus can be defined as the set of points in which the map does not induce a local isomorphism. When the surface is normal this is simply the critical locus of π with its reduced structure. For L general enough in $G(N-2, N)$ the space $P_L(S, 0)$ is either empty or a reduced curve; we call it the **polar curve** associated to $(S, 0)$ defined by L. When $(S, 0)$ is normal, the general polar curve is not empty, see [Tei82, IV].

The definition of polar curves only depends on the surface and the linear space L. We have then a family of curves parametrized by $G(N-2, N)$. This is what we call the family, or linear system, of polar curves on S at 0.

Proposition 2.4.9 *Let $\mu : X \to S$ be a modification of S over 0, with both X and S normal. The modification μ factors through the Nash modification of S if and only if the family of polar curves has no base point by μ; base points of polar curves being defined in a similar way as for hyperplane sections in Sect. 2.4.2.*

For a proof of this proposition see [Spi90, III. Theorem 1.2].

The normalized Nash modification is then the "smallest" normalized modification that removes all base points of the family of polar curves, as the point blowup does for hyperplane sections.

We will now define the exceptional tangents. These objects appear in the description of the set of limits of tangent spaces to a surface at a point. They are important in the thick-thin decomposition of the link of a surface singularity in [BNP14], and in general in the Lipschitz geometry of normal surface singularities.

Unlike the curve case, the set of limits of tangent planes to a surface at a point does not have a straight forward description, but still it is strongly related to the tangent cone. It is easier to describe the **set of limits of tangent hyperplanes.**

A hyperplane $H \subset \mathbb{C}^N$ is tangent (or a limit of tangent hyperplanes) to S at a point $x \in S$ if it contains the tangent space $T_x S$ (or a limit of tangent planes to S at x).

Theorem 2.4.10 *Let* $(S, 0) \subset (\mathbb{C}^N, 0)$ *be the germ of an equidimensional surface singularity. There exists a finite number of lines in the tangent cone $T_0 S$ of S at 0, $l_1, \ldots, l_r$, such that a hyperplane $H \subset \mathbb{C}^N$ is a limit of tangent hyperplanes to S at 0 if and only if, either H is tangent to $T_0 S$ or $l_i \subset H$ for some $i \in \{1, \ldots, r\}$.*

For a proof of this theorem, see [LT88, 2.1.3].

Definition 2.4.11 The lines $l_1, \ldots, l_r$ in Theorem 2.4.10 are called the **exceptional tangents** of the surface S at 0. They are generatrices of the tangent cone $T_0 S$; that is why we will also view them as points in the projective curve associated to $C_{S,0}$, or equivalently, points in the exceptional fiber of the blowup of the origin.

Remark 2.4.12 When the tangent cone $T_0 S$ is not a plane, it is singular and may even have non-isolated singularities. Since it is conical, a tangent plane to it, or a limit of tangent planes, is always obtained as the cone over a tangent line, or over a limit of tangent lines, to the projective tangent cone. Since we understand the limits of tangents to a singular curve, there is no ambiguity in defining the tangent planes, or tangent hyperplanes, to the tangent cone.

The characterization of the exceptional tangents for normal surfaces is explained in [Sno01].

Example 2.4.13 Consider the hypersurface singularity $A_2 \subset \mathbb{C}^3$ defined by the equation $xy - z^3 = 0$. It has an isolated singularity at 0. One can compute by hands the limits of tangent spaces to A_2 at 0. This can be done for instance by taking a parametrization, that exists in this case, of the singularity: $(u, v) \mapsto (u^3, v^3, uv)$. Comparing speeds of convergence to 0 of sequences (u_n, v_n), one gets as limits of tangent planes (equivalently hyperplanes in this case) all the planes containing the z-axis.

The tangent cone of the surface is defined by $xy = 0$; it is a union of two planes: $x = 0$ and $y = 0$. Each of them is tangent to $T_0 S$. The intersection line of these two planes is the only exceptional tangent of S at 0.

In the example above, both planes of the tangent cone are limits of tangent spaces. In general, whenever a two-dimensional plane is a component of the tangent cone, it corresponds to a point in the fiber of the Nash modification over the considered point.

We will establish a relation between exceptional tangents and planes of the tangent cone with base points of some families of curves in some modifications.

Proposition 2.4.14 *Let $(S, 0)$ be a normal surface singularity. Consider the Nash modification v and the blowup of the origin e_0.*

There is a one-to-one correspondence between the exceptional tangents of S at 0 and the base points of the family of polar curves in the modification e_0.

There is a one-to-one correspondence between the planes of the tangent cone of S at 0 and the base points of the family of hyperplane sections in the modification v.

The proof of the first statement can be read in [LT88, 2.2.1], for the second statement see [Sno05, 3.2].

Every point of the exceptional fiber of e_0 is a line in $T_0 S$ and hence a tangent to a curve on the surface. So the first statement claims that the exceptional tangents are the lines of $T_0 S$ that are tangent to almost all the polar curves; that is why we also refer to them as the fixed tangents to the polar curves.

Example 2.4.15 In the A_2 singularity seen in Example 2.4.13, the exceptional fiber of the blowup of the origin is the projective curve associated to $xy = 0$; it is the union of two lines intersecting in one point that corresponds precisely to the exceptional tangent and the base point of the family of polar curves in the blowup. The exceptional fiber of the Nash modification is a line in $\check{\mathbb{P}}^2$, which is the line of hyperplanes containing the z-axis. In this line are two particular points corresponding to the planes $x = 0$ and $y = 0$; the planes of the tangent cone. These are the base points of the hyperplane sections by the Nash modification.

2.4.4   Jung's Method

We will now introduce a classical method for resolution of singularities of normal surfaces: Jung's method (see [KV04] and [Pop11] for more details).

Consider a representative of a normal surface singularity, together with a finite projection $\pi : S \to \mathbb{C}^2$. Call Δ the reduced **discriminant** of π; i.e., the image, with reduced structure, of the critical locus of π. It is a plane curve singularity. We will then process an embedded resolution of Δ as we describe in Theorem 2.3.7. We obtain a map $\rho : X \to \mathbb{C}^2$ such that the total transform $\rho^{-1}(\Delta)$ has only normal crossing singularities. Call ρ' and π' the pull-back of ρ by π and of π by ρ respectively. We obtain the following diagram:

$$
\begin{array}{ccccc}
Z & \xrightarrow{\ n\ } & Y & \xrightarrow{\ \rho'\ } & S \\
 & {\scriptstyle \psi}\searrow & {\scriptstyle \pi'}\downarrow & & \downarrow{\scriptstyle \pi} \\
 & & X & \xrightarrow{\ \rho\ } & \mathbb{C}^2
\end{array}
$$

where $n : Z \to Y$ is the normalization of the space Y obtained as pull-back of S by ρ.

At any point of Z the map $\psi := \pi' \circ n$ is a finite map whose reduced discriminant is contained in the total transform $\rho^{-1}(\Delta)$.

The singularities of Z are then of special type.

Definition 2.4.16 Let S be a representative of a germ of normal surface singularity. If there exists a finite map $S \to \mathbb{C}^2$ whose reduced discriminant has at most normal crossing singularities (i.e., smooth, or union of two smooth curves intersecting transversally), then we say that $(S, 0)$ has a **quasi-ordinary singularity**.

This is a particular type of singularities that has been widely studied. They are each a normalization of a hypersurface of $\mathbb{C}^3$ of type $z^n = x^a y^b$. The resolution of such a singularity is completely determined by the triple (n, a, b). The exceptional fiber of the minimal resolution is a normal crossing curve made of smooth rational curves $E = \cup_{i=1}^r E_i$, such that E_1 and E_r intersect respectively only E_2 and E_{r-1} and E_i intersects only E_{i-1} and E_{i+1} for $1 < i < r - 1$. These are particular cases of the so-called **minimal surface singularities**. For the complete description of the combinatorial aspect, and also the topology of such singularities, we refer to [BPVdV84, III. 5].

Going back to Jung's method, we obtain then a surface Z with finitely many quasi-ordinary singularities, whose resolution is completely described by the normalization of the surface $z^n = x^a y^b$ where n is the degree of π, a and b are the valuation of the defining function of Δ along the corresponding components of the total transform given by the local coordinates x and y respectively, as defined in Definition 2.3.9.

We can then achieve a resolution of the surface $(S, 0)$ by resolving the quasi-ordinary singularities of Z. This leads to a particular resolution of singularities called **Jung's method**.

Note that this resolution need not be the minimal resolution.

We give now two examples where we use Jung's method to obtain a resolution and then the minimal resolution. We do it in the case of surfaces of the form $z^2 = f(x, y)$, for which it is much easier to describe the resolution process. We will use in particular the paper [LW00].

Consider a surface singularity $(S, 0)$ defined by $z^2 = f(x, y)$ with f reduced. Consider on a representative S of $(S, 0)$, the projection to $\mathbb{C}^2$, $\pi : (x, y, z) \mapsto (x, y)$. It is finite and its discriminant Δ, which is reduced in this case, is defined by $f(x, y) = 0$. Apply the minimal embedded resolution ρ of Δ. Call $F = \rho^{-1}(\Delta) = \cup_i F_i$ the decomposition of the total transform of Δ into irreducible components. When F_i is part of the exceptional fiber, recall that the valuation $v_i(f)$ is the vanishing order of the pull-back of f on F_i, see Definition 2.3.9. When F_i is part of the strict transform of Δ, and since f is reduced, by extension we define $v_i(f)$ to be 1. The surface Y of Jung's method, obtained by the pull-back of S by ρ, is locally defined near each point $y \in Y$ by $z^2 = u^{v_i(f)} v^{v_j(f)}$ if $\pi'(y) \in F_i \cap F_j$, or $z^2 = u^{v_i(f)}$ if $\pi'(y)$ is not an intersection point. So the normalization of Y near such a point y is singular if and only if $\pi'(y)$ is an intersection point of F_i and F_j and $v_i(f)$ and $v_j(f)$ are simultaneously odd. If this occurs, then we can apply one extra blowup at the point $\pi'(y)$, creating a new exceptional divisor separating F_i

and F_j for which the valuation of f is $v_i(f) + v_j(f)$ which is odd. This way we can produce a non-necessarily minimal embedded resolution ρ_1 of Δ, for which the normalization of the pull-back, Z, is non-singular.

Now the components of the exceptional fiber on Z have their genus and self-intersection determined by the self-intersection and the valuations of the components of the exceptional fiber on the new resolution of Δ. Call F_i' the irreducible components of $\rho_1^{-1}(\Delta)$.

Proposition 2.4.17 *If $v_{F'}(f)$ is odd, then $D =: \psi^{-1}(F')$ is connected and irreducible and ψ induces an isomorphism between D and F'; moreover $(D.D) = \frac{1}{2}(F'.F')$.*

If $v_{F'}(f)$ is even, then there are two cases:

(i) *If all the components of the total transform $\rho_1^{-1}(\Delta)$ intersecting F' have an even valuation, then $\psi^{-1}(F') = D_1 \cup D_2$ consists of the union of two connected components, each of them being isomorphic to F' and having the same self-intersection as F'.*

(ii) *If F' intersects k irreducible components of the total transform $\rho_1^{-1}(\Delta)$ that have an odd valuation, then $D := \psi^{-1}(F')$ is connected, has genus $\frac{k}{2} - 1$ and has self-intersection $(D.D) = 2(F'.F')$.*

For a proof of this proposition one can see [LW00, Propositions 2.6.1 and 2.7.2]. It has also an extension to the case $z^3 = f(x, y)$ as shown in [LW00, Propositions 3.6.1 and 3.7.1]. However, one can not always ensure the existence of a resolution of the discriminant for which the pulled-back surface has a non-singular normalization. It is already not possible for $z^5 = f(x, y)$.

Example 2.4.18 Let $S \subset \mathbb{C}^3$ be the surface defined by $z^2 = x^3 + y^6$. Consider the projection $\pi : (x, y, z) \mapsto (x, y)$. Its discriminant Δ is reduced and defined by $x^3 + y^6 = 0$. It has three irreducible smooth and tangent components. The embedded resolution of this curve is given by two blowups. After the first one we obtain one exceptional curve of self-intersection -1 on which the three strict transforms intersect transversally in the same point. One extra blowup separates these 3 curves. We obtain the resolution graph of Fig. 2.3.

Now we consider the pull-back of S by the minimal resolution of Δ. Since there are no adjacent components of the total transform with odd valuation, the normalization of the pulled-back surface is smooth.

Applying Proposition 2.4.17, we obtain a dual graph for the resolution $Z \to S$ as in Fig. 2.4, where the values a and $[b]$ refer respectively to the self-intersection

Fig. 2.3 Minimal embedded resolution of $x^3 + y^6 = 0$

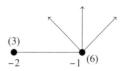

Fig. 2.4 Resolution of $z^2 = x^3 + y^6$ by Jung's method

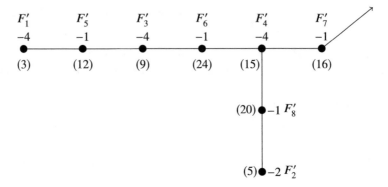

Fig. 2.5 Extended resolution of E_8 curve singularity

and the genus of the corresponding irreducible component of the exceptional fiber. When the genus is 0 we omit it in the notation.

One of the components of the graph is rational with self-intersection -1. The resolution is not minimal. Contracting that component, we obtain a dual graph with one component of genus 1 and self-intersection -1; it corresponds to the minimal resolution of S.

This minimal resolution does not factor through the blowup of the origin. In fact, if we blow up the origin, we obtain a normal surface with one singular point, and one irreducible exceptional curve. A resolution factoring through the blowup needs some exceptional fiber over the singular point.

The hyperplane sections have a base point in the minimal resolution of S which corresponds precisely to the point obtained by blowing down the rational component of self-intersection -1.

Example 2.4.19 We now describe Jung's method for the E_8-surface singularity. Its equation is $z^2 = x^3 + y^5$. Again projecting to (x, y) we obtain a reduced discriminant curve Δ defined by $x^3 + y^5 = 0$. We have already computed the minimal embedded resolution and the dual graph of this curve singularity in Example 2.3.10.

In the description of Jung's method for surfaces of type $z^2 = f(x, y)$ made immediately before Proposition 2.4.17, we saw that the normalized pull-back of the surface singularity will have singular points over each intersection point of two components having both odd valuation. This is the case at each intersection point in the minimal resolution of the discriminant curve. So we blowup once each of these intersection points. We obtain a new resolution of the discriminant with dual graph shown in Fig. 2.5. The self-intersections and valuations on this new dual graph can

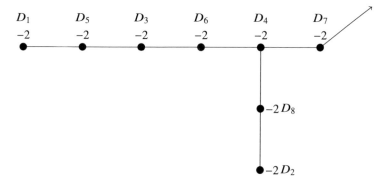

Fig. 2.6 Dual graph of E_8 surface singularity

be computed either by following the explanation we gave after Definition 2.3.9, or by using the following formula of [Lau71, Theorem 2.6]: If we denote the total transform of the discriminant as a cycle in the form $\Sigma_i v_i F_i' + \Delta^*$ where Δ^* denotes the strict transform of the discriminant curve and v_i the valuation of the function $x^3 + y^5$ along each component F_i', then for each i one has $F_i' \cdot (\Sigma_i v_i F_i' + \Delta^*) = 0$. So, knowing the self-intersections one can compute the valuations and vice versa.

The surface Z obtained as normalized pull-back of the original singularity by this extended resolution is non-singular.

Applying the process of Proposition 2.4.17 we obtain a resolution of the E_8 surface singularity with dual graph as in Fig. 2.6.

All components of the exceptional fiber are rational with self-intersection -2. This is the minimal resolution of the surface singularity.

In [Lau71, Section 2], the author gives a full description of Jung's method for surfaces of type $z^n = f(x, y)$. One can find there the results of Proposition 2.4.17, but also how to compute the valuation of the function f along the exceptional components of the obtained resolution. Precisely, in our case $z^2 = x^3 + y^5$, if the valuation v_i of the function defining the discriminant along a component F_i' is odd, then the valuation along the corresponding component in the resolution of the surface is again v_i, when v_i is even the corresponding valuation will be $\frac{v_i}{2}$. These valuations are the valuation of the polar curve defined by the considered projection to $\mathbb{C}^2$ in the resolution of the surface. This is shown in Fig. 2.7, where the value (a) refers to the valuation of the function $x^3 + y^5$.

Note that in this example, the minimal resolution factors through the blowup of the origin. In fact, a simple computation shows that the family of the images of the hyperplane sections in $\mathbb{C}^2$ has no base points after the blowup of the origin of $\mathbb{C}^2$. Since we obtained the minimal resolution of the E_8-surface singularity without blowing-down any component, then there is no base point of the hyperplane sections in the minimal resolution.

70   J. Snoussi

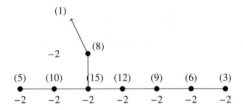

Fig. 2.7 The resolution graph for E_8 with valuations of $x^3 + y^5$

References

[AHV77] J.M. Aroca, H. Hironaka, J.L. Vicente, *Desingularization Theorems*, vol. 30. Memorias de Matemática del Instituto "Jorge Juan" [Mathematical Memoirs of the Jorge Juan Institute] (Consejo Superior de Investigaciones Científicas, Madrid, 1977) 50

[Art66] M. Artin, On isolated rational singularities of surfaces. Amer. J. Math. **88**, 129–136 (1966) 61

[BG80] R.-O. Buchweitz, G.-M. Greuel, The Milnor number and deformations of complex curve singularities. Invent. Math. **58**(3), 241–281 (1980) 58

[BG90] C. Brücker, G.-M. Greuel, Deformationen isolierter Kurvensingularitäten mit eingebetteten Komponenten. Manuscripta Math. **70**(1), 93–114 (1990) 58

[BK86] E. Brieskorn, H. Knörrer, Plane algebraic curves, in *Modern Birkhäuser Classics* (Birkhäuser/Springer Basel AG, Basel, 1986). Translated from the German original by John Stillwell, [2012] reprint of the 1986 edition 55, 57

[BL02] R. Bondil, D.T. Lê, Résolution des singularités de surfaces par éclatements normalisés (multiplicité, multiplicité polaire, et singularités minimales), in *Trends in Singularities*. Trends in Mathematics (Birkhäuser, Basel, 2002), pp. 31–81 61

[BNP14] L. Birbrair, W.D. Neumann, A. Pichon, The thick-thin decomposition and the bilipschitz classification of normal surface singularities. Acta Math. **212**(2), 199–256 (2014) 64

[BPVdV84] W. Barth, C. Peters, A. Van de Ven, Compact complex surfaces, in *Ergebnisse der Mathematik und ihrer Grenzgebiete (3) [Results in Mathematics and Related Areas (3)]*, vol. 4 (Springer, Berlin, 1984) 61, 66

[Cut04] S.D. Cutkosky, Resolution of singularities, in *Graduate Studies in Mathematics*, vol. 63 (American Mathematical Society, Providence, 2004) 50

[dJP00] T. de Jong, G. Pfister, Local analytic geometry, in *Advanced Lectures in Mathematics* (Friedr. Vieweg & Sohn, Braunschweig, 2000). Basic theory and applications 47, 50, 51, 59

[GLS07] G.-M. Greuel, C. Lossen, E. Shustin, *Introduction to Singularities and Deformations*. Springer Monographs in Mathematics (Springer, Berlin, 2007) 50

[Gre17] G.-M. Greuel, Equisingular and equinormalizable deformations of isolated non-normal singularities. Methods Appl. Anal. **24**(2), 215–276 (2017) 51

[Har77] R. Hartshorne, *Algebraic Geometry* (Springer, New York, 1977). Graduate Texts in Mathematics, No. 52 48, 61, 62

[HLOQ00] H. Hauser, J. Lipman, F. Oort, A. Quirós (eds.), *Resolution of Singularities*, vol. 181. Progress in Mathematics (Birkhäuser Verlag, Basel, 2000). A research textbook in tribute to Oscar Zariski, Papers from the Working Week held in Obergurgl, September 7–14, 1997 50

[Hir64] H. Hironaka, Resolution of singularities of an algebraic variety over a field of characteristic zero. I, II. Ann. Math. (2) **79**, 109–203 (1964); ibid. (2), **79**, 205–326 (1964) 50

[KV04] K.-H. Kiyek, J.L. Vicente, Resolution of curve and surface singularities, in *Characteristic Zero, Algebra and Applications*, vol. 4 (Kluwer Academic Publishers, Dordrecht, 2004), xxii+483 pp. 65

[Kol07] J. Kollár, Lectures on resolution of singularities, in *Annals of Mathematics Studies*, vol. 166 (Princeton University Press, Princeton, 2007) 50

[Lau71] H.B. Laufer, *Normal Two-Dimensional Singularities* (Princeton University Press, Princeton; University of Tokyo Press, Tokyo, 1971). Annals of Mathematics Studies, No. 71 69

[LT88] D.T. Lê, B. Teissier, Limites d'espaces tangents en géométrie analytique. Comment. Math. Helv. **63**(4), 540–578 (1988) 64, 65

[LW00] D.T. Lê, C. Weber, Résoudre est un jeu d'enfants. Rev. Semin. Iberoam. Mat. Singul. Tordesillas **3**(1), 3–23 (2000) 66, 67

[Mil68] J. Milnor, *Singular Points of Complex Hypersurfaces*. Annals of Mathematics Studies, No. 61 (Princeton University Press, Princeton; University of Tokyo Press, Tokyo, 1968) 57, 58

[Nar66] R. Narasimhan, *Introduction to the Theory of Analytic Spaces*. Lecture Notes in Mathematics, No. 25 (Springer, Berlin, 1966) 49

[PT69] F. Pham, B. Teissier, Lipschitz fractions of a complex analytic algebra and Zariski saturation, in *Introduction to Lipschitz Geometry of Singularities: Lecture Notes of the International School on Singularity Theory and Lipschitz Geometry, Cuernavaca, June 2018*, ed. by W. Neumann, A. Pichon. Lecture Notes in Mathematics, vol. 2280 (Springer, Cham, 2020), pp. 309–337. https://doi.org/10.1007/978-3-030-61807-0_10 49

[Pop11] P. Popescu-Pampu, Introduction to Jung's method of resolution of singularities, in *Topology of Algebraic Varieties and Singularities*. Contemporary Mathematics, vol. 538 (American Mathematical Society, Providence, 2011), pp. 401–432 65

[Sno01] J. Snoussi, Limites d'espaces tangents à une surface normale. Comment. Math. Helv. **76**(1), 61–88 (2001) 64

[Sno05] J. Snoussi, The Nash modification and hyperplane sections on surfaces. Bull. Braz. Math. Soc. (N.S.) **36**(3), 309–317 (2005) 65

[Spi90] M. Spivakovsky, Sandwiched singularities and desingularization of surfaces by normalized Nash transformations. Ann. Math. (2) **131**(3), 411–491 (1990) 63

[Tei82] B. Teissier, Variétés polaires. II. Multiplicités polaires, sections planes, et conditions de Whitney, in *Algebraic Geometry (La Rábida, 1981)*, vol. 961. Lecture Notes in Mathematics (Springer, Berlin, 1982), pp. 314–491 63

[Wal04] C.T.C. Wall, *Singular Points of Plane Curves*. London Mathematical Society Student Texts, vol. 63 (Cambridge University Press, Cambridge, 2004) 54, 55, 57

[Whi65] H. Whitney, Local properties of analytic varieties, in *Differential and Combinatorial Topology (A Symposium in Honor of Marston Morse)* (Princeton University Press, Princeton, 1965), pp. 205–244 48

[Zar39] O. Zariski, The reduction of the singularities of an algebraic surface. Ann. Math. (2) **40**, 639–689 (1939) 61

[ZS75] O. Zariski, P. Samuel, *Commutative Algebra*, vol. II (Springer, New York, 1975). Reprint of the 1960 edition, Graduate Texts in Mathematics, vol. 29 59

Chapter 3
3-Manifolds and Links of Singularities

Walter D. Neumann

Abstract This chapter gives a brief overview of general 3-manifold topology, and its implications for links of isolated complex surface singularities. It does *not* discuss Lipschitz geometry, but it provides many examples of isolated complex surface singularities on which one can work to find their Lipschitz geometry (e.g., thick-thin decompositions and inner and/or outer bilipschitz classifications).

3.1 Introduction

We start with basic three-dimensional manifold topology, in view of understanding the topology of normal complex surface singularities. The topology of a normal surface singularity determines and is determined by its "link", which is a closed oriented 3-manifold. The **link** of a singularity is defined, for example, in Theorem 1.2 and Definition 3.2.3 of the lecture notes of Anne Pichon in the present volume.

So we start by describing the geometry of closed oriented 3-manifolds. In Sect. 3.2 we give the basics of 3-manifold topology, only considering compact oriented 3-manifolds. In Sect. 3.3 we describe the "JSJ" decomposition of a 3-manifold, which is a minimal decomposition of the 3-manifold into so-called "Seifert fibered" and "hyperbolic" pieces. But hyperbolic pieces do not occur in singularity theory, so from Sect. 3.4 on we only consider Seifert fibered pieces. Seifert fibered manifolds (Seifert manifolds for short) are described in Sects. 3.4 and 3.5 describes the decomposition into Seifert manifolds in terms of "plumbing". Section 3.6 describes how plumbing and minimal good resolution are basically the same thing. Finally Sects. 3.7 and 3.8 describe the "Panorama of classical surface singularities". These classical singularities all come from "Thurston geometries" (see Sect. 3.7), although Thurston geometries have nothing to do with Lipshitz

W. D. Neumann (✉)
Barnard College, Columbia University, New York, NY, USA
e-mail: neumann@math.columbia.edu

© The Author(s), under exclusive license to Springer Nature Switzerland AG 2020
W. Neumann, A. Pichon (eds.), *Introduction to Lipschitz Geometry of Singularities*,
Lecture Notes in Mathematics 2280, https://doi.org/10.1007/978-3-030-61807-0_3

geometry, which is the main interest of most of the book. But they nevertheless provide a plethora of interesting examples of normal complex surface singularities.

3.2 Basics of 3-Manifold Topology

3.2.1 Basics

For us a three-dimensional manifold M (3-*manifold* for short) will always be smooth, compact and oriented. A non-orientable manifold M can always be dealt with by taking its orientation double cover $\tilde{M}$ along with the free $\mathbb{Z}/2$ action on $\tilde{M}$ with quotient M, but we won't need it.

We allow M to have a boundary, but we do not allow 2-spheres S^2 as boundary components (if there are S^2 boundary components we fill them by balls). We also require that the Euler characteristic $\chi(M)$ be 0 (which is automatic if M has empty boundary). Then:

Exercise 3.2.1 Prove that the components of the boundary of M are tori (hint: use Poincaré or Poincaré-Lefschetz duality).

Definition 3.2.2 (Irreducibility, [Kne29]) M is **irreducible** if every embedded S^2 in M bounds a D^3 in M (D^3 is a three-dimensional ball).

Definition 3.2.3 (Connected Sum) Suppose M_1 and M_2 are 3-manifolds other than S^3. Their **connected sum**, denoted $M_1 \# M_2$, is defined as the result of removing the interior of a closed ball from each M_i and then gluing the results together along the resulting S^2 boundaries, with orientation consistent with those of M_1 and M_2. It is a classical fact that $M_1 \# M_2$ is well defined up to oriented diffeomorphism.

Proposition 3.2.4 M *is* **irreducible** *if and only if*

1. $M \ncong S^2 \times S^1$ *(where $\cong$ means oriented diffeomorphism) and*
2. M *is not a non-trivial connected sum* $M_1 \# M_2$.

Exercise 3.2.5 Prove this proposition. (Hint: "only if" is easy, so we just give a hint for "if". Assume M is not a non-trivial connected sum. Suppose there is an essential non-separating S^2. Take a simple path γ that departs this S^2 from one side in M and returns on the other, and let N be a closed regular neighbourhood of $S^2 \cup \gamma$. What are ∂N and $M^3 \setminus N$?)

Definition 3.2.6 A manifold is **prime** if and only if it is either irreducible or it is homeomorphic to $S^2 \times S^1$.

Theorem 3.2.7 (Kneser and Milnor [Kne29, Mil62]) *Any M other than S^3 has a unique "prime factorization" in terms of connected sums. I.e., there exist prime manifolds $M_1, \ldots, M_k$, $k \geq 1$ whose connected sum $M_1 \# \cdots \# M_k$ is diffeomorphic* *to M, and this decomposition of M into prime manifolds is unique up to diffeomorphism and reordering the indices.*

Exercise 3.2.8 Show that if M is irreducible and $M \not\cong (D^2 \times S^1)$ then any boundary torus of M is **essential**, i.e., no embedded circle S^1 in a boundary torus T of M with S^1 not contractible in T can bound an embedded disk in M.

Since our interest is links of surface singularities, the following theorem is useful.

Theorem 3.2.9 (Neumann [Neu81]) *Any link of an isolated surface singularity is irreducible.*

In view of this theorem the manifold M will always be irreducible from now on.

3.3   JSJ Decomposition

JSJ stands for Jaco-Shalen & Johansson ([JS78, Joh79], see also Waldhausen [Wal69], Thurston [Thu82]). It describes a certain decomposition of M into **Seifert fibered** and **hyperbolic** pieces, unique up to isotopy.

We use a simplified version of JSJ decomposition, to fit well with links of surface singularities. This will depend on a simplified version of "Seifert fibered" (but see Remark 3.4.4). So we must define our versions.

Definition 3.3.1 For us, a manifold M is a **Seifert fibered manifold** if it has an action of the circle group $\mathbb{S}^1$ on it, acting with no fixed points.

A manifold M is **hyperbolic** if its interior $M \setminus \partial M$ admits a complete Riemannian metric of constant curvature -1 and has finite volume. This metric is then unique; i.e., two such manifolds which are homeomorphic with each other are isometric with each other.

Recall again that we only consider 3-manifolds which are compact, oriented and irreducible, and with boundary (possibly empty) consisting of tori.

Theorem 3.3.2 (Our Version of JSJ) *A 3-manifold M has a decomposition into Seifert fibered and hyperbolic pieces glued along their torus boundary components, using a least number of tori. This decomposition is then unique up to isotopy.*

A proof can be found in [Neu07] (including some classical 3-manifold theory: Dehn's Lemma, and the Loop and Sphere theorems). See also [NS97] and Remark 3.4.4.

What we call hyperbolic in that theorem was originally called "simple",[1] because the non-Seifert-fibered pieces of the JSJ decomposition were conjectured to be hyperbolic, but this was only finally proved in 2003 by Grigory Perelman (see

---

[1] "M simple" meant any embedded essential torus in M is isotopic to a boundary torus.

[MT07].[2]) This was a very major result which proved the century old Poincaré conjecture, and much more. The Poincaré conjecture states that a 3-manifold with trivial fundamental group is diffeomorphic to the 3-sphere.

Let us return to links of complex surface singularities. Hyperbolic pieces never arise in complex singularity links, so from now on we will only consider irreducible 3-manifolds with no hyperbolic pieces in their JSJ decomposition. Such a 3-manifold M is called a **graph-manifold**, as will be explained when we describe plumbing.

Returning to singularity links, we have in particular:

Theorem 3.3.3 ([Neu81]) *The link of any isolated complex surface singularity has a unique minimal decomposition into Seifert fibered pieces.*

3.4 Seifert Fibered Manifolds

We will say "Seifert manifolds" for short. We first consider the case that M is closed. Recall that for us M is **Seifert fibered** if it has an action of the circle group $\mathbb{S}^1$ on M with no fixed points.

A special case of Seifert fibered manifolds are **lens spaces**, quotients of S^3 by free actions of finite cyclic groups, defined as follows: let $S^3 := \{(z_1, z_2) \subset \mathbb{C}^2, |z_1|^2 + |z_2|^2 = 1\}$ and let p and q be coprime integers with $0 < q < p$; the $\mathbb{Z}/p$-action on (z_1, z_2) generated by $(z_1, z_2) \mapsto (e^{2\pi i/p} z_1, e^{2\pi i q/p} z_2)$ acts freely on S^3 and the quotient is called the lens space $L(p, q)$. We will discuss them later, since these have infinitely many different Seifert fibrations. For now we exclude them.

The orbits of the $\mathbb{S}^1$-action on M are called the **fibers** of M. We then have a map $\pi : M \to M/\mathbb{S}^1 = F^2$ with fibers homeomorphic to S^1. We write

$$\mathbb{S}^1 \to M \to M/\mathbb{S}^1 \text{ and write } F^2 := M/\mathbb{S}^1 .$$

A fiber is **singular** if $\mathbb{S}^1$ does not act faithfully on it, equivalently the group of fixed points of the fiber is a non-trivial cyclic subgroup $C_q \subset \mathbb{S}^1$.

Let $O_1, \ldots, O_s, s \geq 1$, be a collection of disjoint fibers including all the singular ones. Let $T_1, \ldots T_s$ be disjoint $\mathbb{S}^1$-invariant tubular neighborhoods of $O_1, \ldots, O_s$ and $M_0 := M \setminus int(T_1 \cup \cdots \cup T_s)$. Since $M_0 \to M_0/\mathbb{S}^1$ is an S^1-bundle over a connected surface with boundary, it has a section $R \subset M_0$, i.e., a surface $R \subset M_0$ which intersects each fiber of M_0 exactly once.

Let $R_i := R \cap \partial T_i$. Then R_i is homologous in T_i to some multiple $\beta_i O_i$ of the central fiber O_i of T_i. Let α_i be the order of the isotropy subgroup $C_{\alpha_i} \subset S^1$ at O_i. Let g be the genus of the surface F. Then the "unnormalized Seifert invariant" is

---

[2]Perelman put his work on the arXiv but refuses to publish or allow others to publish.

the collection of numbers

$$M(g; (\alpha_1, \beta_1), \ldots, (\alpha_s, \beta_s)),$$

satisfying $g \geq 0$, $\alpha_i \geq 1$, $\gcd(\alpha_i \beta_i) = 1$.

The invariant is not unique: one can add or remove some non-singular fibers (those with $\alpha_i = 1$) and we can change the section $R \subset M_0$ (recall that M is closed).

Theorem 3.4.1 ([NR78]) *If M and M' are Seifert manifolds with invariants*

$$M(g; (\alpha_1, \beta_1), \ldots, (\alpha_s, \beta_s)) \text{ and } M(g; (\alpha'_1, \beta'_1), \ldots, (\alpha'_t, \beta'_t))$$

then M and M' are homeomorphic preserving the fibers if and only if one can re-index the Seifert pairs so that

1. $\alpha_i = \alpha'_i$ for $i = 1, \ldots, k$; $\alpha_i = \alpha'_j = 1$ for $i, j > k$;
2. $\beta_i \equiv \beta'_i \pmod{\alpha_i}$ for $i = 1, \ldots, k$;
3. $\sum_{i=1}^s \frac{\beta_i}{\alpha_i} = \sum_{i=1}^t \frac{\beta'_i}{\alpha'_i}$

Equivalently, given $M(g; (\alpha_1, \beta_1), \ldots, (\alpha_s, \beta_s))$ one can repeatedly add or delete Seifert pairs $(1, 0)$, replace pairs (α_i, β_i), (α_j, β_j) by $(\alpha_i, \beta_i - \alpha_i)$, $(\alpha_j, \beta_j + \alpha_j)$, and reorder the indices.

Exercise 3.4.2 Prove this theorem.

We call $e(M) := -\sum_{i=1}^s \frac{\beta_i}{\alpha_i}$ the **Euler number** of the Seifert fibration, with orientations chosen so that if M is a circle bundle then $e(M)$ is the usual Euler number (or Chern class $c_1(M)$). Note that in general, $e(M)$ is not an integer. (We remark that the choice of orientations goes back to the 1960's, and some authors have preferred the opposite orientation).

The base surface $F^2 = F$ has an **orbifold** structure. The surface Euler characteristic is $\chi(F)$ and its orbifold Euler characteristic is

$$\chi := \chi(F) - \sum_{i=1}^s (1 - \frac{1}{\alpha_1}).$$

If one reverses the orientation of M, either by reversing the orientation of the fibers or of the base, the effect is to replace each (α_i, β_i) by $(\alpha_i, -\beta_i)$, so $e(-M) = -e(M)$, where $-M$ means M with reversed orientation.

As one sees, the "unnormalized Seifert invariant" is more convenient than Seifert's normalized version, which is unique up to ordering of the indices:

$$M(g; (1, b), (\alpha_1, \beta_1), \ldots, (\alpha_k, \beta_k)); \quad \alpha_i > \beta_i > 0 \text{ for } i = 1, \ldots, k.$$

Theorem 3.4.3 ([NR78]) *A Seifert manifold M is the link of a surface singularity iff $e(M) < 0$.*

3.4.1   Seifert Manifolds with Boundary

If M has $t > 0$ boundary components, the simplest classifying invariant has the form $M(g; t; (\alpha_1, \beta_1), \ldots, (\alpha_k, \beta_k))$ with $k \geq 0$ and $\alpha_i > \beta_i > 0$ for each i (but an unnormalized version can be more convenient).

Remark 3.4.4 (Seifert Manifolds over Non-orientable Surfaces) We consider only Seifert manifolds determined by S^1-actions, but Seifert's original definition allowed Seifert fibrations over non-orientable surfaces (which do not have an S^1-action). There is a "geometric" version of JSJ decompositions, mentioned briefly in [Neu07] and [NS97], which allows Seifert fibered manifolds over non-orientable surfaces and which has the advantage of being compatible with taking finite covers of M. This is more of interest to topologists than singularists, but we describe the relationship.

Given a Möbius band Mb, the boundary of a disk neighbourhood of the tangent space of Mb is an oriented 3-manifold fibered over Mb with circle fibers. It is called X, but originally called Q ([Wal67], Waldhausen 1967). X also has the structure of Seifert manifold determined by an S^1-action: its Seifert invariant is $M(0; 1; (1, 2), (1, 2))$. We avoid any Seifert manifold M fibered over a non-orientable surface F by excising either one or two copies of X from M, depending on the parity of the non-orientable genus $g(F) \in \{-1, -2, \ldots\}$.

Exercise 3.4.5 Prove that the two descriptions of X are diffeomorphic.

3.5   Plumbing, Plumbing Graphs, Dual Resolution Graphs

First we give a brief preview. As we already said, a 3-manifold M with only Seifert fibered pieces is called a **graph-manifold**. The JSJ decomposition of such a 3-manifold M can be refined into smaller pieces by what is called "plumbing". If M has the topology of the link of an isolated surface singularity, then the topological picture of M via a suitable plumbing and the dual graph of the minimal good resolution of a corresponding singularity are simply views of the same object. See also Sect. 3.5 of the lecture notes of Haydée Aguilar-Cabrera and José Luis Cisneros-Molina in the present volume.

3.5.1 Plumbing

Let F be an oriented real surface or complex curve and E a D^2-bundle E over F. It's Euler number (or first Chern class) $e(E \to F)$ has many definitions. The most geometric is given by the intersection number of $F \subset E$ with a nearby copy $F' \subset E$ intersecting transversally (each intersection point is counted as 1 or -1 depending on orientation).

Let $\xi_1 \colon E_1 \to F_1$ and $\xi_2 \colon E_2 \to F_2$ be two such bundles. Choose embedded disks $D_1^2 \subset F_1$ and $D_2^2 \subset F_2$. Let $f_i \colon D_i^2 \times D^2 \to E_i|_{D_i^2}$ be trivializations of the restricted bundles $E_i|_{D_i^2}$. To **plumb** we take the disjoint union of E_1 and E_2 and then identify the points $f_1(x, y)$ with $f_2(y, x)$ for each $(x, y) \in D^2 \times D^2$. We can perform additional plumbings of bundles $\xi_j \colon E_j \to F_j$, $\xi_k \colon E_k \to F_k$ in the same way, so long as we make sure that for each plumbing the embedded disks we use in the surfaces F_j, F_k do not overlap.

Our result is a four-dimensional manifold with boundary. We give here a schematic picture of a pair of plumbings (imagine them as real versions of complex objects).

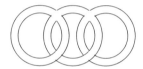

This is "solid plumbing". The boundary of the solid plumbed manifold gives 3-*manifold plumbing*, or simply **plumbing**. The boundary of the solid plumbing is the result of gluing $\partial E_1 \setminus int(E_2)$ and $\partial E_2 \setminus int(E_1)$ along their boundaries $S^1 \times S^1$, again using $(x, y) \to (y, x)$ to exchange the two circles, and again we can do additional plumbing. The result consists of gluing together circle bundles over surfaces which have the interior of disks removed.

3.5.2 Constructing S^1-Bundles over Surfaces

An S^1-bundle M of Euler number e over S^2 can be constructed as

$$M = (D^2 \times \mathbb{S}^1) \cup_{H_e} (D^2 \times \mathbb{S}^1) \text{ with } H_e = \begin{pmatrix} -1 & 0 \\ -e & 1 \end{pmatrix} \colon \mathbb{S}^1 \times \mathbb{S}^1 \to \mathbb{S}^1 \times \mathbb{S}^1,$$

where we write $\mathbb{S}^1$ additively as $\mathbb{R}/\mathbb{Z}$. The minus signs reflect the fact that the base circle $S^1 \times \{1\}$ inherits opposite orientations from each side.

Instead of D^2 on the left (or right) we can replace D^2 by an orientable surface F_0 consisting of a closed surface F with the interior of a disk removed. This gives an S^1-bundle M of Euler number e over F.

3.5.3 Seifert Manifolds via Plumbing

Denote $J = \begin{pmatrix} 0 & 1 \\ 1 & 0 \end{pmatrix}$ and consider a collection of plumbed sequences of S^1 bundles over spheres, where A represents an annulus obtained by puncturing D^2:

$$(D^2 \times S^1 \cup_{H_{e_{i1}}} A \times S^1) \cup_J \cdots \cup_J (A \times S^1 \cup_{H_{e_{is_i}}} D^2 \times S^1), \quad i = 1, \ldots, m.$$

We then plumb an S^1-bundle of Euler number e_0 over a surface F of genus g onto the start of each of these sequences to obtain a Seifert fibered manifold M. The following weighted tree is the plumbing tree for M ([g] is generally omitted if $g = 0$).

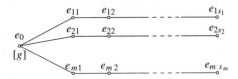

The Seifert invariant of this Seifert manifold is $M(g; (1, -e_0), (\alpha_1, \beta_1), \ldots, (\alpha_m, \beta_m))$ where α_i/β_i is the continued fraction

$$e_{i1} - \cfrac{1}{e_{i2} - \cfrac{1}{e_{i3} - \cfrac{1}{\ddots - \cfrac{1}{e_{is_i}}}}}$$

for each $i = 1, \ldots, m$.

For example the Seifert invariant of the link of the singularity E_8:

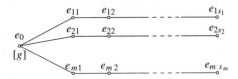

is $M(0; (1, 2), (5, -4), (3, -2), (2, -1)) \cong M(0; (1, -1), (5, 1), (3, 1), (2, 1))$. (As remarked in Sect. 3.4, some people prefer the opposite orientation and then the Seifert invariant for the link of E_8 is $M(0; (1, -2), (5, 4), (3, 2), (2, 1))$.)

3.5.4 General Plumbing

Using plumbing as described above we can construct general plumbing graphs such as the ones with diagrams in Sect. 3.8. We always use $J = \begin{pmatrix} 0 & 1 \\ 1 & 0 \end{pmatrix}$ in these constructions, as in Sect. 3.5.3.

For general plumbing (see [Neu81]) one also uses $-J := \begin{pmatrix} 0 & -1 \\ -1 & 0 \end{pmatrix}$ as well as J, but for singularity links we don't need this so we don't use it. In fact a general plumbed manifold always has a double cover of it which does not use $-J$.

3.6 Resolution and Plumbing Graphs

Let Γ be a plumbing graph with vertices $v_1, \ldots, v_r$ and Euler number weights e_i and genus weights g_i. We obtain an intersection form $A(\Gamma) = (a_{ij})$ with $a_{ij} = e_i$ if $i = j$ and a_{ij} is the number of edges connecting v_i to v_j in Γ if $i \neq j$.

If $A(\Gamma) = (a_{ij})$ is negative definite, then the plumbing graph Γ corresponds to a good resolution of an isolated surface singularity. (See Anne Pichon's notes in the present volume for resolution graphs of surface singularities. A "good" resolution means that the exceptional curves of the resolution are smooth curves intersecting each other transversally with only normal crossings).

Conversely, if the graph Γ is a graph of a minimal good resolution of an isolated surface singularity then $A(\Gamma)$ is negative definite. So we can talk just in terms of good resolution graphs rather than plumbing graphs.

We summarize some important properties of isolated complex surface singularities.

Theorem 3.6.1 ([Neu81])

1. *Isolated complex surface singularity links are irreducible 3-manifolds.*
2. *The oriented homeomorphism type of the link determines the minimal good resolution of the singularity.*
3. *If M is a singularity link then $-M$ (i.e., reversed orientation) is a singularity link if and only if M is a lens space or a torus bundle over a circle whose monodromy has trace ≥ 3.*

3.7 Relationship with Thurston Geometries

We describe briefly the relationship with the Thurston geometries. These geometries have nothing to do with the Lipschitz geometry of singularities (despite Lipschitz geometry being the main topic of the workshop). For the definition of inner and outer Lipschitz geometry, see the first paragraph of the introduction of the lecture notes of Lev Birbrair and Andrei Gabrielov, or Definition 7.2.7 of the lecture notes of Anne Pichon, or Definition 3.3.1 of the lecture notes of Maria Aparecida Soares Ruas (all in the present volume).

3.7.1   The Thurston Geometries

A Thurston geometry is a simply-connected 3-manifold with a Riemannian metric which is homogeneous and complete, and which is the universal cover of some finite volume locally homogeneous 3-manifold.

There are 8 such geometries. Six of them, $\mathbb{S}^3$, $\mathbb{N}il$, $\mathbb{P}SL$, $\mathbb{S}^2 \times \mathbb{E}^1$, $\mathbb{E}^3$ and $\mathbb{H}^2 \times \mathbb{E}^1$, are the universal covers of Seifert manifolds. The two others are called $\mathbb{S}ol$ and $\mathbb{H}^3$. Each of them is unique up to a finite dimensional deformation of its homogeneous complete Riemannian metric, e.g., one-dimensional for $\mathbb{S}^3$, $\mathbb{S}^2 \times \mathbb{E}$, $\mathbb{H}^3$ or $\mathbb{H}^2 \times \mathbb{E}^1$ (which one can normalize by taking curvature 1 resp. -1 for $\mathbb{S}^3$ and $\mathbb{S}^2$ resp. $\mathbb{H}^3$ and $\mathbb{H}^2$).

Given a Seifert fibered manifold with $M \to F = F^2$, the base surface F has extra structure as an **orbifold** F. If M is closed there are two invariants associated with this situation (see paragraph after Theorem 3.4.1):

* the Euler number of the Seifert fibration $e(M \to F)$,
* and the orbifold Euler characteristic χ of F, as described earlier.

The relevant geometry is determined by these as:

	$\chi > 0$	$\chi = 0$	$\chi < 0$
$e \neq 0$	$\mathbb{S}^3$	$\mathbb{N}il$	$\mathbb{P}SL$
$e = 0$	$\mathbb{S}^2 \times \mathbb{E}^1$	$\mathbb{E}^3$	$\mathbb{H}^2 \times \mathbb{E}^1$

We have already said that a Seifert manifold is a singularity link iff $e < 0$, and that hyperbolic manifolds never arise in singularity links, so the singularity links which are locally homogeneous finite volume 3-manifolds can only come from four of the Thurston geometries: $\mathbb{S}^3$, $\mathbb{N}il$, $\mathbb{P}SL$ and $\mathbb{S}ol$. They do indeed occur for all four of these geometries, and they include many classical singularities, which we will now describe.

3.8   The Panorama of Classical Singularities

These are the families of "classical singularities" listed below, which are determined by their connections to the 3-manifold geometries listed above. All of them are complex normal surface singularities.

3.8.1   ADE

These are the Du-Val singularities. Their resolution graphs and Seifert invariants are as follows.

$A_n : x^2 + y^2 + z^{n+1} = 0$ $L(n, n-1) = M(0; (1, n))$

$D_n : x^2 + y^2 z + z^{n-1} = 0$ $M(0; (1, -1), (2, 1), (2, 1), (n-2, 1))$

$E_6 : x^2 + y^3 + z^4 = 0$ $M(0; (1, -1), (2, 1), (3, 1), (3, 1))$

$E_7 : x^2 + y^3 + yz^3 = 0$ $M(0; (1, -1), (2, 1), (3, 1), (4, 1))$

$E_8 : x^2 + y^3 + z^5 = 0$ $M(0; (1, -1), (2, 1), (3, 1), (5, 1))$

3.8.2   Hirzebruch-Jung Singularities

Those are the normal surface singularities whose links are lens spaces. The resolution graph is a string of rational curves. As mentioned earlier, a lens space has infinitely many different Seifert invariants; they are described in [JN83] and the recent papers [GL18] and [Web18] give a clearer and more detailed version.

Sections 3.8.1 and 3.8.2 belong to the Thurston geometry S^3.

3.8.3   Quasihomogeneous Surface Singularities

They are singularities with a unique $\mathbb{S}^1$ action with no fixed point outside the origin. Their links are Seifert fibered manifolds. The resolution graph has a central node connecting to strings of rational curves (see Sect. 3.5.3).

Section 3.8.3 belongs to the Thurston geometries S^3, $\mathbb{N}il$ and $\mathbb{P}SL$, thus overlapping with 3.8.1.

3.8.4   Hirzebruch Cusp Singularities

There are two types of Hirzebruch cusp singularities ("cusp singularities" for short). A **cusp singularity** usually means one whose minimal resolution graph is a cycle of vertices representing rational curves E_i. We denote the resolution graph by $[-e_1, \ldots, -e_k]$ (this notation is well-defined up to cyclic permutation and reversal). If $k > 1$ then $-e_i$ is the self-intersection number $E_i \cdot E_i$; but if $k = 1$ then $E_1 \cdot E_1 = -e_1 + 2$, since E_1 intersects itself normally in one point. The e_i's satisfy $e_i \geq 2$ for all i and some $e_j \geq 3$.

The cusp singularity link is a torus bundle over S^1 with monodromy

$$A = \begin{pmatrix} 0 & -1 \\ 1 & e_k \end{pmatrix} \cdots \begin{pmatrix} 0 & -1 \\ 1 & e_1 \end{pmatrix}.$$

This matrix A has trace ≥ 3. Conversely, every torus bundle over S^1 with monodromy matrix A having trace ≥ 3 is the link of a cusp singularity (such a matrix is conjugate to a matrix as above and the singularity is the corresponding cusp). The fundamental group of the singularity link is the semi-direct product $\mathbb{Z}^2 \rtimes \mathbb{Z}$ where a generator of $\mathbb{Z}$ acts on $\mathbb{Z}^2$ by the matrix A.

The second type of Hirzebruch cusp singularity is closely related to first. It is one with a resolution of graph the form:

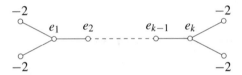

with $k \geq 2$, $e_i \leq -2$ for each i and at least one $e_i \leq -3$. It has a double cover with singularity link of the first type, namely it has resolution graph

$$2 + 2e_1 \overset{e_2 \quad e_3}{\diagup} \text{-} \text{-} \text{-} \text{-} \text{-} \overset{e_{k-1}}{\diagup} 2 + 2e_k$$
$$\underset{e_2 \quad e_3}{\diagdown} \text{-} \text{-} \text{-} \text{-} \underset{e_{k-1}}{\diagdown}$$

Note that when we speak of a "finite cover of an isolated singularity" we mean that there is an analytic map $(\tilde{X}, 0) \to (X, 0)$ which is a finite covering map on the complement of the point 0 in $(X, 0)$.

Section 3.8.4 belongs to the Thurston geometry $\mathcal{S}ol$.

3.8.5 Further Comments

The singularities of the Panorama are very special. All of them except quasihomogeneous surface singularities are "taut", i.e., analytically determined by their topological type (some quasihomogeneous surface singularities are also taut). Taut normal surface singularities were classified by Henry Laufer in 1973 ([Lau73]).

Of course for taut singularities the topology determines the Lipschitz geometry since it determines the analytic type. But for quasihomogeneous surface singularities which are not taut the topology does not necessarily determine the Lipschitz geometry.

A classic example of Briançon-Speder in the Panorama is the family of singularities

$$X_t = \{(x, y, z) \subset \mathbb{C}^3 : x^5 + z^{15} + y^7 z + t x y^6 = 0\},$$

with $(X_t, 0)$ depending on the parameter t. Each X_t is quasihomogeneous since it has an S^1 action $(e^{3\theta i} x, e^{2\theta i} y, e^{\theta i} z)$, $\theta \in [0, 2\pi]$ preserving X_t. The topology does not change as t changes, but the inner Lipschitz geometry (and hence also the outer Lipschitz geometry) changes very radically when t becomes 0. For details see, e.g., its discussion in [BNP14, Example 15.7] or Example 3.21 in Anne Pichon's lecture notes of the present volume. The same holds for other Briançon-Speder families.

A striking property of the Panorama of singularities is that any finite cover of one of the singularities of the Panorama which is unramified outside the singular point is a singularity of the Panorama.

Exercise 3.8.1 Prove this.

The Lipschitz geometry of a finite cover of a singularity $(X, 0)$ rarely has a relationship with the Lipschitz geometry of $(X, 0)$, so the Panorama provides a plethora of examples whose singularities have Lipschitz geometry which have not yet been described.

References

[BNP14] L. Birbrair, W.D. Neumann, A. Pichon, The thick-thin decomposition and the bilipschitz classification of normal surface singularities. Acta Math. **212**(2), 199–256 (2014) 85

[GL18] H. Geiges, C. Lange, Seifert fibrations of lens spaces. Abh. Math. Semin. Univ. Hambg. **88**(1), 1–22 (2018) 83

[JN83] M. Jankins, W.D. Neumann, *Lectures on Seifert Manifolds*, vol. 2. Brandeis Lecture Notes (Brandeis University, Waltham, 1983) 83

[Joh79] K. Johannson, *Homotopy Equivalences of 3-Manifolds with Boundaries*, vol. 761. Lecture Notes in Mathematics (Springer, Berlin, 1979) 75

[JS78] W. Jaco, P.B. Shalen, A new decomposition theorem for irreducible sufficiently-large 3-manifolds, in *Algebraic and Geometric Topology (Proceedings of symposia in pure mathematics, Stanford University, Stanford, Calif., 1976), Part 2*. Proceedings of symposia in pure mathematics, vol. XXXII (American Mathematical Society, Providence, 1978), pp. 71–84 75

[Kne29] H. Kneser, Geschlossene flächen in dreidimensionalen mannigfaltigkeite. Jahresbericht der Deutschen Mathematiker Vereinigung **38**, 248–260 (1929) 74

[Lau73] H.B. Laufer, Taut two-dimensional singularities. Math. Ann. **205**, 131–164 (1973) 84

[Mil62] J. Milnor, A unique decomposition theorem for 3-manifolds. Am. J. Math. **84**, 1–7 (1962) 74

[MT07] J. Morgan, G. Tian, *Ricci Flow and the Poincaré Conjecture*, vol. 3. Clay Mathematics Monographs (American Mathematical Society, Providence; Clay Mathematics Institute, Cambridge, 2007) 76

[Neu81] W.D. Neumann, A calculus for plumbing applied to the topology of complex surface singularities and degenerating complex curves. Trans. Amer. Math. Soc. **268**(2), 299–344 (1981) 75, 76, 81

[Neu07] W.D. Neumann, Graph 3-manifolds, splice diagrams, singularities, in *Singularity Theory* (World Scientific Publishing, Hackensack, 2007), pp. 787–817 75, 78

[NR78] W.D. Neumann, F. Raymond, Seifert manifolds, plumbing, μ-invariant and orientation reversing maps, in *Algebraic and Geometric Topology (Proceedings of Symposia, University of California, Santa Barbara, Calif., 1977)*, vol. 664. Lecture Notes in Mathematics (Springer, Berlin, 1978), pp. 163–196 77, 78

[NS97] W.D. Neumann, G.A. Swarup, Canonical decompositions of 3-manifolds. Geom. Topol. **1**, 21–40 (1997) 75, 78

[Thu82] W.P. Thurston, Three-dimensional manifolds, Kleinian groups and hyperbolic geometry. Bull. Amer. Math. Soc. (N.S.) **6**(3), 357–381 (1982) 75

[Wal67] F. Waldhausen, Eine Klasse von 3-dimensionalen Mannigfaltigkeiten. I, II. Invent. Math. **3**, 308–333 (1967); ibid. **4**, 87–117 (1967) 78

[Wal69] F. Waldhausen, On the determination of some 3-manifolds by their fundamental groups alone, in *Proceedings of International Symposium in Topology, Hercy-Novi, Yugoslavia, 1968, Belgrad* (1969), pp. 331–332 75

[Web18] C. Weber, Lens spaces among 3-manifolds and quotient surface singularities. Rev. R. Acad. Cienc. Exactas Fís. Nat. Ser. A Mat. RACSAM **112**(3), 893–914 (2018) 83

Chapter 4
Stratifications, Equisingularity and Triangulation

David Trotman

Abstract This text is based on 3 lectures given in Cuernavaca in June 2018 about stratifications of real and complex analytic varieties and subanalytic and definable sets. The first lecture contained an introduction to Whitney stratifications, Kuo-Verdier stratifications and Mostowski's Lipschitz stratifications. The second lecture concerned equisingularity along strata of a regular stratification for the different regularity conditions: Whitney, Kuo-Verdier, and Lipschitz, including thus the Thom-Mather first isotopy theorem and its variants. (Equisingularity means continuity along each stratum of the local geometry at the points of the closures of the adjacent strata.) A short discussion follows of equisingularity for complex analytic sets including Zariski's problem about topological invariance of the multiplicity of complex hypersurfaces and its bilipschitz counterparts. In the real subanalytic (or definable) case we mention that equimultiplicity along a stratum translates as continuity of the density at points on the stratum, and quote the relevant results of Comte and Valette generalising Hironaka's 1969 theorem that complex analytic Whitney stratifications are equimultiple along strata. The third lecture provided further evidence of the tameness of Whitney stratified sets and of Thom maps, by describing triangulation theorems in the different categories, and including definable and Lipschitz versions. While on the subject of Thom maps we indicate examples of their use in complex equisingularity theory and in the definition of Bekka's (c)-regularity. Some very new results are described as well as old ones.

D. Trotman (✉)
Aix Marseille Univ, CNRS, Centrale Marseille, I2M, Marseille, France
e-mail: david.trotman@univ-amu.fr

© The Author(s), under exclusive license to Springer Nature Switzerland AG 2020
W. Neumann, A. Pichon (eds.), *Introduction to Lipschitz Geometry of Singularities*,
Lecture Notes in Mathematics 2280, https://doi.org/10.1007/978-3-030-61807-0_4

4.1 Stratifications

Consider some singular spaces which are real algebraic varieties.

(i) Let V be the curve $\{y^2 = x^2 + x^3\}$. Then V has a double point singularity at the origin in $\mathbb{R}^2$ (Fig. 4.1).

(ii) Let V be the curve $\{y^2 = x^3\}$. Here V has a cusp singularity at the origin in $\mathbb{R}^2$.

(iii) Let V be the surface $\{z^2 = x^2 + y^2\}$ in $\mathbb{R}^3$. This is a cone with an isolated singularity at the origin.

(iv) Let V be the variety $\{z(x^2 + (y+z)^2) = 0\}$ in $\mathbb{R}^3$. This is the union of a plane P and a transverse line ℓ (Fig. 4.2).

In each of these four examples the singular set of the variety V is a point. However in Example (iv) the regular part of V is not equidimensional—both 1 and 2 occur as local dimensions. In the other examples the regular part is equidimensional.

Now we give an example of a surface whose singular set is a line.

(v) Let V be $\{y^2 = t^2x^2 - x^3\}$ in $\mathbb{R}^3$. Then the singular set of V is the line $< Ot >$ (Fig. 4.3).

Fig. 4.1 $y^2 = x^2 + x^3$

Fig. 4.2
$z(x^2 + (y+z)^2) = 0$

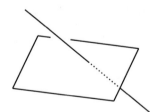

Fig. 4.3 $y^2 = t^2x^2 - x^3$

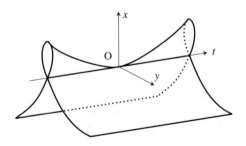

4.1.1   Whitney's Conditions (a) and (b)

We will "stratify" our singular spaces X (closed subsets of some $\mathbb{R}^n$) by expressing them as a union of smooth manifolds defined by means of a filtration by closed subsets:

$$X = X^d \supseteq X^{d-1} \supseteq \cdots \supseteq X^1 \supseteq X^0 \supseteq X^{-1} = \emptyset$$

where each difference $X^j - X^{j-1}$ is either a smooth manifold of dimension j, or is empty. Each connected component of $X^j - X^{j-1}$ is called a **stratum** of dimension j.

In Example (iv) the natural filtration can be either

$$V \supset \ell \supset \{0\} \supset \emptyset$$

or

$$V \supset \ell \supset \emptyset = \emptyset.$$

Because the intersection point 0 is different from other points on the line ℓ we like to take the first filtration. The natural 1-dimensional stratum is thus $\ell \setminus \{0\}$. Also, in Example (v) the natural 1-dimensional strata are the two components of $< Ot > \setminus\{0\}$, because 0 is a different point. The local topology of V at points of the t-axis changes as we pass through $t = 0$.

Question How can we formalise this difference?

Whitney (§19 in [Whi65a]; §8 in [Whi65b]) defined two **regularity** conditions (a) and (b).

Let X, Y be two strata (disjoint smooth submanifolds of $\mathbb{R}^n$) and let $y_0 \in Y \cap \overline{X} \setminus X$. Then **condition** (a) **holds for** (X, Y) **at** y_0 if given any sequence of points $x_i \in X$ tending to y_0, such that the tangent spaces $T_{x_i} X$ tend to τ in the appropriate grassmannian, then $T_0 Y \subseteq \tau$

If we stratify Example (iv) without removing the point 0 from the line ℓ, then Whitney's condition (a) fails to hold for the pair of strata $(P - \{0\}, \ell)$ at $0 \in \ell$, where P is the plane $\{z = 0\}$.

Look now at Example (v) (the Whitney cusp). We can stratify V by the filtration

$$V \supset < Ot > \supset \emptyset$$

and then Whitney's condition (a) holds for $(V \setminus < Ot >, < Ot >)$ at all points. So we need to impose more regularity so that the point $\{0\}$ becomes a stratum: the local topology of slices $\{t = constant\} \cap V$ changes at $t = 0$.

We say that **condition** (*b*) **holds for** (X, Y) at $y_0 \in Y \cap (\overline{X} - X)$ if given sequences $x_i \in X$ and $y_i \in Y$ both tending to y_0, such that $T_{x_i} X$ tends to τ and $y_i x_i / \|y_i x_i\|$ tends to λ, then $\lambda \in \tau$.

Look at Example (v). A sequence on $V \cap \{y = 0\} = \{x(t^2 + x) = 0\}$, i.e. $x = -t^2$, has $\lambda = (1 : 0 : 0)$ and $\tau = (1, 0, 0)^{\perp}$ (the (t, y)-plane), so that $\lambda \notin \tau$, and condition (*b*) fails to hold.

Definition 4.1.1 A locally finite stratification of a closed set $Z \subseteq \mathbb{R}^n$ is called a **Whitney stratification** if every adjacent pair of strata satisfy condition (b) of Whitney.

Lemma 4.1.2 *Condition* (*b*) *implies condition* (*a*).

The proof is an exercise.

Theorem 4.1.3 (Theorem 2.B.1 in [Tho69], Corollary 10.5 in [Mat12]) *A Whitney stratification automatically satisfies the* **frontier condition**, *i.e., whenever a stratum Y intersects the closure of a stratum X, then Y is contained in the closure of X.*

Remark 4.1.4 In Example (iv) the stratification

$$V \supset \ell \supset \emptyset$$

does not satisfy the frontier condition. In Example (v), stratifying by

$$V \supset < Ot >$$

there are 4 strata of dimension 2 (recall that the strata are the connected components of $V^2 \setminus V^1$).

Let $X_1 = \{V \cap \{x \le 0\} \cap \{y \le 0\}\}$, $X_2 = \{V \cap \{x \le 0\} \cap \{y \ge 0\}\}$, $X_3 = \{V \cap \{t \le 0\} \cap \{x \ge 0\}\}$ and $X_4 = \{V \cap \{t \ge 0\} \cap \{x \ge 0\}\}$. These are the 4 strata of our stratification (Fig. 4.4) We see that $Y \cap X_3 \ne \emptyset$, but that Y is not a subset of $\overline{X_3}$, and similarly for X_4, so that the frontier condition fails for (X_3, Y) and for (X_4, Y).

Fig. 4.4 The strata of V

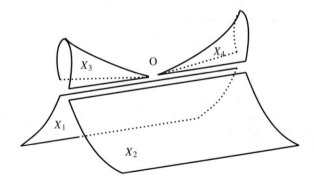

However the frontier condition holds for the pairs of adjacent strata (X_1, Y) and (X_2, Y).

Theorem 4.1.5 ([Whi65a], Theorem 19.2) *Every analytic variety V (real or complex) admits a Whitney stratification.*

In fact this is also true for more general sets: for semialgebraic sets (Łojasiewicz [Loj65], Thom [Tho65], Wall [Wal75], Kaloshin [Kal05]), more generally for subanalytic sets (Hironaka [Hir73], Hardt [Har75], Verdier [Ver76], Denkowska, Wachta and Stasica [DWS85]), and even more generally for definable sets in o-minimal structures (Loi [Loi98], van den Dries and Miller [vdDM96], Nguyen, Trivedi and Trotman [NTT14], and Halupczok [Hal14a, Hal14b]).

One says a regularity condition is **generic** if every variety (or semialgebraic set, etc.) admits a stratification such that every pair of adjacent strata satisfy the regularity condition.

So Whitney's condition (a) and Whitney's condition (b) are generic.

[The term "generic" arises as follows. To prove existence of a regular stratification one proves that for an adjacent pair of strata (X, Y),

$$\{y \in Y \subset \overline{X} - X | (X, Y) \text{ is regular at } y\}$$

is generic in Y in the Baire sense of containing a countable intersection of open dense subsets, so that its complement can be added to a closed set lower in the filtration than $\overline{Y}$.]

Theorem 4.1.6 *Both (a) and (b) are C^1 invariants, i.e. given an (a)-regular (resp. (b)-regular) stratification of $Z \subset \mathbb{R}^n$ and a C^1 diffeomorphism $\phi : \mathbb{R}^n \longrightarrow \mathbb{R}^n$ then $\phi(Z)$ inherits an (a)-regular (resp. (b)-regular) stratification.*

The previous result follows at once from the following characterizations of (a) and (b).

Let $\phi : (U, U \cap Y, y) \longrightarrow (\mathbb{R}^n, \mathbb{R}^m \times 0^{n-m}, 0)$ be a C^1 chart for Y as a submanifold of $\mathbb{R}^n$. Let $\pi_\phi = \phi^{-1} \circ \pi_m \circ \phi : U \longrightarrow U \cap Y$ where $\pi_m : \mathbb{R}^n \longrightarrow \mathbb{R}^m \times 0^{n-m}$ is projection onto the first m coordinates, and let $\rho_\phi = \rho_m \circ \phi : U \longrightarrow [0, \infty)$ where $\rho_m : \mathbb{R}^n \longrightarrow [0, \infty)$ is defined by $\rho_m(x_1, \ldots, x_n) = \Sigma_{i=m+1}^n x_i^2$.

First we characterize (a)-regularity.

Theorem 4.1.7 (Theorem A in [Tro79]) *A pair of adjacent strata (X, Y) is (a)-regular at $y \in Y \Longleftrightarrow$ for every C^1 foliation $\mathcal{F}$ transverse to Y at y, there is a neighbourhood of y in which $\mathcal{F}$ is transverse to $X \Longleftrightarrow$ for every C^1 chart (U, ϕ) for Y at y, there exists a neighbourhood V of y, $V \subset U$, such that the retraction $\pi_\phi|_{V \cap X}$ is a submersion.*

Next we characterize (b)-regularity.

Theorem 4.1.8 (Theorem B in [Tro79]) *A pair of adjacent strata (X, Y) is (b)-regular at $y \in Y \Longleftrightarrow$ for every C^1 chart (U, ϕ) for Y at y, there is a neighbourhood V of y, $V \subset U$, such that $(\pi_\phi, \rho_\phi)|_{V \cap X}$ is a submersion.*

4.1.2   The Kuo-Verdier Condition (w)

A natural idea is to seek stronger generic regularity conditions. Now Whitney's condition (a) says that

$$\text{dist}(T_x X, T_{y_0} Y) \longrightarrow 0 \text{ as } x \to y_0.$$

We can quantify this convergence in the stronger Kuo-Verdier condition [Ver76]

$$(w) \qquad \text{dist}(T_x X, T_{y_0} Y) = O(\|x - \pi_Y(x)\|) = O(\text{dist}(x, Y))$$

i.e., there exists $C > 0$, and a neighborhood U of y_0 in $\mathbb{R}^n$ such that

$$\text{dist}(T_x X, T_{y_0} Y) \leq C \|x - \pi_Y(x)\| \ \forall x \in U \cap X.$$

Here π_Y denotes a C^1 submersive retraction from a tubular neighbourhood of Y onto Y.

Theorem 4.1.9 *Condition (w) is generic, i.e. (w)-regular stratifications exist in the various classes of sets.*

See Verdier (Théorème 2.2 in [Ver76]), Denkowska and Wachta [DW87] or Łojasiewicz, Stasica and Wachta [LSW86] in the subanalytic case, and Tà Lê Loi [Loi98] for definable sets.

In Brodersen and Trotman ([BT79], Proposition 2) it was shown that condition (w) can be characterized by lifting of vector fields. Precisely, (w) holds for (X, Y) at $y_0 \in Y$ if and only if every vector field v_Y on Y extends in a neighborhood U of y_0 to a vector field v_X on X which is **rugose**: $\exists C > 0$ such that

$$\forall x \in U \cap X, \forall y \in U \cap Y, \|v_X(x) - v_Y(y)\| \leq C \|x - y\|.$$

Remark 4.1.10 The stratified vector field on $X \cup Y$ is weakly Lipschitz. For it to be Lipschitz one would need to impose the condition that

$$\forall x \in U \cap X, \forall x' \in U \cap X, \|v_X(x) - v_X(x')\| \leq C \|x - x'\|.$$

Theorem 4.1.11

(1) For semi algebraic sets (also for subanalytic sets, and for definable sets in o-minimal structures), (w) implies (b).

(2) For complex analytic stratifications, $(w) \Longleftrightarrow (b)$.

For (1) in the subanalytic case see Kuo [Kuo71] or Verdier (Théorème 1.5 in [Ver76]). The definable case is due to Loi [Loi98]. (2) is due to Teissier (Théorème 1.2 in Chapter V of [Tei82]).

Example

(vi) Let $V = \{y^4 = t^4 x + x^3\} \subset \mathbb{R}^3$, and stratify by $V \supset < Ot > \supset \emptyset$. This satisfies
   (*b*) but not (*w*). In fact V is a C^1 submanifold of $\mathbb{R}^3$, as proved in my thesis
   (Example 7.1 in [Tro77]). This shows that (*w*) is not a C^1 invariant.

   One can check easily that condition (*w*) is a C^2 invariant. In fact it is a $C^{1+\epsilon}$
invariant where the ϵ refers to a Hölder property of the first derivative. This fact is
useful in proofs that (*w*) is a generic condition.

4.1.3 Mostowski's Lipschitz Stratifications

Mostowski [Mos85] introduced in 1985 a very strong regularity condition for complex analytic varieties and proved genericity. Then Parusiński successively proved
genericity of Mostowski's Lipschitz condition for real analytic varieties [Par88b],
for semi-analytic sets [Par88a] and finally for subanalytic sets [Par94]. Recently, N.
Nguyen and Valette [NV16] proved genericity of Mostowski's Lipschitz condition
for definable sets in polynomially bounded o-minimal structures.

   Mostowski's original condition is rather technical and takes long to write down,
so we will give an equivalent version due to Parusiński (Proposition 1.5 of [Par88a]).

Definition 4.1.12 A stratification Σ of a set Z defined by

$$Z = Z^d \supset Z^{d-1} \supset \cdots \supset Z^0 \supset Z^{-1} = \emptyset$$

is said to be a Lipschitz stratification (or satisfy condition (L)) if there exists a
constant $K > 0$ such that for every subset $W \subset Z$ such that

$$Z^{j-1} \subseteq W \subseteq Z^j$$

for some $j \doteq \ell, \ldots, d$ where ℓ is the lowest dimension of a stratum of Z, each
Lipschitz Σ-compatible vector field on W with Lipschitz constant L which is
bounded on $W \cap Z^\ell$ by a constant $C > 0$, can be extended to a Lipschitz Σ-
compatible vector field on Z with Lipschitz constant $K(L + C)$.

Proposition 4.1.13 *Every Lipschitz stratification satisfies condition* (*w*).

   This proposition is actually an immediate consequence of Mostowki's original
definition [Mos85].

   In fact, so far the Lipschitz condition is the strongest generic regularity condition
on stratifications of definable sets.

4.1.4 Applications of Whitney (a)-regularity

We have been describing successively stronger regularity conditions. So, why should one study the rather weak Whitney (a)-regular stratifications? One reason is because in singularity theory and dynamical systems (in classification problems and in the study of stability) one often uses that transversality to a Whitney stratification is an open condition. And in fact one can show the following equivalence, which gives another characterisation of (a)-regularity and hence another proof that (a) is a C^1 invariant.

Theorem 4.1.14 (Theorem 1.1 in [Tro77, Tro79]) *Given a stratification Σ of a closed subset Z of a smooth manifold M, Σ is Whitney (a)-regular $\Leftrightarrow \{f : N \longrightarrow M | f$ is transverse to $\Sigma\}$ is an open set of $C^1(N, M)$ in the strong C^1 topology, for all C^1 manifolds N.*

Recently, Trivedi gave holomorphic versions of this theorem for Stein manifolds N, M [Tri13].

Another application of Whitney (a)-regularity is the following.

Theorem 4.1.15 (Kuo, Li, and Trotman [KTL89]) *Given a stratum X of an (a)-regular stratification of a subset Z of $\mathbb{R}^n$, then for all $x \in X$ and for every pair of Lipschitz transversals M_1, M_2 to X at x (a Lipschitz transversal is defined to be the graph of a Lipschitz map $N_x X \to T_x X$), there is a homeomorphism*

$$(M_1, Z \cap M_1, x) \longrightarrow (M_2, Z \cap M_2, x).$$

These results justify the study and verification of (a)-regularity.

4.2 Equisingularity

We have seen in the examples how Whitney (b)-regularity allows us to distinguish points where the local topology changes. This is in fact a general property.

4.2.1 Topological Equisingularity

Theorem 4.2.1 (Thom-Mather: Théorème 2.B.1 in [Tho69] and Proposition 11.1 in [Mat12]) *A Whitney (b)-regular stratification (of a closed subset Z of a manifold M) is locally topologically trivial along each stratum.*

This means more precisely that for every point x in a stratum X there is a neighbourhood U of x in M, a stratified set L, and a stratified homeomorphism

$$h : (U, U \cap Z, U \cap Z, x) \longrightarrow (U \cap X) \times (\mathbb{R}^k, cL, \star)$$

such that $p_1 \circ h = \pi_X$, where cL denotes the cone on L with vertex $\star$.

The proof of this theorem, known as the Thom-Mather first isotopy theorem, is by integration of a continuous stratified controlled vector field v on Z: for each stratum X, there is a lift of v_X to a vector field v_Y on neighbouring strata Y such that $\pi_{X\star}v_Y = \pi_X$ and $\rho_{X\star}v_Y = 0$ (these two conditions state that v_Y is a lift of v_X and that v_Y is tangent to the level hypersurfaces of ρ_Y).

In particular the isotopy theorem states that the local topological type of Z at points of a stratum X is locally constant, hence constant, as X is connected.

Remark 4.2.2 That the lifted stratified vector field $v_X \cup v_Y$ in the Thom-Mather isotopy theorem can be chosen to be continuous was first independently proved by Shiota (Lemma I.1.5 in [Shi97]) and du Plessis [dP99]. A much stronger statement was recently proved as part of Whitney's fibering conjecture (Conjecture 9.2 in [Whi65b]). From the statement of the Thom-Mather theorem one can see that h defines a foliation by leaves $h^{-1}(p)$ for $p \in cL$, each diffeomorphic to $U \cap X$. In the complex holomorphic case Whitney conjectured that the leaves be holomorphic and that their tangents vary continuously as we take the limit for points on a stratum Y tending to an adjacent stratum X. This was proved by Parusiński and Paunescu (Theorem 7.6 in [PP17]) in the real and complex algebraic and analytic cases, using a hypothesis of a stratification which is Zariski equisingular (in a generic sense), stronger than (w)-regularity. In 2018 Parusiński has announced that this generic Zariski equisingularity implies the Lipschitz regularity of Mostowski for families of hypersurfaces in $\mathbb{C}^3$.

With the hypothesis of (b)-regularity (in fact with the even weaker (c)-regularity defined in Lecture III), Whitney's fibering conjecture was proved in the smooth case in 2017 by Murolo, du Plessis and Trotman (Theorem 7 in [MdPT17]): the leaves of $\{h^{-1}(p)\}_{p\in cL}$ form a $C^{0,1}$ foliation.

We saw above that a Kuo-Verdier (w)-regular stratification admits locally rugose vector fields tangent to strata. These may be integrated to provide a local rugose trivialization.

Theorem 4.2.3 (Verdier: Théorème 4.14 in [Ver76]) *Every (w)-regular stratification is locally rugosely trivial along strata.*

This is to say that a homeomorphism defining a trivialization (almost) as in the Thom-Mather theorem can be chosen to be rugose. This requires two clarifications. Firstly the homeomorphism of the Thom-Mather theorem is in fact already rugose because it is controlled—h can be chosen to respect the level hyper surfaces of the control function ρ_X. Secondly in Verdier's theorem [Ver76] the homeomorphism is not in general with the product of $U \cap X$ and a cone, but rather with a normal slice— see the counterexample using the topologist's sine curve below (Example 4.2.10).

Because of the definition we gave above of a Lipschitz stratification (of Mostowski) it is no surprise that there is also a local trivialization theorem for Lipschitz stratifications.

Theorem 4.2.4 (Mostowski [Mos85], Parusiński (Theorem 1.6 in [Par94]))
Every Lipschitz stratification is locally bilipschitz trivial along strata.

Corollary 4.2.5 *Every semialgebraic/subanalytic/definable subset of $\mathbb{R}^n$ admits a locally bilipschitz trivial stratification.*

Remark 4.2.6 Here "definable" must be taken in a polynomially bounded o-minimal structure: this means that every definable function $f : \mathbb{R}^n \longrightarrow \mathbb{R}$ satisfies $|f(x)| \leq C||x||^k$, for some $C > 0$ and some positive integer k, in a neighbourhood of infinity (i.e. outside some compact set $K \subset \mathbb{R}^n$).

Example 4.2.7 (Parusiński) Let $X(t)$ be $< Ox > \cup\{(x, x^t, t)|x > 0\} \subset \mathbb{R}^3$. Then the Lipschitz types of the $X(t)$ are all distinct for $t > 1$. Hence there is no locally bilipschitz trivial stratification of $\bigcup X(t)$, thus no Lipschitz stratification.

This example is definable in any o-minimal structure which is not polynomially bounded ($x^t = exp(t\log x)$). Recall the theorem of C. Miller.

Theorem 4.2.8 (Miller [Mil94]) *An o-minimal structure is not polynomially bounded if and only if the exponential function is definable in the structure.*

Remark 4.2.9 When working outside of the class of definable sets, in the local triviality theorems for (w)-regular and Lipschitz stratifications we must replace cL by a normal slice F (not necessarily a cone), as shown by the following example.

Example 4.2.10 Let $Z = \overline{\{y = sin(1/x), x \neq 0\}} \subset \mathbb{R}^2$, the topologist's sine curve. If $Y = (-1, 1) \times 0$ and $X = Z - Y$, with $(-1, 0)$ and $(1, 0)$ the 0-strata, then we obtain a (w)-regular stratification and a Lipschitz stratification, but not a (b)-regular stratification. The stratification is locally topologically trivial indeed locally bilipschitz trivial along Y but is not locally topologically conical. It is clear that Z is not definable in an o-minimal structure because the x-axis intersects Z in an infinite number of connected components.

Although local bilipschitz triviality is in general strictly weaker than the Lipschitz property of Mostowski, there exist (w)-regular stratified sets which are not locally bilipschitz trivial.

Example 4.2.11 (Koike) Let $Z = \{y^2 = t^2x^2 - x^3, x \leq 0\}$. This is obtained by removing the "upper half" of the Whitney cusp $V = \{y^2 = t^2x^2 - x^3\}$ (Fig. 4.5).

Because the slices $t = $ constant vary between half of a double point, with a nonzero angle between the two branches, and a cusp, with zero angle between the 2

Fig. 4.5 Example 4.2.11

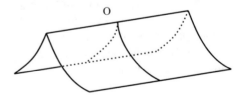

branches, one sees easily that these two types of slices are not bilipschitz equivalent. However the following calculation shows that (w)-regularity holds.

$$d(< 0t >, T_p X) = || < (0, 0, 1), \frac{grad_p F}{||grad_p F||} ||$$

$$= \frac{2|tx^2|}{\sqrt{(3x^2 - 2xt^2)^2 + 4y^2 + 4t^2 x^4}}$$

$$\leq \frac{2|tx^2|}{2y} \leq |x| \leq \sqrt{x^2 + y^2} = ||p - \pi(p)||.$$

Thus (w) holds.

4.2.2 Some Complex Equisingularity and Real Analogues

We have seen that (b)-regularity implies the constance of the local topological type of a stratified set along each stratum. For families of complex plane curves defined by

$$F : \mathbb{C}^2 \times \mathbb{C} \longrightarrow \mathbb{C}, 0$$

$(F^{-1}(0), 0 \times \mathbb{C})$ is (b)-regular if and only if the local topological type of $F_t^{-1}(0)$ is constant as t varies, where (z, t) are the coordinates of $\mathbb{C}^2 \times \mathbb{C}$. However this equivalence does not extend to higher dimensions as shown by the following celebrated example.

Example 4.2.12 (Briançon and Speder [BS75]) Let $F(x, y, z, t) = x^3 + txy^3 + y^4 z + z^9$. Then $(F^{-1}(0), 0 \times \mathbb{C})$ is not (b)-regular at $(0, 0, 0, 0)$, but the local topological type at $(0, 0, 0, t)$ of $F_t^{-1}(0)$ is constant.

The theory of equisingularity aims at comparing different notions of regularity on stratifications, in particular of analytic varieties (where much work has been done in particular by Zariski, Teissier, and Gaffney).

A basic invariant in algebraic geometry is the multiplicity $m_0(V)$ at a point 0 of a variety V in $\mathbb{C}^n$. An informal definition of $m_0(V)$ is the number of points near 0 in $P \cap V$ for a generic plane P of dimension equal to the codimension of V, passing near 0.

A relation with stratifications is given by a theorem of Hironaka.

Theorem 4.2.13 (Hironaka 1969 (Corollary 6.2 in [Hir69])) *Given a complex analytic Whitney (b)-regular stratification of a complex analytic variety V, the multiplicity of V at points of V is constant on strata.*

Thus (b) implies equimultiplicity.

The proof is by integration of a vector field, and works for subanalytic sets, interpreted as (b) implying normal pseudo flatness (this is equivalent to equimultiplicity in the complex case), as shown in a paper of mine with Orro (Proposition 5.2 in [OT02]). (One defines the normal cone of a stratified set along a stratum X by taking limits on X of orthogonal secant vectors from Y to the set and then normal pseudo flatness means that the associated projection of the normal cone to X is open.)

4.2.2.1 Zariski's Problem

In 1971, Zariski stated the following problem (Question A in [Zar71]): Given analytic functions $f, g : \mathbb{C}^{n+1}, 0 \longrightarrow \mathbb{C}, 0$ and a germ at 0 of a homeomorphism h of $\mathbb{C}^{n+1}$ sending $f^{-1}(0)$ onto $g^{-1}(0)$, does $m_0(f^{-1}(0)) = m_0(g^{-1}(0))$?

As this school concerns the Lipschitz geometry of singularities I will mention some results about Zariski's problem when the homeomorphism h is assumed to be bilipschitz.

Theorem 4.2.14 (Fernandes and Sampaio [FS16]) *Zariski's problem has a positive answer if $n = 2$ and h is bilipschitz.*

Theorem 4.2.15 (Risler and Trotman [RT97]) *Zariski's problem has a positive answer if h is bilipschitz and $f = g \circ h$, for all n.*

In 2018 it was announced by Birbrair, Fernandes, Sampaio, and Verbitsky [BFSV18] that for the non hypersurface case there are infinitely many counterexamples to the bilipschitz invariance of the multiplicity with the dimension of the varieties being at least 3.

For normal complex surfaces (possibly embedded in higher dimensions), Neumann and Pichon [NP12] have proved that the multiplicity is an outer bilipschitz invariant.

Theorem 4.2.16 (Comte [Com98]) *Zariski's problem has a positive answer for complex analytic germs if h is bilipschitz with Lipschitz constants (of h and h^{-1}) sufficiently close to 1.*

More precisely if X_1 and X_2 are complex analytic germs of dimension d in $\mathbb{C}^n$ and there exist constants $C > 0, C' > 0$ such that

$$(1/C')||x - y|| \leq ||h(x) - h(y)|| \leq C||x - y||$$

for all x, y near 0 in X_1 for a bilipschitz homeomorphism $h : X_1, 0 \longrightarrow X_2, 0$ and

$$1 \leq CC' \leq (1 + \frac{1}{M})^{\frac{1}{2d}}$$

where $M = max(m_0(X_1), m_0(X_2))$, then $m_0(X_1) = m_0(X_2)$.

The proof uses a characterization of the multiplicity as the density, originally due to Lelong [Lel57].

Definition 4.2.17 The **density** of a set X at $p \in X$ is defined as the limit as r tends to 0 of the volume of the intersection of X with the ball of radius r centred at p divided by the volume of the intersection of a plane through p of the same dimension as X with the ball of radius r centred at p.

Corollary 4.2.18 (Comte [Com98]) *In a bilipschitz trivial family of complex analytic germs (defined by a Lipschitz isotopy) the multiplicity is constant.*

While on the topic of equimultiplicity and stratifications one should mention the important characterization due to Teissier.

Theorem 4.2.19 (Teissier: Théorème 1.2 in Chapter V of [Tei82]) *A complex analytic stratification of a complex analytic variety is Whitney (b)-regular $\Longleftrightarrow$ the multiplicities of the local polar varieties are constant on strata.*

Here the local polar varieties at a point of the variety are the closures of the critical sets of the restrictions to strata, whose closure contains the point, of locally defined projections to general linear subspaces of dimensions lying between two and the dimension of the variety (see section 3.2 in [FT]).

There are real analogues of these complex results involving what are known as Lipschitz-Killing invariants on strata of a definable stratification, due to Comte and Merle [CM08] and Nguyen and Valette [NV18]. These generalize another real analogue of Hironaka's theorem stated above, due to Comte (who proved in 2000 the partial result (Théorème 0.4 of [Com00]) of continuity of the density along strata of a (w)-regular subanalytic stratification) and G. Valette.

Theorem 4.2.20 (Valette [Val08]) *The density is a Lipschitz function along strata of a (w)-regular subanalytic stratification, and a continuous function along strata of a (b)-regular subanalytic stratification.*

Part of the proof of Teissier's Theorem 4.2.19 above involves studying how equi-singularity is preserved after taking generic plane sections of different dimensions. Precisely, let $Y \subset \overline{X} - X$.

Definition 4.2.21 Consider a plane $P \supset Y$, then $(X \cap P, Y)$ is a stratified pair. If E is an equisingularity condition, such as (b) or (w), etc., then one says that the pair (X, Y) is $E^\star$-regular at $0 \in Y$ if for all $k, 0 \leq k \leq n - m$ there exists an open dense set of planes P of codimension k such that P is transverse to X near 0 and $(X \cap P, Y)$ is E-regular at 0.

If we abbreviate local topological triviality by $(T.T.)$ then Teissier proved in the complex case that (b) implies $(T.T.^\star)$, a strengthening of the Thom-Mather theorem, while the converse, that $(T.T.^\star)$ implies (b), was proved by Lê and Teissier (see Théorème 5.3.1 in [LT83]). This is thus a converse to Teissier's strengthened Thom-Mather theorem. Next we give some results concerning the * condition in the subanalytic case.

Theorem 4.2.22 *For subanalytic stratifications, (w) implies $(w^\star)$, the Lipschitz property (L) implies $(L^\star)$ and, when Y has dimension 1, (b) implies $(b^\star)$.*

The first and third implications are proved by Navarro and Trotman (Theorem 3.14 in [NAT81]), and the second implication is proved by Juniati, Trotman and Valette (Corollary 2.9 in [JTV03]). It is unknown if (b) implies $(b^\star)$ when the dimension of Y is greater than 1 for subanalytic stratifications, but in the complex case this follows from the implication that (w) implies $(w^\star)$ since (b) and (w) are equivalent (Teissier: Théorème 1.2 in Chapter V of [Tei82]). Probably the implications in the previous theorem are valid for definable sets in polynomially bounded o-minimal structures. A counterexample to the third implication in the non polynomially bounded case is given in a paper by myself and Valette (in section 4 of [TV17]). Another such example is given in a paper by myself and L. Wilson [TW06].

Example 4.2.23 (Trotman and Wilson [TW06]) Let $f(x, z) = z - \frac{z \log(x + \sqrt{x^2 + z^2})}{\log z}$, $z > 0$. Then let S_f be the closure of the graph of f in $\mathbb{R}^3$. Then $(S_f - < Ox >, < 0x >)$ is (b)-regular, but $(b^\star)$ fails. Also (w) fails to hold, and normal pseudo flatness fails.

Example 4.2.24 (Trotman and Valette (section 4 in [TV17])) Let

$$g(x, z) = z^{x^2 + 1} = \exp((x^2 + 1) \log z), z > 0.$$

Let S_g be the closure of the graph of g in $\mathbb{R}^3$ (Fig. 4.6). Then $(S_g - < 0x >, < 0x >)$ is (b)-regular, but $(b^\star)$ fails. Also normal pseudo flatness and (w) fail to hold.

Consider the convex hull K_g of S_g and the half-plane $\{y = 0, z > 0\}$. The density of K_g is not continuous along $0x$ at 0. This is a counterexample to a possible generalization of the Comte-Valette theorem 2.20 [Com00, Val08] to non polynomially bounded o-minimal structures. Note too that the bilipschitz type of K_g varies continuously along $0x$.

Fig. 4.6 Example 4.2.24

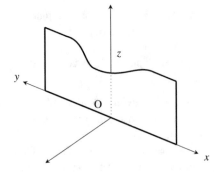

These two examples provide examples of definable sets in *any* non polynomially bounded o-minimal structure, because in such a structure the exponential function and its logarithm inverse are definable by Miller's dichotomy [Mil94] stated above.

These examples prevent definable extensions to the following theorem.

Theorem 4.2.25 (Pawłucki (Theorem 1.1 in [Paw85])) *Let X, Y be (locally) connected subanalytic strata in $\mathbb{R}^n$, $Y \subset \overline{X} - X$, such that $dim X = dim Y + 1$. Then (X, Y) is (b)-regular $\Longleftrightarrow X \cup Y$ is a C^1 manifold-with-boundary.*

One implication is just the C^1 invariance of (b). The other is more delicate.

In 2017 with Valette (Corollary 3.11 in [TV17]) I proved that Pawłucki's characterization is valid for definable sets in polynomially bounded o-minimal structures.

The two examples above show this fails in non polynomially bounded o-minimal structures.

4.3 Triangulation of Stratified Sets and Maps

While stratifications can be thought of as a more efficient alternative to triangulations, as there are less strata in a stratification into manifolds than simplexes of maximal dimension in a triangulation, it remains the case that triangulations of sets (and maps) are useful for calculating homology and cohomology. In this section we present results concerning sets and maps, rather incomplete, but which may serve as an introduction to the theory.

4.3.1 Triangulation of Sets

Theorem 4.3.1 (Hironaka [Hir75]) *Every semialgebraic set S is triangulable: there exists a polyhedron K and a semialgebraic homeomorphism $\phi : K \longrightarrow S$. Moreover given a finite family $\{S_j\}_{j=1,...,m}$ of semialgebraic subsets of S, we can choose $K = \{\sigma_i\}_{I=1,...,p}$ (the simplexes) and ϕ such that each S_j is the union of some of the $\phi(\sigma_i^o)$.*

The proof applies also to the case of subanalytic sets.

Corollary 4.3.2 *A Whitney stratified semialgebraic set S admits a triangulation such that every stratum is a union of (images of) open simplexes.*

There are similar results in the smooth category.

Theorem 4.3.3 (Goresky [Gor78]) *Every Whitney stratified set is triangulable, such that strata are unions of open simplexes.*

Conjecture 4.3.4 (Thom) Every Whitney stratified set admits a Whitney triangulation, i.e. a triangulation such that the refined stratification defined by the open simplexes is itself (b)-regular.

As a partial answer to Thom's conjecture we have the following result of Shiota in the semialgebraic case.

Theorem 4.3.5 (Shiota [Shi05]) *Every semialgebraic set S admits a semialgebraic Whitney triangulation, i.e. the open simplexes $\phi(\sigma_i^o)$ form the strata of a Whitney stratification, and this may be chosen to be compatible with a finite set of semialgebraic subsets of S.*

Shiota's theorem was improved and extended by Malgorzata Czapla in her thesis.

Theorem 4.3.6 (Czapla [Cza12]) *Every definable set S admits a definable C^2 (w)-regular triangulation, compatible with a finite number of definable subsets of S. Moreover the triangulation $\phi : |K| \longrightarrow S$ is a locally Lipschitz mapping.*

So Czapla improves on Shiota's theorem in two ways: (w)-regularity and definability. The main tool of Czapla is a bilipschitz triviality theorem of Valette for definable families, itself an improvement of a celebrated theorem of Hardt.

A continuous semialgebraic mapping $p : A \longrightarrow \mathbb{R}^k$ where $A \subset \mathbb{R}^n$ is semialgebraic, is said to be **semialgebraically trivial** over a semialgebraic subset $B \subset \mathbb{R}^k$ if there is a semialgebraic set F and a semialgebraic homeomorphism $h : p^{-1}(B) \longrightarrow B \times F$ such that $p_1 \circ h = p$. Then h is called a **semialgebraic trivialization of p over B**. We say h is compatible with $C \subset A$ if there exists a semialgebraic set $G \subset F$ such that $h(C \cap p^{-1}(B)) = B \times G$.

Theorem 4.3.7 (Hardt's Semialgebraic Triviality [Har80]) *Let $A \subset \mathbb{R}^n$ be a semialgebraic set and $p : A \longrightarrow \mathbb{R}^k$ a continuous semialgebraic mapping. Then there is a finite semialgebraic partition of $\mathbb{R}^k$ into $B_1, \ldots, B_m$ such that p is semi algebraically trivial over each B_i. Moreover if $C_1, \ldots, C_q$ are semialgebraic subsets of A we can assure that each trivialization*

$$h_i : p^{-1}(B_i) \longrightarrow B_i \times F_i$$

is compatible with all C_j.

In particular if $b, b' \in B_i$, then $p^{-1}(b)$ and $p^{-1}(b')$ are semialgebraically homeomorphic. One can take $F_i = p^{-1}(b_i)$, $b_i \in B_i$ and set $h_i(x) = (x, b_i)$ for all $x \in p^{-1}(b_i)$.

There is a definable version of Hardt's triviality theorem too, given by Coste in his Pisa notes on semialgebraic geometry [Cos00]. We now consider a further improvement, a definable bilipschitz triviality theorem due to G. Valette [Val05a].

Fix a polynomially bounded o-minimal structure over $\mathbb{R}$ (take semialgebraic sets if preferred). Let $A \subset \mathbb{R}^n \times \mathbb{R}^p$ be a definable set, considered as a family of definable subsets of $\mathbb{R}^n$ parametrized by $\mathbb{R}^p$. For $U \subset \mathbb{R}^p$, let

$$A_U = \{q = (x, t) \in \mathbb{R}^n \times \mathbb{R}^p | q \in A, t \in U\}$$

and for $t \in \mathbb{R}^p$, let $A_t = \{x \in \mathbb{R}^n | q = (x, t) \in A\}$, the fibre of A at t.

Definition 4.3.8 A is said to be **definably bilipschitz trivial along** $U \subseteq \mathbb{R}^p$ if there exists $t_0 \in U$ and a definable homeomorphism $h : A_{t_0} \times U \longrightarrow A_U$ mapping (x, t) to $h(x, t) = (h_t(x), t)$ together with a definable continuous function $C : U \longrightarrow \mathbb{R}$ such that for all $x, x' \in A_{t_0}$ and all $t \in U$,

$$|h_t(x) - h_t(x')| \le C(t)|x - x'|$$

and for all $x, x' \in A_t$, and all $t \in U$,

$$|h_t^{-1}(x) - h_t^{-1}(x')| \le C(t)|x - x'|.$$

Theorem 4.3.9 (Valette [Val05a, Val05b]) *Let A be a definable subset of $\mathbb{R}^n \times \mathbb{R}^p$ in some polynomially bounded structure over $\mathbb{R}$. Then there exists a definable partition of $\mathbb{R}^p$ such that the family A is definably bilipschitz trivial along each element of the partition.*

Notes The Mostowski-Parusiński condition (L) of section 1.3 together with the definable existence theorem of Nguyen and Valette [NV16] gives a local bilipschitz trivialization h. Here we have definability of h as well. There is also better control of the Lipschitz constants of the bilipschitz trivialization here.

As in the case of Hardt's theorem for topological types we can deduce from Valette's theorem bounds on the number of Lipschitz types of sets given as zeros of polynomials of bounded degree.

To prove his theorem, Valette proves a preparation theorem, and uses ultrafilters as in Coste's account of the definable Hardt triviality theorem (cf. Coste's Pisa notes on o-minimal geometry [Cos00]).

4.3.2 Thom Maps and the (a_f) Condition

We will describe a class of stratified maps which are triangulable.

Definition 4.3.10 Let Z be a closed subset of $\mathbb{R}^n$ (or $\mathbb{C}^n$) with a stratification Σ. Let $f : \mathbb{R}^n \longrightarrow \mathbb{R}^p$ be a C^1 map. Then Σ is said to satisfy (a_f) if each $f|_X$, for

X a stratum of Σ, is of constant rank (depending on X), and for sequences $x_i \in X$ tending to y in a stratum Y of Σ,

$$lim_{x_i \to y} T_{x_i}(f^{-1}(f(x_i))) \supseteq T_y(f^{-1}(f(y))).$$

When further,

$$\text{dist}(T_x(f^{-1}(f(x))), T_y(f^{-1}(f(y))) \leq C||x - \pi_Y(x)||$$

for some $C > 0$ and x in a neighbourhood U of y in $\mathbb{R}^n$ (or $\mathbb{C}^n$), we say that Σ satisfies the (w_f) condition.

Theorem 4.3.11 (Loi [Loi98]) *For polynomially bounded o-minimal structures, every definable function $f : \mathbb{R}^n \longrightarrow \mathbb{R}$ admits a stratification such that (w_f) holds.*

For (a_f) this is true in any o-minimal structure. In the complex case the result is due to Henry, Merle, and Sabbah [HMS84].

Definition 4.3.12 Let $f : \mathbb{R}^n \longrightarrow \mathbb{R}^m$ be a C^1 map. If there exist Whitney stratifications Σ of $Z \subset \mathbb{R}^n \subset \mathbb{R}^m$ such that f maps each stratum X of Σ to a stratum X' of Σ', such that $f|_X$ is a submersion onto X', Σ satisfies (a_f), and each $f|_X$ is proper, then one says that f is a **Thom map** .

Thom maps have nice properties.

Theorem 4.3.13 (Shiota [Shi00]) *If Z, W are respectively closed subsets of $\mathbb{R}^n$ and $\mathbb{R}^m$ and $f : Z \longrightarrow W$ is a proper C^∞ Thom map, then f is triangulable, i.e. there exist polyhedra P, Q and homeomorphisms $\phi : Z \longrightarrow P, \psi : W \longrightarrow Q$ such that $\psi \circ f \circ \phi^{-1} : P \longrightarrow Q$ is piecewise linear.*

For non-proper maps there is still a theorem.

Theorem 4.3.14 (Shiota [Shi10]) *Nonproper semialgebraic C^1 Thom maps between closed semialgebraic subsets are triangulable , i.e. there exist finite simplicial complexes K, L and semialgebraic (resp. definable) C^0 embeddings $\phi : Z \longrightarrow |K|, \psi : W \longrightarrow |L|$ such that $\phi(Z)$ and $\psi(W)$ are unions of open simplexes of K, L and $\psi \circ f \circ \phi^{-1} : \phi(Z) \longrightarrow \psi(W)$ can be extended to a simplicial map $K \longrightarrow L$.*

When the target space is of dimension > 1, the transform of the map by suitable blowing-ups of the target space becomes (a_f) stratifiable (see Sabbah [Sab83]) and locally triangulable (see Teissier [Tei89]). Note that maps not satisfying (a_f) may not be triangulable. For example the blowup of a point in $\mathbb{R}^2$ does not satisfy (a_f) and is not triangulable. Any 2-simplex attached to the exceptional fibre (a projective line) is mapped to a 1-simplex by linearity. One sees that (a_f) fails because outside the origin the fibres of points are just points and the limit of a point cannot contain a line as the tangent space of the exceptional fibre. Thom [Tho69] called maps satisfying (a_f) maps **"sans éclatement"**, i.e. without blowing-up, so that this example is in some sense a paradigm.

Analogous to the characterization of (a)-regularity by the openness of the set of maps transverse to a stratification, we have a similar result for (a_f)-maps.

Theorem 4.3.15 (Trivedi-Trotman [TT14]) *Let N, P be C^1 manifolds. Let $f :$ $N \longrightarrow P$ be a C^1 map of constant rank on the strata of a stratification Σ of a closed subset Z of N. Let $\mathcal{F}$ denote the foliations of strata X of Σ induced by the fibers of $f|_X$. The following are equivalent:*

(1) Σ is (a_f)-regular;
(2) for any C^1 manifold M, $\{g \in C^1(M, N) : g$ is transverse to $\mathcal{F}\}$ is open in the strong C^1 topology;
(3) $\{g \in C^1(N, N) : g$ is transverse to $\mathcal{F}\}$ is open in the strong C^1 topology.

The (a_f) condition has a particular role in equisingularity of families of complex hypersurfaces.

Let $F = \mathbb{C}^{n+1} \times \mathbb{C}, O \times \mathbb{C} \longrightarrow \mathbb{C}, 0$ ba an analytic function such that the singular locus of $F^{-1}(0)$ is $0 \times \mathbb{C}$. Let $F_t(z) = F(z, t)$.

Theorem 4.3.16 (Lê and Saito [LS73], Teissier (Remarque 3.10 in [Tei73])) . *The following conditions are equivalent:*

(1) $\mu(F_t)$ is constant as t varies,
(2) (a_f) holds for the stratification $(F^{-1}(0) - 0 \times \mathbb{C}, 0 \times \mathbb{C})$,
(3) $\lim_{(z,t) \to (0,0} \frac{|\partial F/\partial t|}{|grad F|} = 0.$

Corollary 4.3.17 *If $F(z, t) = g(z) + th(z)$ has $\mu(F_t)$ constant, then F_t is equimultiple along $0 \times \mathbb{C}$.*

This simple consequence of the previous theorem should be linked to a striking result of Parusiński.

Theorem 4.3.18 (Parusiński [Par99]) *With the same hypotheses as in the previous corollary, the topological type of $F_t^{-1}(0)$ is constant as t varies.*

This in turn should make us think again of an important general result.

Theorem 4.3.19 (Lê-Ramanujam [LR76]) *If $n \neq 2$ and $\mu(F_t)$ is constant, then the topological type of $F_t^{-1}(0)$ is constant.*

Question What happens when $n = 2$?

Remark 4.3.20 There are at least 3 different definitions of (b_f)-regularity, due to Thom (in section IIIB of [Tho69]), Henry-Merle (Definition 9.1.1 of [HM87]), and Nakai (see §1 of [Nak00]). Their properties have not been studied beyond the original papers so far as I know. And no work has been done on a possible (L_f).

4.3.2.1   (c)-regularity

The notion of Thom map has been used by Karim Bekka to define a new regularity condition called (c).

Definition 4.3.21 One says that a stratification Σ of a closed set Z in a manifold M is (c)-**regular** if for each stratum X of Σ there is a neighbourhood U_X of X in M and a C^1 function $\rho_X : U_X \longrightarrow [0, 1)$ such that $X = \rho_X^{-1}(0)$ and ρ_X is a Thom map for Σ.

One shows fairly easily that $(b) \implies (c) \implies (a)$. Note that U_X is a neighbourhood of the whole of X and not just of a point of X.

Moreover, by a careful analysis of the proof of the Thom-Mather isotopy theorem, Bekka showed:

Theorem 4.3.22 (Bekka (see §3 in [Bek91])) *Every (c)-regular stratification is locally topologically trivial along strata (and conical).*

Thus, as for Whitney (b)-regular stratified sets (Z, Σ), for every point x in a stratum X there is a neighbourhood U of x in M, a stratified set L and a homeomorphism

$$h : (U, U \cap Z, U \cap X) \longrightarrow (U \cap X) \times (\mathbb{R}^k, cL, \star)$$

given by $h(z) = (\pi_X(z), \rho_X(z), \theta(z))$ where cL is the cone on L with vertex $\star$. As for (b)-regularity, fix the values of ρ_X and θ, then $\{z | \rho_X(z) = \rho, \theta(z) = \theta\}$ is a **leaf** diffeomorphic to $U \cap X$.

Theorem 4.3.23 (Murolo-du Plessis-Trotman (Theorem 7 in [MdPT17])) *Given a (c)-regular stratified set we can choose h such that the tangent spaces to the leaves vary continuously on U, in particular as points tend to X.*

Again we may fix just θ. Then $\{z | \theta(z) = \theta\}$ is a **wing**, a C^0 manifold with boundary $U \cap X$ and smooth interior. Then one can choose h so that the tangent spaces to the wings vary continuously and each wing is itself (c)-regular (Theorem 8 in [MdPT17]).

Question What can one say in the semialgebraic or subanalytic cases? Note that the Parusiński-Paunescu theorem (Theorem 7.6 of [PP17]) is only for the algebraic and analytic cases.

References

[Bek91] K. Bekka, C-régularité et trivialité topologique, in *Singularity Theory and Its Applications, Part I (Coventry, 1988/1989)*. Lecture Notes in Mathematical, vol. 1462 (Springer, Berlin, 1991), pp. 42–62 106

[BFSV18] L. Birbrair, A. Fernandes, J.E. Sampaio, M. Verbitsky, Multiplicity of singularities is not a bi-lipschitz invariant. Math. Annalen **377**, 115–121 (2020) 98

[BS75] J. Briançon, J.-P. Speder, La trivialité topologique n'implique pas les conditions de Whitney. C. R. Acad. Sci. Paris Sér. A-B **280**(6), Aiii, A365–A367 (1975) 97

[BT79] H. Brodersen, D. Trotman, Whitney (*b*)-regularity is weaker than Kuo's ratio test for real algebraic stratifications. Math. Scand. **45**(1), 27–34 (1979) 92

[Com98] G. Comte, Multiplicity of complex analytic sets and bi-Lipschitz maps, in *Real analytic and Algebraic Singularities (Nagoya/Sapporo/Hachioji, 1996)*. Pitman Research Notes in Mathematics Series, vol. 381 (Longman, Harlow, 1998), pp. 182–188 98, 99

[Com00] G. Comte, Équisingularité réelle: nombres de Lelong et images polaires. Ann. Sci. École Norm. Sup. (4) **33**(6), 757–788 (2000) 99, 100

[CM08] G. Comte, M. Merle, Équisingularité réelle. II. Invariants locaux et conditions de régularité. Ann. Sci. Éc. Norm. Supér. (4) **41**(2), 221–269 (2008) 99

[Cos00] M. Coste, *An Introduction to O-minimal Geometry*, (Dottorato di ricerca in matematica/Università di Pisa, Dipartimento di Matematica. Istituti editoriali e poligrafici internazionali, Pisa, 2000) 102, 103

[Cza12] M. Czapla, Definable triangulations with regularity conditions. Geom. Topol. **16**(4), 2067–2095 (2012) 102

[DW87] Z. Denkowska, K. Wachta, Une construction de la stratification sous-analytique avec la condition (w). Bull. Polish Acad. Sci. Math. **35**(7–8), 401–405 (1987) 92

[DWS85] Z. Denkowska, K. Wachta, J. Stasica, Stratification des ensembles sous-analytiques avec les propriétés (A) et (B) de Whitney. Univ. Iagel. Acta Math. **25**, 183–188 (1985) 91

[dP99] A. du Plessis, Continuous controlled vector fields, in *Singularity theory (Liverpool, 1996)*. London Mathematical Society Lecture Note Series, vol. 263 (Cambridge University, Cambridge, 1999), pp. 189–197 95

[FS16] A. Fernandes, J.E. Sampaio, Multiplicity of analytic hypersurface singularities under bi-Lipschitz homeomorphisms. J. Topol. **9**(3), 927–933 (2016) 98

[FT] A.G. Flores, B. Teissier, Local polar varieties in the geometric study of singularities. Ann. Fac. Sci. Toulouse Math. **27**(6), 679–775 (2018) 99

[Gor78] M. Goresky, Triangulation of stratified objects. Proc. Am. Math. Soc. **72**(1), 193–200 (1978) 101

[Hal14a] I. Halupczok, Non-Archimedean Whitney stratifications. Proc. London Math. Soc. **109**(3), 1304–1362 (2014) 91

[Hal14b] I. Halupczok, Stratifications in valued fields, in *Valuation Theory in Interaction, EMS Series Congr. Report, European Mathematical Society, Zurich* (2014), pp. 288–296 91

[Har75] R. Hardt, Stratification of real analytic mappings and images. Invent. Math. **28**, 193–208 (1975) 91

[Har80] R. Hardt, Semi-algebraic local-triviality in semi-algebraic mappings. Am. J. Math. **102**(2), 291–302 (1980) 102

[HM87] J.-P. Henry, M. Merle, Conditions de régularité et éclatements. Ann. Inst. Fourier (Grenoble) **37**(3), 159–190 (1987) 105

[HMS84] J.-P. Henry, M. Merle, C. Sabbah, Sur la condition de Thom stricte pour un morphisme analytique complexe. Ann. Sci. École Norm. Sup. (4) **17**(2), 227–268 (1984) 104

[Hir69] H. Hironaka, Normal cones in analytic Whitney stratifications. Inst. Hautes Études Sci. Publ. Math. **36**, 127–138 (1969) 97

[Hir73] H. Hironaka, Subanalytic sets, in *Number Theory, Algebraic Geometry and Commutative Algebra, in Honor of Yasuo Akizuki* (1973), pp. 453–493 91

[Hir75] H. Hironaka, Triangulations of algebraic sets, in *Algebraic Geometry (Proceedings of the Symposium Pure Mathematical, Vol. 29, Humboldt State University, Arcata, Califorinia, 1974)* (American Mathematical Society, Providence, 1975), pp. 165–185 101

[JTV03] D. Juniati, D. Trotman, G. Valette, Lipschitz stratifications and generic wings. J. London Math. Soc. (2) **68**(1), 133–147 (2003) 100

[Kal05] V.Y. Kaloshin, A geometric proof of the existence of Whitney stratifications. Mosc. Math. J. **5**(1), 125–133 (2005) 91

[Kuo71] T.C. Kuo, The ratio test for analytic Whitney stratifications, in *Proceedings of Liverpool Singularities-Symposium, I (1969/70)*. Lecture Notes in Mathematics, vol. 192 (Springer, Berlin, 1971), pp. 141–149 92

[KTL89] T.C. Kuo, P.X. Li, D.J.A. Trotman, Blowing-up and Whitney (a)-regularity. Canad. Math. Bull. **32**(4), 482–485 (1989) 94

[LR76] D.T. Lê, C.P. Ramanujam, The invariance of Milnor's number implies the invariance of the topological type. Am. J. Math. **98**(1), 67–78 (1976) 105

[LS73] D.T. Lê, K. Saito, La constance du nombre de Milnor donne des bonnes stratifications. C. R. Acad. Sci. Paris Sér. A-B **277**, A793–A795 (1973) 105

[LT83] D.T. Lê, B. Teissier, Cycles evanescents, sections planes et conditions de Whitney. II, in *Singularities, Part 2 (Arcata, Calif., 1981)*. Proceedings of the Symposium Pure Mathematical, vol. 40 (American Mathematical Society, Providence, 1983), pp. 65–103 99

[Lel57] P. Lelong, Intégration sur un ensemble analytique complexe. Bull. Soc. Math. France **85**, 239–262 (1957) 99

[Loi98] T.L. Loi, Verdier and strict Thom stratifications in o-minimal structures. Illinois J. Math. **42**(2), 347–356 (1998) 91, 92, 104

[Loj65] S. Łojasiewicz, Ensembles semi-analytiques, *I.H.E.S.notes* (1965) 91

[LSW86] S. Łojasiewicz, J. Stasica, K. Wachta, Stratifications sous-analytiques. Condition de Verdier. Bull. Polish Acad. Sci. Math. **34**(9–10), 531–539 (1987/1986) 92

[Mat12] J. Mather, Notes on topological stability. Bull. Am. Math. Soc. (N.S.) **49**(4), 475–506 (2012) 90, 94

[Mil94] C. Miller, Exponentiation is hard to avoid. Proc. Am. Math. Soc. **122**(1), 257–259 (1994) 96, 101

[Mos85] T. Mostowski, Lipschitz equisingularity. Dissertationes Math. (Rozprawy Mat.) **243**, (1985) 93, 96

[MdPT17] C. Murolo, A. du Plessis, D.J.A. Trotman, On the smooth Whitney fibering conjecture, in *hal preprint hal- 01571382v1* (2017) 95, 106

[Nak00] I. Nakai, Elementary topology of stratified mappings, in *Singularities—Sapporo 1998*. Advantage of Studies Pure Mathematical, vol. 29 (Kinokuniya, Tokyo, 2000), pp. 221–243 105

[NAT81] V. Navarro Aznar, D.J.A. Trotman, Whitney regularity and generic wings. Ann. Inst. Fourier (Grenoble) **31**(2), v, 87–111 (1981 100

[NP12] W.D. Neumann, A. Pichon, Lipschitz geometry of complex surfaces: analytic invariants and equisingularity (2012). arXiv preprint:1211.4897 98

[NV16] N. Nguyen, G. Valette, Lipschitz stratifications in o-minimal structures. Ann. Sci. Éc. Norm. Supér. (4) **49**(2), 399–421 (2016) 93, 103

[NV18] N. Nguyen, G. Valette, Whitney stratifications and the continuity of local Lipschitz-Killing curvatures. Ann. Inst. Fourier (Grenoble) **68**(5), 2253–2276 (2018) 99

[NTT14] N. Nguyen, S. Trivedi, D. Trotman, A geometric proof of the existence of definable Whitney stratifications. Illinois J. Math. **58**(2), 381–389 (2014) 91

[OT02] P. Orro, D. Trotman, Cône normal et régularités de Kuo-Verdier. Bull. Soc. Math. France **130**(1), 71–85 (2002) 98

[Par88a] A. Parusiński, Lipschitz properties of semi-analytic sets. Ann. Inst. Fourier (Grenoble) **38**(4), 189–213 (1988) 93

[Par88b] A. Parusiński, Lipschitz stratification of real analytic sets, in *Singularities (Warsaw, 1985)*. Banach Center Publicationvol. 20 (PWN, Warsaw, 1988), pp. 323–333 93

[Par94] A. Parusiński, Lipschitz stratification of subanalytic sets. Ann. Sci. École Norm. Sup. (4) **27**(6), 661–696 (1994) 93, 96

[Par99] A. Parusiński, Topological triviality of μ-constant deformations of type $f(x) + tg(x)$. Bull. London Math. Soc. **31**(6), 686–692 (1999) 105

[PP17] A. Parusiński, L. Păunescu, Arc-wise analytic stratification, Whitney fibering conjecture and Zariski equisingularity. Adv. Math. **309**, 254–305 (2017) 95, 106

[Paw85] W. Pawłucki, Quasiregular boundary and Stokes's formula for a subanalytic leaf, in *Seminar on Deformations (Łódź/Warsaw, 1982/84)*.. Lecture Notes in Mathematical, vol. 1165 (Springer, Berlin, 1985), pp. 235–252 101

[RT97] J.-J. Risler, D. Trotman, Bi-Lipschitz invariance of the multiplicity. Bull. London Math. Soc. **29**(2), 200–204 (1997) 98

[Sab83] C. Sabbah, Morphismes sans éclatements et cycles évanescents, in *Analysis and Topology on Singular Spaces II, III, Astérisque*. Society Mathematical France, vol. 101–102 (1983), pp. 286–319 104

[Shi97] M. Shiota, *Geometry of Subanalytic and Semialgebraic Sets*. Progress in Mathematics, vol. 150 (Birkhäuser Boston, Inc., Boston, 1997) 95

[Shi00] M. Shiota, Thom's conjecture on triangulations of maps. Topology **39**(2), 383–399 (2000) 104

[Shi05] M. Shiota, Whitney triangulations of semialgebraic sets. Ann. Polon. Math. **87**, 237–246 (2005) 102

[Shi10] M. Shiota, Triangulations of non-proper semialgebraic Thom maps, in *The Japanese-Australian Workshop on Real and Complex Singularities—JARCS III*. Proceedings of the Centre Mathematical Application Australia National University, vol. 43 (Australia National University, Canberra, 2010), pp. 127–140 104

[Tei73] B. Teissier, Cycles évanescents, sections planes et conditions de Whitney, in *Singularités à Cargèse (Rencontre Singularités Géom. Anal., Institut Etudes Scient. Cargèse, 1972)*. Astérisque, Nos. 7 et 8 (1973), pp. 285–362 105

[Tei82] B. Teissier, Variétés polaires. II. Multiplicités polaires, sections planes, et conditions de Whitney, in *Algebraic geometry (La Rábida, 1981)*. Lecture Notes in Mathematical, vol. 961 (Springer, Berlin, 1982), pp. 314–491 92, 99, 100

[Tei89] B. Teissier, Sur la triangulation des morphismes sous-analytiques. Publ. Math. de l'IHES **70**, 169–198 (1989) 104

[Tho69] R. Thom, Ensembles et morphismes stratifiés. Bull. Am. Math. Soc. **75**, 240–284 (1969) 90, 94, 104, 105

[Tho65] R. Thom, Propriétés différentielles locales des ensembles analytiques (d'après H. Whitney), in *Séminaire Bourbaki*, vol. 9 (Société Mathématique de France, Paris, 1995), Exp. No. 281, 69–80 91

[Tri13] S. Trivedi, Stratified transversality of holomorphic maps. Internat. J. Math. **24**(13), 1350106, 12 (2013) 94

[TT14] S. Trivedi, D. Trotman, Detecting Thom faults in stratified mappings. Kodai Math. J. **37**(2), 341–354 (2014) 105

[Tro77] D. Trotman, *Whitney Stratifications: Faults and Detectors* (Warwick University, Warwick, 1977) 93, 94

[Tro79] D.J.A. Trotman, Geometric versions of Whitney regularity for smooth stratifications. Ann. Sci. École Norm. Sup. (4) **12**(4), 453–463 (1979) 91

[Tro79] D.J.A. Trotman, Stability of transversality to a stratification implies Whitney (*a*)-regularity. Invent. Math. **50**(3), 273–277 (1978/1979) 94

[TV17] D. Trotman, G. Valette, On the local geometry of definably stratified sets, in *Ordered Algebraic Structures and Related Topics*. Contemporary Mathematics, vol. 697 (American Mathematical Society, Providence, 2017), pp. 349–366 100, 101

[TW06] D. Trotman, L. Wilson, (*r*) does not imply (*n*) or (*npf*) for definable sets in non polynomially bounded o-minimal structures, in *Singularity Theory and Its Applications*. Advanced Studies in Pure Mathematics, vol. 43 (Mathematical Society Japan, Tokyo, 2006), pp. 463–475 100

[Val05a] G. Valette, A bilipschitz version of Hardt's theorem. C. R. Math. Acad. Sci. Paris **340**(12), 895–900 (2005) 102, 103

[Val05b] G. Valette, Lipschitz triangulations. Illinois J. Math. **49**(3), 953–979 (2005) 103

[Val08] G. Valette, Volume, Whitney conditions and Lelong number. Ann. Polon. Math. **93**(1), 1–16 (2008) 99, 100 [vdDM96] L. van den Dries, C. Miller, Geometric categories and o-minimal structures. Duke Math. J. **84**(2), 497–540 (1996) 91

[Ver76] J.-L. Verdier, Stratifications de Whitney et théorème de Bertini-Sard. Invent. Math. **36**, 295–312 (1976) 91, 92, 95

[Wal75] C.T.C. Wall, Regular stratifications, in *Dynamical Systems—Warwick 1974 (Proc. Sympos. Application Topology and Dynamical Systems, University Warwick, Coventry, 1973/1974; Presented to E. C. Zeeman on his Fiftieth Birthday).* Lecture Notes in Mathematics, vol. 468, (1975), pp. 332–344 91

[Whi65a] H. Whitney, Tangents to an analytic variety. Ann. of Math. (2) **81**, 496–549 (1965) 89, 91

[Whi65b] H. Whitney, Local properties of analytic varieties, in *Differential and Combinatorial Topology (A Symposium in Honor of Marston Morse)* (Princeton University Press, Princeton, 1965), pp. 205–244 89, 95

[Zar71] O. Zariski, Some open questions in the theory of singularities. Bull. Am. Math. Soc. **77**, 481–491 (1971) 98

Chapter 5
Basics on Lipschitz Geometry

Maria Aparecida Soares Ruas

Abstract In this course we introduce the main tools to study the Lipschitz geometry of real and complex singular sets and mappings: the notions of semialgebraic sets and mappings and basic notions of Lipschitz geometry. The course then focuses on the real setting, presenting the outer Lipschitz classification of semialgebraic curves, the inner classification of semialgebraic surfaces, the bi- Lipschitz invariance of the tangent cone, and ending with a presentation of several results on Lipschitz geometry of function germs.

5.1 Introduction

These are the notes of the course *Basics on Lipschitz Geometry* taught at the first week of the International School on Singularities and Lipschitz Geometry, in Cuernavaca, Mexico, 11th to 22nd of June, 2019.

Lipschitz geometry is the study of those properties of metric spaces which are left invariant by bi-Lipschitz homeomorphisms. It is the geometry of a topological space based on the notion of length. Lipschitz singularity theory deals with classification of singular points of functions and mappings with respect to bi-Lipschitz equivalences.

Denoting by M the category of metric spaces the Lipschitz mappings are the morphisms of M and the bi-Lipschitz homeomorphisms are the isomorphisms of this category.

In this course we discuss basic properties of Lipschitz geometry of real and complex singular sets and introduce the framework for the Lipschitz theory of singularities.

By **singular set** we mean a compact semialgebraic or subanalytic (or definable in some o-minimal structure) set in $\mathbb{R}^n$ which is not a smooth submanifold of $\mathbb{R}^n$.

M. A. S. Ruas (✉)
Instituto de Ciências Matemáticas e de Computação - USP, São Carlos, SP, Brazil
e-mail: maasruas@icmc.usp.br

© The Author(s), under exclusive license to Springer Nature Switzerland AG 2020
W. Neumann, A. Pichon (eds.), *Introduction to Lipschitz Geometry of Singularities*,
Lecture Notes in Mathematics 2280, https://doi.org/10.1007/978-3-030-61807-0_5

In many situations we reduce our investigation to a small neighborhood of a singular point. In this case, the language of set germs and map germs will be useful.

What is the motivation to study Lipschitz geometry and Lipschitz singularity theory? The topological structure of real and complex analytic sets are well known and, in many cases, a complete topological classification of singular sets is given.

The following are well known results on the topology of singularities:

Theorem 5.1.1 (Topological Conical Structure, [Mil68]) *In a small neighborhood of a singular point, any analytic set is homeomorphic to a cone over the link.*

Theorem 5.1.2 (Topological Classification of Function Germs, [Kin78, Per85]) *Let $f, g : (\mathbb{C}^n, 0) \to (\mathbb{C}, 0)$ analytic function germs with isolated singularity, $X_f = f^{-1}(0)$, $X_g = g^{-1}(0)$. Then X_f is ambient topologically equivalent to X_g if and only if f is topologically right equivalent to g or to $\bar{g}$, where $\bar{g}$ is the conjugate of g.*

See also Part 1 of Anne Pichon's notes [Pic19] and Section 3 of Haydée Aguilar-Cabrera and José Luis Cisneros notes [ACCM19] in the present volume.

We shall see that these two results do not remain true if we replace "homeomorphism" by "bi-Lipschitz homeomorphism". In fact, Lev Birbrair and Alexandre Fernandes proved a decade ago in [BF08a] that an algebraic set is not always bi-Lipschitz equivalent to a cone. On the other hand, it follows from a result of Jean-Pierre Henry and Adam Parusiński [HP03] that the bi-Lipschitz classification of analytic functions has moduli. Furthermore, in a unique bi-Lipschitz class of real or complex hypersurfaces there are infinitely many bi-Lipschitz classes of functions $f : (\mathbb{K}^n, 0) \to (\mathbb{K}, 0)$ with $X = f^{-1}(0)$.

The foundations of Lipschitz theory were the work of Frédéric Pham and Bernard Teissier [PT69] on bi-Lipschitz classification of complex curves (published at the end of the present volume), and Tadeusz Mostowski's result on the existence of a locally finite Lipschitz stratification of complex analytic sets in [Mos85].

Lipschitz geometry is now at its golden age. The notes of the courses of this school will be a guide to students and researchers who want to proceed in this area.

The contents of these notes are organized in five sections:

1. Semi-algebraic sets and mappings
2. Basic notions in Lipschitz geometry
3. Lipschitz geometry of semialgebraic curves and surfaces
4. Bi-Lipschitz invariance of the tangent cone
5. Lipschitz theory of singularity.

In these notes we restrict our presentation to semialgebraic sets in Euclidean spaces. Most results can be extended to the class of subanalytic sets. Both, semialgebraic and subanalytic sets are examples of o-minimal structures. An Appendix on definable sets in o-minimal structures by Nhan Nguyen is presented at the end of the notes.

I would like to thank the organizers of the International School on Singularities and Lipschitz Geometry for the invitation to teach the course Basics on Lipschitz

Geometry and to make available these lecture notes. I am also grateful to Lev Birbrair and Alexandre Fernandes for very helpful discussions on the material presented here. It is a pleasure to thank Nhan Nguyen for writing the Appendix on o-minimal structures. Thanks are also due to Saurabh Trivedi and Nhan Nguyen for reading and making some corrections to these notes and to Pedro Benedini Riul for the help with the figures.

I am also grateful to the referee for his/her careful reading and suggestions that improved very much the presentation of these notes.

5.2 Semialgebraic Sets and Mappings

In this section we define semialgebraic sets and study their properties. Semialgebraic geometry is the study of sets of real solutions of polynomial equations and polynomial inequalities. Semialgebraic sets have a simple topology. In a small neighborhood of a singular point, a semialgebraic set is topologically equivalent to a cone over the link. Moreover, the homeomorphism can be chosen to be semialgebraic.

Semialgebraic sets are examples of o-minimal structures. For a brief introduction to definable sets in o-minimal structures see the Appendix.

5.2.1 Definitions and Basic Properties

We start reviewing the notion of algebraic sets.

Definition 5.2.1 A set $X \subset \mathbb{R}^n$ is said to be an **algebraic set** if there are polynomial functions $f_i : \mathbb{R}^n \to \mathbb{R}$, $i = 1, \ldots, k$ such that

$$X = \{x \in \mathbb{R}^n \mid f_1(x) = 0, \ldots, f_k(x) = 0\}.$$

In a similar way we can define complex algebraic sets, replacing $\mathbb{R}$ by $\mathbb{C}$ in the above definition.

Remark 5.2.2 In the real setting, we can say that X is algebraic if there exists a polynomial function $f : \mathbb{R}^n \to \mathbb{R}$ such that $X = \{x \in \mathbb{R}^n \mid f(x) = 0\}$ by taking $f = f_1^2 + \cdots + f_k^2$.

Example 5.2.3 $S^{n-1} = \{(x_1, \ldots, x_n) \in \mathbb{R}^n \mid x_1^2 + \cdots + x_n^2 = 1\}$ is an algebraic set in $\mathbb{R}^n$. In fact, let $f(x_1, \ldots, x_n) = \sum_{i=1}^{n} x_i^2 - 1$ then $S^{n-1} = f^{-1}(0)$.

Proposition 5.2.4 *Let X, Y be algebraic sets in $\mathbb{R}^n$. Then $X \cup Y$ and $X \cap Y$ are algebraic sets.*

Proposition 5.2.5 *Let* $Y \subset \mathbb{R}^p$ *be an algebraic set and* $F : \mathbb{R}^n \to \mathbb{R}^p$ *a polynomial mapping. Then* $F^{-1}(Y) \subset \mathbb{R}^n$ *is an algebraic set.*

At this point a natural question is whether the image of an algebraic set under a polynomial mapping is an algebraic set. This is not always true, as we can see in the following example.

Example 5.2.6 Let $S^1 = \{(x, y) \mid x^2 + y^2 = 1\} \subset \mathbb{R}^2$ and $\pi : \mathbb{R}^2 \to \mathbb{R}, \pi(x, y) = x$. We know that S^1 is an algebraic set but $\pi(S^1) = [-1, 1]$ is not algebraic.

This example motivates the introduction of a larger class of objects called **semialgebraic** sets. We start with some notations.

Let $f, g : (X, a) \to (\mathbb{R}, 0)$ be germs of non-negative continuous functions defined on a subset X in $\mathbb{R}^n, a \in X$. We denote $f \lesssim g$ when there exists a neighborhood U of a in X and a constant K such that $f(x) \leq Kg(x), \forall x \in U$. We denote $f \approx g$ when $f \lesssim g$ and $g \lesssim f$. Also, we denote $f \prec\prec g$ when $\lim_{x \to a} \frac{f(x)}{g(x)} = 0$ and $f \succ\succ g$ when $\lim_{x \to a} \frac{g(x)}{f(x)} = 0$.

Definition 5.2.7 A subset X of $\mathbb{R}^n$ is said to be **semialgebraic** if there exist polynomial functions $f_{i,j}, g_{i,j} : \mathbb{R}^n \to \mathbb{R}$, $1 \leq i \leq p$, $1 \leq j \leq q$ such that

$$X = \cup_{i=1}^{p} \cap_{j=1}^{q} \{x \in \mathbb{R}^n \mid f_{i,j}(x) = 0; \ g_{i,j}(x) > 0\}.$$

Proposition 5.2.8 *The following properties hold:*

1. *An algebraic set is semialgebraic.*
2. *In* $\mathbb{R}$*, semialgebraic sets are finite unions of intervals and points.*
3. *If* $X, Y \subset \mathbb{R}^n$ *are semialgebraic sets then* $X \cup Y$, $X - Y$, *and* $X \cap Y$ *are semialgebraic.*
4. *If* $X \subset \mathbb{R}^n$ *and* $Y \subset \mathbb{R}^p$ *are semialgebraic sets then* $X \times Y \subset \mathbb{R}^n \times \mathbb{R}^p$ *is a semialgebraic set.*

Theorem 5.2.9 (Tarski-Seidenberg) *Let* $\pi : \mathbb{R}^n \times \mathbb{R}^p \to \mathbb{R}^p$ *be the canonical projection. If* $X \subset \mathbb{R}^n \times \mathbb{R}^p$ *is semialgebraic then* $\pi(X)$ *is also semialgebraic.*

Definition 5.2.10 Let $X \subset \mathbb{R}^n$ be a semialgebraic set. A mapping $F : X \to \mathbb{R}^p$ is said to be **semialgebraic** if its graph,

$$\text{graph}(F) = \{(x, y) \in \mathbb{R}^n \times \mathbb{R}^p \mid y = F(x)\}$$

is a semialgebraic subset of $\mathbb{R}^n \times \mathbb{R}^p$.

The following results are consequences of Tarski-Seidenberg Theorem.

Corollary 5.2.11 *If* $X \subset \mathbb{R}^n$ *is semialgebraic and* $F : \mathbb{R}^n \to \mathbb{R}^p$ *is a semialgebraic map then* $F(X)$ *is a semialgebraic set of* $\mathbb{R}^p$.

Corollary 5.2.12 *If* $X \subset \mathbb{R}^n$ *is semialgebraic then the closure* $\overline{X}$ *and the interior* X° *of* X *are also semialgebraic.*

Fig. 5.1 Whitney umbrella

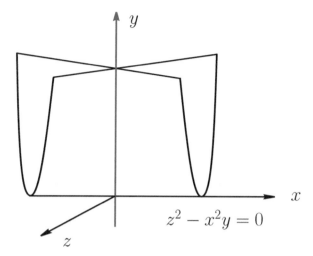

$$z^2 - x^2 y = 0$$

Exercise 5.2.13 Prove Corollaries 5.2.11 and 5.2.12.

Example 5.2.14 (The Whitney Umbrella (Cross-Cap)) Let $f(x, y, z) = z^2 - x^2 y$. Then $X = f^{-1}(0)$ is an algebraic set in $\mathbb{R}^3$ called Whitney umbrella. The image of the map $\Phi : \mathbb{R}^2 \to \mathbb{R}^3$ given by $\Phi(u, v) = (u, v^2, y)$ is a parametrization of the semialgebraic set $\{f = 0, y \geq 0\}$ (Fig. 5.1).

5.2.2 Semialgebraic Arcs and Curve Selection Lemma

Let $f : [0, \alpha) \to \mathbb{R}$ be a continuous semialgebraic function with $f(0) = 0$. Up to reducing α, the function f is a converging Puiseux series

$$f(t) = \sum_{k \geq 1} a_k t^{k/p}, \tag{5.1}$$

where p is a positive integer (see [Wal04, BCR98]).

The **order** of such an f is defined as $\mathrm{ord}(f) = \frac{q}{p}$ where $q = \min\{k \,|\, a_k \neq 0\}$.

Exercise 5.2.15 Verify that the function f is C^1 if and only if $\mathrm{ord}(f) \geq 1$.

Definition 5.2.16 A **semialgebraic arc** at $x_0 \in \mathbb{R}^n$ is the image $\gamma([0, \epsilon))$ of a C^1-semialgebraic map $\gamma : [0, \epsilon) \to \mathbb{R}^n$, $\gamma(0) = x_0$.

It follows that a semialgebraic arc at $x_0 \in \mathbb{R}^n$ is a converging Puiseux power series of the form

$$[0, \epsilon) \to \mathbb{R}^n \tag{5.2}$$

$$t \mapsto \gamma(t) = x_0 + \sum_{k \geq 1} u_k t^{k/p} \tag{5.3}$$

for p a positive integer and $u_k \in \mathbb{R}^n$ (see [BCR98, GBGPPP17].)

Remark 5.2.17 A semialgebraic arc admits a reparametrization $s \to t(s)$ into a converging Puiseux series $s \to \gamma(s)$ such that $\|\gamma(s) - x_0\| \approx s$ so that

$$\gamma(s) = x_0 + sv(s) = x_0 + s \left[\frac{u_0}{\|u_0\|} + \sum_{\ell \geq 1} s^{\ell/r} v_\ell \right]$$

for r a positive integer and $v_l \in \mathbb{R}^n$. Notice that $\lim_{s \to 0} \frac{\gamma(s) - x_0}{\|\gamma(s) - x_0\|} = \frac{u_0}{\|u_0\|}$.

Definition 5.2.18 The **tangent cone** to the image of γ at x_0 is the half line $\mathbb{R}_{\geq 0} u_0$. We denote it $C_0(\gamma)$.

The Curve Selection Lemma stated and proved by Milnor in [Mil68] is a fundamental result in the local structure of semialgebraic sets. The first versions of this lemma appeared for the first time around the middle of last century in the works of Bruhat and Cartan [BC57] (see also Wallace [Wal58] and Moreira and Ruas [MR09]).

Lemma 5.2.19 (Curve Selection Lemma) *Let V be a semialgebraic set in $\mathbb{R}^n$ and $x_0 \in \bar{V} \setminus V$. Then there exists a continuous semialgebraic mapping $p : [0, \epsilon) \to \mathbb{R}^n$, with $p(0) = x_0$ and $p(t) \in V$ for every $t > 0$.*

Remark 5.2.20 The parametrization of the curve in this lemma can be assumed to be analytic, see [BCR98].

See Theorem 5.7.8 in the Appendix for the general version of the Curve Selection lemma for definable sets in o-minimal structures.

5.2.3 The Conic Structure Theorem

We denote by $\overline{B}(x_0, \epsilon)$ (resp. $S(X_0, \epsilon)$) the closed ball (resp. the sphere) with center at x_0 and radius ϵ. Given a topological space Z the cone of vertex x_0 and base Z is the set

$$\text{Cone}(Z, x_0) = \{y = (1 - t)x_0 + tx, x \in Z, 0 \leq t \leq 1\}.$$

Let X be a semialgebraic subset of $\mathbb{R}^n$ and $x_0 \in X$. Let $\bar{B}(x_0, \epsilon)$ (respectively $S(x_0, \epsilon)$) be the closed ball (respectively the sphere) with center in x_0 and radius ϵ and let $K(x_0, X) = X \cap S(x_0, \epsilon)$ for a sufficiently small $\epsilon > 0$. The next result, called the Conic Structure Theorem, shows that in a neighborhood of a point x_0 of a semialgebraic set X, the intersection of X with a ball is semialgebraically homeomorphic to the cone of the intersection of X with the sphere (see [BR90, BCR98]).

Theorem 5.2.21 (Conical Structure) *For $\epsilon > 0$, sufficiently small, there is a semialgebraic homeomorphism*

$$h : \bar{B}(x_0, \epsilon) \cap X \to \mathrm{Cone}(K(x_0, X))$$

such that $\|h(x) - x_0\| = \|x - x_0\|$ and $h|_{K(x_0,X)} = \mathrm{Id}$.

The proof of Theorem 5.2.21 follows by applying Hardt's semialgebraic trivialization Theorem to the (continuous, semialgebraic) distance function in $\mathbb{R}^n$, $\rho(x) = \|x - x_0\|$. The reader can see the statement and the proof of Hardt's Theorem in [BR90] or [BCR98], for instance. Hardt's Theorem for definable sets in o-minimal structures is Theorem 5.7.11 in the Appendix.

For every $\epsilon > 0$ sufficiently small it follows that we can associate to the semialgebraic distance function $\rho : X \to (0, \epsilon]$, the semialgebraic homeomorphism

$$\{x \in \mathbb{R}^n; 0 < \|x - x_0\| \leq \epsilon\} \to (0, \epsilon] \times S(x_0, \epsilon)$$

$$x \mapsto (\|x - x_0\|, \tilde{h}(x))$$

where $\tilde{h}(x) = x_0 + \frac{\epsilon}{\|x - x_0\|}(x - x_0)$. This homeomorphism is compatible with X and satisfies the condition $\tilde{h}|_{S(x_0,\epsilon)} = \mathrm{Id}$. Then, the homeomorphism

$$h(x) = x_0 + \frac{\|x - x_0\|}{\epsilon}(\tilde{h}(x) - x_0),$$

gives the result.

Corollary 5.2.22 *For $\epsilon > 0$, sufficiently small the semialgebraic topological type of the set $K(x_0, X) = X \cap S(x_0, \epsilon)$ does not depend on ϵ.*

It follows from this corollary that the set $K(x_0, X)$ is a topological invariant of X at x_0. We call it the **link** at x_0 in X. When x_0 and X are clear from the context we denote $K(x_0, X)$ simply by K.

Remark 5.2.23 The Conic Structure Theorem for complex algebraic hypersurfaces with isolated singularities was proved by Milnor [Mil68], Theorem 2.10. We recommend to the reader to compare the results of this section with the notes by Haydée Aguilar-Cabrera and José Luis Cisneros-Molina, Geometric viewpoint of Milnor's Fibration Theorem, [ACCM19], published in the present volume.

5.3   Basic Concepts on Lipschitz Geometry

In this section we review the basic definitions and properties of metric structures on topological spaces. The notion of Lipschitz normal embedding plays a central role in Lipschitz geometry and classification of singular spaces.

5.3.1 Inner and Outer Metrics: Lipschitz Normal Embeddings

Let X and Y be metric spaces.

Definition 5.3.1 A map $f : X \to Y$ is **Lipschitz** if there exists a positive constant $k \in \mathbb{R}$ such that for every $x_1, x_2 \in X$, we have

$$d_Y(f(x_2), f(x_1)) \leq kd_X(x_2, x_1),$$

where d_Y and d_X are the metrics in Y and X, respectively.

A bi-Lipschitz homeomorphism between two metric spaces X and Y is a Lipschitz homeomorphism $h : X \to Y$ whose inverse $h^{-1} : Y \to X$ is also Lipschitz. Bi-Lipschitz maps are the isomorphisms in the category of metric spaces.

Let X be a subset of $\mathbb{R}^n$. There are two natural metrics defined on X, the **outer metric**

$$d_{\text{out}}(x, y) = \|x - y\|$$

which is the induced Euclidean metric on X and the **inner metric** or (**intrinsic metric**)

$$d_{\text{in}}(x, y) = \inf_{\gamma \in \Gamma(x,y)} l(\gamma)$$

where $\Gamma(x, y)$ is the set of rectifiable arcs $\gamma : [0, 1] \to X$ with $\gamma(0) = x$ and $\gamma(1) = y$ and $l(\gamma)$ is the length of γ.

Is clear that $d_{\text{out}}(x, y) \leq d_{\text{in}}(x, y)$ but the converse does not hold in general.

Definition 5.3.2 We say that X is **Lipschitz normally embedded** (**LNE**) if there exists $k > 0$ such that for all $x, y \in X$, we have

$$d_{\text{in}}(x, y) \leq kd_{\text{out}}(x, y). \tag{5.4}$$

Any convex subset of $\mathbb{K}^n$ is a trivial example of Lipschitz normally embedded set. For an example of a set which is not Lipschitz normally embedded consider the (real or complex) curve defined by $x^3 - y^2 = 0$ (see Example 1.8 in [Pic19]). Then

$$d_{\text{out}}((t^2, t^3), (t^2, -t^3)) = 2|t|^3$$

but,

$$d_{\text{in}}((t^2, t^3), (t^2, -t^3)) = 2|t|^2 + o(t^2).$$

Then

$$\lim_{t \to 0} \frac{d_{\text{in}}((t^2, t^3), (t^2, -t^3))}{d_{\text{out}}((t^2, t^3), (t^2, -t^3))} = +\infty$$

and hence there cannot exist a constant k satisfying (5.4).

Definition 5.3.3 A space X is **locally Lipschitz normally embedded** at $x \in X$ if there is an open neighborhood U in X such that U is Lipschitz normally embedded. We say that X is locally normally embedded if this condition holds at each $x \in X$.

In his thesis Sampaio [Sam15, Example 4.4.7] (see also Sampaio [Sam20]) proves that the cone $z^3 = x^3 + y^3$ in $\mathbb{R}^3$ is Lipschitz normally embedded. It is possible to generalize his proof to the complex case and to higher dimensions.

The next proposition extends Sampaio's result for metric cones whose links are also Lipschitz normally embedded.

Proposition 5.3.4 (Proposition 2.8, [KPR18]) *Let $X \subset \mathbb{K}^n$ be the semialgebraic cone over the link $L \subset S$ with vertex at the origin in $\mathbb{K}^n$, where $S = S^{n-1}$ if $\mathbb{K} = \mathbb{R}$ and $S = S^{2n-1}$ if $\mathbb{K} = \mathbb{C}$. Then, the following conditions hold:*

1. *If L is Lipschitz normally embedded then X is Lipschitz normally embedded.*
2. *If X is Lipschitz normally embedded and L is compact then each connected component of L is Lipschitz normally embedded.*

Proof We will prove 1. Since L is Lipschitz normally embedded with Lipschitz constant K_L, the same is true for the scaled version $rL = \{x \in X \mid \|x\| = r\}$ where $r \in \mathbb{R}^+$.

Let $x, y \in X$. We first assume $0 \le \|x\| \le \|y\|$. If $x = 0$, then $d_{\text{in}}^X(x, y) = d_{\text{out}}(x, y)$ since the straight line through 0 and y is in X, because X is a cone with vertex at the origin.

If $\|x\| = \|y\| = r$, then x and y are both in rL and hence

$$d_{\text{in}}^X(x, y) \le d_{\text{in}}^{rL}(x, y) \le K_L d_{\text{out}}(x, y).$$

Now if, $0 < \|x\| < \|y\|$, we let $y' = \frac{y}{\|y\|}\|x\|$. Then, $d_{\text{in}}^X(y, y') = d_{\text{out}}(y, y')$ since both lie in the same straight line through the origin. Let $r = \|x\|$. Then $x, y' \in rL$ and like before $d_{\text{in}}^X(x, y') \le K_L d_{\text{out}}(x, y')$. Now y' is the closest point to y in rL. Hence all of rL lies on the other side of the hyperplane through y' orthogonal to the line yy'. So, in the triangle $yy'x$, the angle at y' is $> \frac{\pi}{2}$. Then $d_{\text{out}}(y, x) \ge d_{\text{out}}(y, y')$ and $d_{\text{out}}(y, x) \ge d_{\text{out}}(y', x)$ and we obtain

$$d_{\text{in}}^X(x, y) \le d_{\text{in}}^X(x, y') + d_{\text{in}}^X(y', y)$$

$$K_L d_{\text{out}}(x, y') + d_{\text{out}}(y', y)$$

$$\le (K_L + 1) d_{\text{out}}(x, y),$$

and X is Lipschitz normally embedded.

We leave the proof of item (2) to the reader.   □

Corollary 5.3.5 *Let* $(X, 0)$ *be an algebraic hypersurface in* $\mathbb{K}^n$ ($\mathbb{K} = \mathbb{R}$ *or* $\mathbb{C}$) *defined by a homogeneous polynomial with isolated singularity. Then* X *is Lipschitz normally embedded.*

Example 5.3.6 The D_4-singularity $X = f^{-1}(0)$ with $f(x, y) = x^2y + y^3$ is Lipschitz normally embedded.

The notion of (Lipschitz) normally embedded subsets was defined by L. Birbrair and T. Mostowski in [BM00]. Their main result is the Normal Embedding theorem stated below.

Theorem 5.3.7 (Normal Embedding Theorem) *Let* $X \subset \mathbb{R}^n$ *be a compact semialgebraic set equipped with the inner metric. Then there is a Lipschitz normally embedded semialgebraic set* $\widetilde{X} \subset \mathbb{R}^m$ *which is bi-Lipschitz homeomorphic to* (X, d_{in}).

Exercise 5.3.8 (β-Horn Surface) The semialgebraic surface in $\mathbb{R}^3$ defined by

$$X = \{(x, y, z) | (x^2 + y^2)^q = z^{2p}, z \geq 0\}$$

with $\beta = \frac{p}{q} \geq 1$ is Lipschitz normally embedded.

Example 5.3.9 (E_8–Singularity) Let $f : \mathbb{R}^3 \to \mathbb{R}$, $f(x, y, z) = x^2 - y^3 + z^5$ and $X = \{(x, y, z) \in \mathbb{R}^3 | f(x, y, z) = 0\}$.

The polynomial f is the normal form of the real E_8 singularity.

We outline here the argument of A. Fernandes (personal communication) to prove that E_8 is not Lipschitz normally embedded.

In fact, let $\gamma^{\pm}(t) = (\pm t^{\frac{3}{2}}, t, 0)$, $t \in \mathbb{R}$ be the two branches of the curve $X \cap \{z = 0\}$, and let β be the curve defined by $X \cap \{x = 0\}$. The common unit tangent vector to γ^+ and γ^- at $t = 0$ is $(0, 1, 0)$, while the tangent to the curve β at the origin is $(0,0,1)$, see Fig. 5.2.

Now for any fixed t, if C is any arc in X from $\gamma^-(t)$ to $\gamma^+(t)$ then C always intersect β. We denote by $Q = (0, \bar{y}, \bar{z})$ this intersection point (we can assume Q is unique).

We have $d_{in}(\gamma^-(t), \gamma^+(t)) = \inf_C l(C)$ where $l(C) = l(C^-) + l(C^+)$ and C^- and C^+ are the arcs from $\gamma^-(t)$ and $\gamma^+(t)$, to Q. Then it is easy to verify that:

$$l(C) \geq \|\gamma^-(t) - Q\| + \|\gamma^+(t) - Q\| = 2\sqrt{t^3 + (t - \bar{y})^2 + \bar{z}^5} \approx t,$$

Fig. 5.2 E_8 is not Lipschitz normally embedded. **(a)** $x^2 - y^3 + z^5 = 0$. **(b)** $d_{in}(\gamma^-(t), \gamma^+(t)) = \inf_C l(C)$

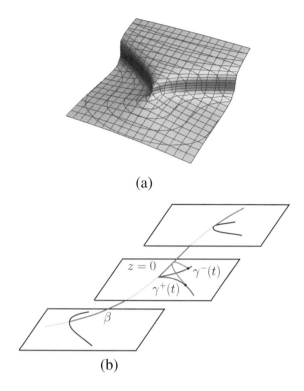

(a)

(b)

for $\bar{y}$, $\bar{z}$ sufficiently small. The argument now is similar to the previous argument to show that the cusp $x^2 - y^3 = 0$ is not locally normally embedded at the origin, that is,

$$\frac{d_{in}(\gamma^-(t), \gamma^+(t))}{d_{out}(\gamma^-(t), \gamma^+(t))} \to \infty$$

when $t \to 0$, and the result follows.

5.3.2 Existence of Extension of Lipschitz Mappings

We finish this section with a few theorems about existence of extension of Lipschitz mappings.

The Theorem of Kirszbraun considers the problem of extending Lipschitz functions. It says that a $c-$Lipschitz function $f : Y \to \mathbb{R}$ can be extended to a $c-$Lipschitz function $f : X \to \mathbb{R}$, where Y is a metric subspace of X. In 1934, E. McShane in [McS34] found a new proof with an explicit formula for the extension.

We state here a weaker version of Kirszbraun's Theorem including McShane's formula in the statement.

Theorem 5.3.10 (Kirszbraun's Theorem) *Let X be a subset of $\mathbb{R}^n$ equipped with the outer metric and $f : X \to \mathbb{R}$ be a c-Lipschitz map. Then*

$$F(z) = \sup_{x \in X}(f(z) - c\|x - z\|), \; z \in \mathbb{R}^n$$

is a c-Lipschitz extension of f.

We recall a classical criterion to verify if a given semialgebraic function is Lipschitz.

Lemma 5.3.11 *Let U be a convex open subset of $\mathbb{R}^n$ and $f : U \to \mathbb{R}$ a continuous semialgebraic function. If $Df(x)$ exists and $\|Df(x)\| \leq M$ except for a finite number of points in U, then F is Lipschitz with Lipschitz constant also bounded by M.*

The following proposition proved in [FKK98] justifies the usefulness of working with semialgebraic bi-Lipschitz homeomorphisms.

Proposition 5.3.12 *Let $h : (\mathbb{R}^n, 0) \to (\mathbb{R}^n, 0)$ be a semialgebraic homeomorphism. Then h is bi-Lipschitz if and only if for every pair of analytic curves $\alpha_1, \alpha_2 : [0, \delta) \to (\mathbb{R}^n, 0)$ we have*

$$\mathrm{ord}_t \|h(\alpha_1(t)) - h(\alpha_2(t))\| = \mathrm{ord}_t \|\alpha_1(t) - \alpha_2(t)\|$$

5.4 Lipschitz Geometry of Real Curves and Surfaces

We present here the main ingredients we need to classify real semialgebraic sets of dimensions 1 and 2. The results are due to L. Birbrair and A. Fernandes, [Bir99, BF00] and [Fer03].

Definition 5.4.1 Let X, Y be semialgebraic sets in $\mathbb{R}^n$ and d_X and d_Y chosen metrics in X and Y respectively. We say that X, Y are **bi-Lipschitz equivalent** if there exists a bi-Lipschitz homeomorphism $h : (X, d_X) \to (Y, d_Y)$ such that $h(X) = Y$. If h is semialgebraic we say that X, Y are semialgebraically bi-Lipschitz equivalent.

Similarly, we say that the germs (X, x_0) and (Y, y_0) are bi-Lipschitz equivalent if there exists a germ of a bi-Lipschitz homeomorphism $h : (X, x_0) \to (Y, y_0)$ satisfying the conditions of the definition above.

In these notes we distinguish three kinds of bi-Lipschitz equivalence:

(1) **inner bi-Lipschitz equivalence** when d_X and d_Y are inner metrics in X and Y, respectively,

Fig. 5.3 Semialgebraic arc

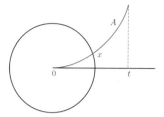

(2) **outer bi-Lipschitz equivalence** when d_X and d_Y are outer metrics, that is, the Euclidean metric induced in X and Y, respectively,

(3) **ambient bi-Lipschitz equivalence** when there exists a germ of a bi-Lipschitz homeomorphism $h : (\mathbb{R}^n, x_0) \to (\mathbb{R}^n, y_0)$ such that $h(X) = Y$.

5.4.1   Lipschitz Geometry of Semialgebraic Curves

As we have seen in Sect. 5.2 a semialgebraic arc A is the image $\gamma([0, \epsilon))$ of a C^1-semialgebraic map $\gamma : [0, \epsilon) \to \mathbb{R}^n$, $\gamma(0) = x_0$. It has a Puiseux series expansion

$$\gamma(t) = x_0 + \sum_{k \geq 1} u_k t^{k/p}$$

where p is a positive integer and $u_k \in \mathbb{R}^n$ (Fig. 5.3).

Remark 5.4.2   Notice that every arc is Lipschitz normally embedded.

Proposition 5.4.3   *Every arc in $\mathbb{R}^n$ is bi-Lipschitz outer equivalent to a segment.*

Corollary 5.4.4   *If $(X, 0)$ is the germ of a semialgebraic curve in $\mathbb{R}^n$, then $(X, 0)$ is inner bi-Lipschitz equivalent to a star with r segments, where r is the number of half-branches $(X_i, 0)$ of $(X, 0)$.*

Proof   Let v_i denote the vectors defining the segments of the star $L = \cup_{i \in r} L_i$, $\|v_i\| = 1$ (Fig. 5.4).
    We define

$$\Phi_i : X_i \to L_i$$

$$x \mapsto |x| v_i$$

and

$$\Phi : (X, 0) \to L$$

$$x \mapsto \Phi_i(x), \text{ if } x \in X_i.$$

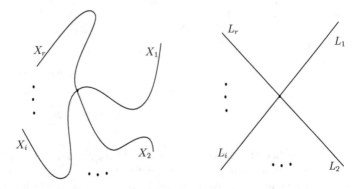

Fig. 5.4 Inner bi-Lipschitz geometry of a semialgebraic curve

Then, denoting by x_i points in X_i, we have

$$d_{in}(\Phi(x_i), \Phi(x_j)) \le d_{in}(\Phi(x_i), 0) + d_{in}(0, \Phi(x_j))$$
$$\le \lambda(d_{in}(x_i, 0) + d_{in}(x_j, 0))$$
$$= \lambda d_{in}(x_i, x_j)$$

The contact order between arcs that we define now is an important invariant for the bi-Lipschitz equivalence of semialgebraic curves.

Definition 5.4.5 Let A and B be arcs, we define the **contact order between A and B** as

$$t_{ord}(A, B) = ord_r dist(A \cap S(0, r), B \cap S(0, r)),$$

where $S(0, r)$ is the sphere of center 0 and radius r in $\mathbb{R}^n$.

Proposition 5.4.6 *Let $A = \gamma([0, \epsilon))$ and $B = \beta([0, \epsilon))$ be two semialgebraic arcs at $x_0 = 0$ in $\mathbb{R}^n$ such that their tangent cones are distinct (see Definition 5.2.18). Then $t_{ord}(A, B) = 1$.*

Proof From Remark 5.2.17, we can reparametrize A and B to get

$$\gamma(s) = s \left(\frac{u_0}{\|u_0\|} + \sum_{\ell \ge 1} s^{\ell/r} u_\ell \right),$$

$$\beta(s) = s \left(\frac{w_0}{\|w_0\|} + \sum_{\ell \ge 1} s^{\ell/r} w_\ell \right),$$

where u_0, w_0, u_l, w_l are vectors in $\mathbb{R}^n$. The hypothesis imply that u_0 and w_0 are not parallel.

Then $\text{ord}(\text{dist}(\gamma(s), \beta(s))) = 1$. $\qquad\qquad\square$

Example 5.4.7 Let $x^2 - y^3 = 0$ and $x^4 - y^7 = 0$ be two algebraic curves in $\mathbb{R}^2$. Then the contact order between the branches $A : x = y^{\frac{3}{2}}$ and $B : x = y^{\frac{7}{4}}$, $y \geq 0$ is

$$\text{t}_{\text{ord}}(A, B) = \text{ord}(\text{dist}((y^{\frac{3}{2}}, y), (y^{\frac{7}{4}}, y))) = \text{ord}(|y|^{\frac{3}{2}}|1 - y^{\frac{1}{4}}|) = \frac{3}{2}$$

Example 5.4.8 Let $\gamma, \beta : [0, \epsilon) \to \mathbb{R}^3$, given by

$$\gamma(t) = (t, t^{\frac{4}{3}}, t^{\frac{5}{2}} + t^{\frac{7}{4}}),$$

and

$$\beta(t) = (t, t^{\frac{5}{2}}, t^{\frac{7}{3}} + t^{\frac{8}{5}}).$$

Then $\text{t}_{\text{ord}}(\gamma(t), \beta(t)) = \frac{4}{3}$.

Exercise 5.4.9 Let $\sigma : (\mathbb{R}^n, 0) \to (\mathbb{R}^n, 0)$ be the germ of a semialgebraic diffeomorphism. Let A and B be semialgebraic arcs in $(\mathbb{R}^n, 0)$. Prove that $\text{t}_{\text{ord}}(A, B) = \text{t}_{\text{ord}}(\sigma(A), \sigma(B))$.

In the next proposition we prove that $\text{t}_{\text{ord}}(A, B)$ is a Lipschitz invariant with respect to the outer metric. We define the following auxiliary relation.

Definition 5.4.10

$$\tilde{\text{t}}_{\text{ord}}(A, B) = \text{ord}_r \text{dist}(A \setminus B(0, r), B \setminus B(0, r)).$$

Lemma 5.4.11 (Comparison Lemma [BF00]) $\text{t}_{\text{ord}}(A, B) = \tilde{\text{t}}_{\text{ord}}(A, B)$.

Proposition 5.4.12 $\text{t}_{\text{ord}}(A, B)$ *is a Lipschitz invariant with respect to the outer metric.*

Proof Let σ be a bi-Lipschitz map, $\sigma(A) = A_1$, $\sigma(B) = B_1$. Then,

$$d_{out}(A \cap S(0, r), B \cap S(0, r)) = |x(r) - y(r)| \geq k|\sigma(x(r)) - \sigma(y(r))|$$

$$\geq k d_{out}(A_1 \setminus B(0, \frac{1}{\lambda}r), B_1 \setminus B(0, \frac{1}{\lambda}r)),$$

where we assume $\frac{r}{\lambda} = \|\sigma(x(r))\| \leq \|\sigma(y(r))\|$. Hence,

$$\text{t}_{\text{ord}}(A, B) \geq \tilde{\text{t}}_{\text{ord}}(A_1, B_1)$$

By the Comparison Lemma we have $\tilde{t}_{\text{ord}}(A_1, B_1) = t_{\text{ord}}(A_1, B_1)$, hence it follows that $t_{\text{ord}}(A, B) \geq t_{\text{ord}}(A_1, B_1)$. The converse follows similarly. $\square$

We are now ready to state and prove the classification theorems for semialgebraic curves. We organize the proofs in three parts.

First, in Theorem 5.4.13, we prove that the contact between the pairs of branches is a complete invariant for the outer bi-Lipschitz equivalence of curves in $\mathbb{R}^n$.

In Lemmas 5.4.15, 5.4.17 and Proposition 5.4.19 we give the main steps of the classification of plane curves with respect to the ambient bi-Lipschitz equivalence.

Finally, in Theorem 5.4.20 we state and prove the classification of curves in $\mathbb{R}^n$ with respect to ambient bi-Lipschitz equivalence.

Theorem 5.4.13 *Let $(X, 0)$ and $(Y, 0)$ be semialgebraic curves with branches $X_1, \ldots, X_l$ and $Y_1, \ldots, Y_s$ respectively. Then, $(X, 0)$ is outer bi-Lipschitz equivalent to $(Y, 0)$ if and only if $l = s$ and there is a permutation σ of $\{1, \ldots, l\}$ such that*

$$t_{\text{ord}}(X_i, X_j) = t_{\text{ord}}(Y_{\sigma_i}, Y_{\sigma_j}) \, i, j \in \{1, 2, \ldots, l\}.$$

Proof We know that the number of branches is a topological invariant of semi-algebraic curves (Theorem 5.2.21). Let $F : (X, 0) \rightarrow (Y, 0)$ be a bi-Lipschitz homeomorphism with respect to the outer metric, such that $F(X_i) = Y_{\sigma_i}$. Then, from Proposition 5.4.12 it follows that $t_{\text{ord}}(X_i, X_j) = t_{\text{ord}}(Y_{\sigma_i}, Y_{\sigma_j})$. To prove the converse, for $x \in X_i$ we define $F(x) = Y_{\sigma_i} \cap S(0, |x|)$. We have already seen that $F|_{X_i} : X_i \rightarrow Y_{\sigma_i}$ is outer bi-Lipschitz. To complete the proof we show that $F : X \rightarrow Y$ is bi-Lipschitz. In fact, supposing the contrary, we would have $x_i(r) \in X_i$, $x_j(r) \in X_j$ such that $\|x_i(r) - x_j(r)\| \ll \|F(x_i(r)) - F(x_j(r))\|$ and this would imply $t_{\text{ord}}(X_i, X_j) \neq t_{\text{ord}}(Y_{\sigma_i}, Y_{\sigma_j})$, a contradiction. $\square$

Remark 5.4.14 We write $t_{\text{ord}}(X, x) = t_{\text{ord}}(Y, y)$ to indicate that there is a bijection $\sigma : I \rightarrow J$ such that $t_{\text{ord}}(X_i, X_j) = t_{\text{ord}}(Y_{\sigma(i)}, Y_{\sigma(j)})$ for each pair $i \neq j \in I$.

We now want to prove that the contact between branches is a complete invariant for the (semialgebraic) bi-Lipschitz ambient equivalence between X and Y in $\mathbb{R}^n$.

We first deal with plane curves. Let $f : [0, \delta] \rightarrow \mathbb{R}$, $A_f = \{(x, y) \in \mathbb{R}^2 | x \geq 0, \, y = f(x)\}$. When $f(x) = x^\alpha$ we denote $A_f = A_\alpha$. Interchanging the axis x and y, if necessary, we can assume $\alpha \geq 1$.

Lemma 5.4.15 *Let $f : [0, \delta] \rightarrow \mathbb{R}$, be a semialgebraic function with Newton-Puiseux decomposition in $x = 0$ given by $f(x) = x^\alpha h(x)$, with $h(0) > 0$ and $\alpha \geq 1$. Let $F : (\mathbb{R}^2, 0) \rightarrow (\mathbb{R}^2, 0)$ be the semialgebraic map-germ defined by*

$$F(x, y) = \begin{cases} (x, \frac{y}{h(x)}) & \text{if } (x, y) \in Q \text{ and } 0 \leq y \leq f(x), \\ (x, y - f(x) + x^\alpha) & \text{if } (x, y) \in Q \text{ and } 0 \leq f(x) \leq y, \\ (x, y) & \text{if } (x, y) \notin Q. \end{cases}$$

where $Q = [0, \delta] \times [0, \epsilon]$ with $\epsilon = \max_{x \in [0,\delta]} f(x)$.

  Then F is a semialgebraic bi-Lipschitz map-germ such that $F(A_f, 0) = (A_\alpha, 0)$ and it is the identity in the complement of Q.

Proof F clearly defines a semialgebraic homeomorphism such that $(F(A_f), 0) = (A_\alpha, 0)$. To show that F is bi-Lipschitz in Q it is enough to show its derivative is bounded.   □

Remark 5.4.16 As a consequence of Lemma 5.4.15 it follows that if $f, g : [0, \delta] \to \mathbb{R}$ are semialgebraic functions with the same first Newton-Puiseux exponent α in $x = 0, \alpha \geq 1$ then there is a semialgebraic bi-Lipschitz $F : (\mathbb{R}^2, 0) \to (\mathbb{R}^2, 0)$ such that $F(A_f) = A_g$. That is, A_f and A_g are ambient outer bi-Lipschitz equivalent.

Lemma 5.4.17 *Let $f_i, g_i : [0, \delta] \to \mathbb{R}$ be semialgebraic functions with $f_i(0) = g_i(0) = 0$ and $0 \leq f_i(x), g_i(x) \leq \epsilon$ for $i = i, \ldots, m$. Let us suppose that their first Newton-Puiseux exponent are bigger than or equal to one. Let $X = \cup_i \mathrm{graph}(f_i)$ and $Y = \cup_i \mathrm{graph}(g_i)$. If $\mathrm{t_{ord}}(X, 0) = \mathrm{t_{ord}}(Y, 0)$, then there exists a semialgebraic bi-Lipschitz map-germ $F : (\mathbb{R}^2, 0) \to (\mathbb{R}^2, 0)$ such that $F(X) = Y$, and F is the identity in the complement of $Q = [0, \delta] \times [0, \epsilon]$.*

Proof We can assume $f_m \leq f_{m-1} \leq \cdots \leq f_1$ and $g_m \leq g_{m-1} \leq \cdots \leq g_1$ and prove the result by induction on m.

  If $m = 2$ the result follows from Lemma 5.4.15. Applying Lemma 5.4.15 again, we can assume $f_m = g_m = \lambda$ where $\lambda(x) = x^\alpha h(x), h(0) > 0$. Let $\widetilde{X} = \cup_{i=1}^{m-1} \mathrm{graph}(\widetilde{f_i})$, $\widetilde{Y} = \cup_{i=1}^{m-1} \mathrm{graph}(\widetilde{g_i})$, where $\widetilde{f_i} = f_i - \lambda$ and $\widetilde{g_i} = g_i - \lambda$ for $i = 1, \ldots, m - 1$.

  By induction hypothesis, there exists a semialgebraic bi-Lipschitz map-germ $\widetilde{F} : (\mathbb{R}^2, 0) \to (\mathbb{R}^2, 0)$ such that $\widetilde{F}(\widetilde{X}) = \widetilde{Y}$, and $\widetilde{F}$ is the identity outside Q. We define $\sigma : (\mathbb{R}^2, 0) \to (\mathbb{R}^2, 0)$ by $\sigma(x, y) = (x, y - \lambda(x))$. Then, $F : (\mathbb{R}^2, 0) \to (\mathbb{R}^2, 0)$ defined by $F = \sigma^{-1} \circ \widetilde{F} \circ \sigma$ is a semialgebraic bi-Lipschitz map-germ such that $F(X, 0) = (Y, 0)$ and is the identity in the complement of Q.   □

Remark 5.4.18 Let $c > 0$ and $K = \{(x, y) \in \mathbb{R}^n | 0 \leq x, 0 \leq y \leq cx\}$. Then $\Phi : (\mathbb{R}^2, 0) \to (\mathbb{R}^2, 0)$ defined by $\Phi(x, y) = (cx - y, y)$ is a semialgebraic bi-Lipschitz map such that $\phi(K) = Q$. So, in the statement of Lemma 5.4.17 we can take the map-germ $F : (\mathbb{R}^2, 0) \to (\mathbb{R}^2, 0)$ such that $F(X) = Y$ and F is the identity outside the cone K.

Proposition 5.4.19 *Let $X, Y \subset \mathbb{R}^2$ be semialgebraic curves and $x \in X, y \in Y$. If $\mathrm{t_{ord}}(X, x) = \mathrm{t_{ord}}(Y, y)$, then there exists a semialgebraic bi-Lipschitz map-germ $F : (\mathbb{R}^2, x) \to (\mathbb{R}^2, y)$ such that $F(X) = Y$.*

Proof We take $x = y = 0$ in $\mathbb{R}^2$. If $\mathrm{t_{ord}}(X, 0) = \mathrm{t_{ord}}(Y, 0)$, it follows from Theorem 5.4.13 that the number r of half lines tangent to $(X, 0)$ is the same as number of the ones tangent to $(Y, 0)$. Then we can separate the branches of $(X, 0)$, respectively $(Y, 0)$, in cones $K_1^X, \ldots, K_r^X$, respectively $K_1^Y, \ldots, K_r^Y$, that intersect only at the origin, and such that two branches of $(X, 0)$, respectively $(Y, 0)$, belong to the same cone if and only if they have the same unit tangent vectors at the origin.

Since $t_{ord}(X, 0) = t_{ord}(Y, 0)$, we can enumerate the cones in such a way that

$$t_{ord}(X \cap K_i^X, 0) = t_{ord}(Y \cap K_i^Y, 0)$$

for each $i = 1, \ldots, r$.

From Lemma 5.4.17, for each $i = 1, \ldots, r$, there is a semialgebraic bi-Lipschitz map-germ

$$F_i : (\mathbb{R}^2, 0) \to (\mathbb{R}^2, 0)$$

such that $F_i(X \cap K_i^X, 0) = (Y \cap K_i^Y, 0)$ and F_i is the identity in the complement of the cone K_i^X.

We define $F : (\mathbb{R}^2, 0) \to (\mathbb{R}^2, 0)$ as $F(x, y) = F_i(x, y)$ if $(x, y) \in K_i^X$ for some i and $F(x, y) = (x, y)$ otherwise.   □

Theorem 5.4.20 *Let (X, x) and (Y, y) be germs of semialgebraic curves in $\mathbb{R}^n$ with branches $X = \cup_{i \in I} X_i$ and $Y = \cup_{i \in I} Y_i$. If $\sigma : I \to J$ is a bijection such that $t_{ord}(X_i, X_j) = t_{ord}(Y_i, Y_j)$, for each pair $i \neq j \in I$, then there exists a semialgebraic bi-Lipschitz $F : (\mathbb{R}^n, x) \to (\mathbb{R}^n, y)$ such that $F(X) = Y$.*

Proof We prove by induction on $n \geq 2$. The case $n = 2$ follows from Lemma 5.4.17 and Proposition 5.4.19. Let $n \geq 2$ and (X, x) and (Y, y) be semialgebraic curves in $\mathbb{R}^{n+1}$ such that $t_{ord}(X, x) = t_{ord}(Y, y)$. We assume by the induction hypothesis that the theorem holds for semialgebraic curves in $\mathbb{R}^n$. Let us assume with no loss of generality that $x = y = 0 \in \mathbb{R}^{n+1}$.

Let $\Pi : (\mathbb{R}^{n+1}, 0) \to (\mathbb{R}^n, 0)$ be a orthogonal projection such that its kernel does not contain the tangent cones of X and Y. Then Π induces semialgebraic outer bi-Lipschitz homeomorphisms $(X, 0) \simeq (\Pi(X), 0)$ and $(Y, 0) \simeq (\Pi(Y), 0)$. Hence by Theorem 5.4.13, $t_{ord}(\Pi(X), 0) = t_{ord}(X, 0)$ and $t_{ord}(\Pi(Y), 0) = t_{ord}(Y, 0)$. Moreover, the semialgebraic bi-Lipschitz map-germs $f_1 : (\Pi(X), 0) \to (\mathbb{R}, 0)$ and $f_2 : (\Pi(Y), 0) \to (\mathbb{R}, 0)$ with $f_1 = \Pi^{-1}|_{\Pi(X)}$ and $f_2 = \Pi^{-1}|_{\Pi(Y)}$ are such that $\text{graph}(f_1(\Pi(X))) = X$ and $\text{graph}(f_2(\Pi(Y))) = Y$.

By Kirszbraun's Theorem (Theorem 5.3.10), there are Lipschitz semialgebraic extensions $\Phi_{1,}, \Phi_2 : (\mathbb{R}^n, 0) \to (\mathbb{R}, 0)$ of f_1, f_2, respectively. Notice that these extensions are semialgebraic.

Let $F_i, F_2 : (\mathbb{R}^{n+1}, 0) \to (\mathbb{R}^{n+1}, 0)$ be defined by

$$F_i(z_1, \ldots, z_n, z_{n+1}) = (z_{1,}, \ldots, z_n, z_{n+1} - \Phi_i(z_1, \ldots, z_n)),$$

for $i = 1, 2$. Then, there are semialgebraic curves $\widetilde{X}$ and $\widetilde{Y}$ in $\mathbb{R}^n$ such that $(F_1(X), 0) = (\widetilde{X} \times 0, 0)$ and $(F_2(Y), 0) = (\widetilde{Y} \times 0, 0)$.

As F_i are germs of semialgebraic bi-Lipschitz homeomorphisms, it follows that

$$t_{ord}(\widetilde{X}, 0) = t_{ord}(X, 0) = t_{ord}(Y, 0) = t_{ord}(\widetilde{Y}, 0).$$

By the induction hypothesis, there exists a bi-Lipschitz semialgebraic map-germ $\widetilde{F} : (\mathbb{R}^n, 0) \to (\mathbb{R}^n, 0)$ such that $\widetilde{F}(\widetilde{X}) = \widetilde{Y}$.

Finally, let $F_3 : (\mathbb{R}^{n+1}, 0) \to (\mathbb{R}^{n+1}, 0)$ given by

$$F_3(z_1, \ldots, z_n, z_{n+1}) = (\widetilde{F}(z_1, \ldots, z_n), z_{n+1}).$$

Then $F = F_2^{-1} \circ F_3 \circ F_1$ satisfies the required conditions and $F(X) = Y$.

5.4.2 Lipschitz Geometry of Real Surfaces

A semialgebraic surface $X \subset \mathbb{R}^n$ is a semialgebraic set of pure dimension 2. In this section we study the inner local classification of surfaces with respect to bi-Lipschitz equivalence.

The classification of real semialgebraic and subanalytic surfaces was investigated by several authors independently, using different notations and terminology.

We first discuss this classification for surfaces with isolated singularity based on [Bir99] and [Val05]. A summary of the proof of the classification for general semialgebraic surfaces (Theorem 8.1 in [Bir99]) is presented in Sect. 5.4.2.1.

Definition 5.4.21 A standard β–Hölder triangle T_β, $\beta \in [1, \infty) \cap \mathbb{Q}$, is the semialgebraic set in $\mathbb{R}^2$ defined by $\{(x, y) \in \mathbb{R}^2 | 0 \le y \le x^\beta, 0 \le x \le 1\}$ (Fig. 5.5).

Proposition 5.4.22 *The β–Hölder triangle T_β is Lipschitz normally embedded.*

Proof Let $p, q \in T_\beta$ and let $\overline{pq}$ be the segment joining these two points. There are two possibilities

(1) $\overline{pq}$ is contained in T_β.
(2) $\overline{pq}$ intersects the curve $y = x^\beta, 0 \le x \le 1$ in two points.   □

If (1) holds, it follows that $d_{in}(p, q) = d_{out}(p, q)$ (Fig. 5.5).

If (2) holds, let p', q' be the two points of intersection of $\overline{pq}$ and the curve $y = x^\beta, 0 \le x \le 1$. We can write $p' = (x_0, x_0^\beta)$ and $q' = (x_1, x_1^\beta)$ where $0 \le x_0 < x_1 \le 1$. Then

$$d_{in}(p', q') \le |x_1 - x_0| + |x_1^\beta - x_0^\beta| \le 2d_{out}(p', q')$$

Fig. 5.5 β-Hölder triangle

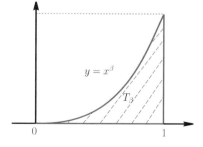

$$y = x^\beta$$

$$T_\beta$$

$$0 \qquad\qquad\qquad 1$$

and hence

$$d_{in}(p,q) \leq d_{in}(p, p') + d_{in}(p', q') + d_{in}(q', q)$$
$$\leq d_{out}(p, p') + d_{in}(p', q') + d_{out}(q', q)$$
$$\leq 2d_{out}(p, q).$$

Remark 5.4.23 Let $\beta_1 \neq \beta_2$. Then the germs T_{β_1} and T_{β_2} are not bi-Lipschitz equivalent.

Definition 5.4.24 A semialgebraic subset X of $\mathbb{R}^n$ is called β−Hölder triangle with principal vertex $a \in X$ if the germ (X, a) is (semialgebraically) bi-Lipschitz equivalent to $(T_\beta, 0)$.

Remark 5.4.25 We consider in X the inner metric. As T_β is Lipschitz normally embedded, we may take any metric on $(T_\beta, 0)$.

Definition 5.4.26 A standard β-horn is the semialgebraic set $H_\beta \subset \mathbb{R}^3$ defined by

$$H_\beta = \{(x_1, x_2, y) \in \mathbb{R}^3 | (x_1^2 + x_2^2)^q = y^{2p}, y \geq 0\},$$

where $\beta = \frac{p}{q} \geq 1$ (here p and q are natural numbers.) (Fig. 5.6)

Theorem 5.4.27 *Let $(X, 0)$ be the germ at $x = 0$ of a semialgebraic surface with isolated singularity at 0, and such that the link $X \cap S(0, \epsilon)$ is connected for all small ϵ. Then X is inner bi-Lipschitz equivalent to a β−Horn for some $\beta \in \mathbb{Q}$, $\beta \geq 1$.*

A semialgebraic triangle in a semialgebraic germ of surface (X, a) is a β−Hölder triangle $(\mathcal{T}, a)$ with vertex at a.

The proof of Theorem 5.4.27 will follow from the following lemmas. The first one follows from Theorem 2.2 in [Val05].

Fig. 5.6 β−horn

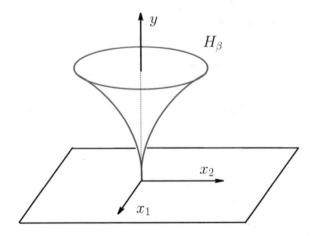

Fig. 5.7 Partition in triangles

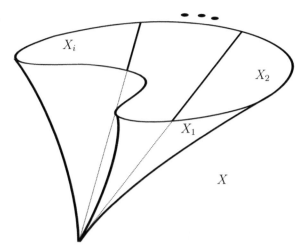

Lemma 5.4.28 *There exists a partition of X in semialgebraic triangles X_i such that*

1. $X_i \cap X_j$ *is an arc.*
2. *For any i, there exists a generic map $p : X_i \to \mathbb{P} \subset \mathbb{R}^n$ such that $p|_{X_i}$ is a bi-Lipschitz map-germ with respect to the inner metric, where $\mathbb{P}$ is a 2-plane in $\mathbb{R}^n$.*

Notice that $p(X_i)$ is a semialgebraic triangle in $\mathbb{P} \simeq \mathbb{R}^2$, bounded by 2 arcs γ_1^i and γ_2^i whose order of contact is $\mathrm{t_{ord}}(\gamma_1^i, \gamma_2^i) = \beta_i$ (Fig. 5.7).

Lemma 5.4.29 *The germ $p(X_i)$ at the origin in $\mathbb{R}^2$ is bi-Lipschitz equivalent to a standard Hölder triangle T_β, $\beta = \mathrm{t_{ord}}(\gamma_1, \gamma_2)$.*

Exercise 5.4.30 Prove Lemma 5.4.29.

Lemma 5.4.31 (Non-archimedean Lemma) *Let (T_1, β_1) and (T_2, β_2) be standard Hölder triangles of exponents β_1 and β_2 and let X be the union $T_1 \cup T_2$ such that $T_1 \cap T_2$ contains an arc. Then X is bi-Lipschitz equivalent to a Hölder triangle T_β, with $\beta = \min\{\beta_1, \beta_2\}$ (Fig. 5.8).*

Proof Let us assume $\beta_1 \leq \beta_2$ and

$$X = \{(x, y)\mid 0 \leq x \leq 1, -x^{\beta_2} \leq y \leq x^{\beta_1}\}.$$

We define $h : [0, 1] \times [-1, 1] \to Q$ as $h(x, y) = (x, y + x^{\beta_2})$. Then

$$h(x, -x^{\beta_2}) = (x, 0),$$

$$h(x, 0) = (x, x^{\beta_2}),$$

$$h(x, -x^{\beta_1}) = (x, x^{\beta_1}(1 + x^{\beta_2 - \beta_1})).$$

Fig. 5.8 Union of triangles

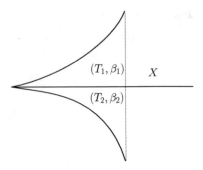

Hence X is inner bi-Lipschitz equivalent to T_β, $\beta = \min\{\beta_1, \beta_2\}$.

Proof of Theorem 5.4.27 By hypothesis, the link $K = X \cap S(0, \epsilon)$ of X is homeomorphic to S^1. Then we can decompose K in connected arcs $\lambda_1, \lambda_2, \ldots, \lambda_s$ such that they determine triangles $(T_1, \beta_1) \ldots (T_s, \beta_s)$ satisfying the condition of Lemma 5.4.28. More precisely, each triangle T_i is determined by the arc λ_i of the link K and arcs at the origin, γ_1^i and γ_2^i such that $t_{\text{ord}}(\gamma_1^i, \gamma_2^i) = \beta_i$. For each pair of adjacent triangles (T_i, β_i), (T_{i+1}, β_{i+1}) we can apply Lemma 5.4.31, to reduce the pair to a Hölder triangle with $\beta = \min\{\beta_1, \beta_2\}$. In this way, the decomposition of X can be reduced to two triangles with common arcs γ_1, γ_2 and $t_{\text{ord}}(\gamma_1, \gamma_2) = \beta$.

5.4.2.1   The General Case

The link of a semialgebraic surface is a compact semialgebraic subset of dimension 1 of the sphere. Each one dimensional connected component of this set is homeomorphic to a finite union of one dimensional spheres S^1's and closed intervals, to which we can naturally associate a graph.

To prove the bi-Lipschitz Classification Theorem for general semialgebraic surfaces, Lev Birbrair introduced in [Bir99] a new bi-Lipschitz invariant, the Hölder Complex. His construction can be summarized as follows:

Let Γ be a finite graph. We denote by E_Γ the set of edges and by V_Γ be the set of vertices of Γ.

Definition 5.4.32 A *Hölder Complex* is a pair (Γ, β), where $\beta : E_\Gamma \rightarrow \mathbb{Q}$ such that $\beta(g) \geq 1$, for every $g \in E_\Gamma$. Two Hölder Complexes (Γ_1, β_1) and (Γ_2, β_2) are called combinatorially equivalent if there exists a graph isomorphism $i : \Gamma_1 \rightarrow \Gamma_2$ such that, for every $g \in E_{\Gamma_1}$, we have $\beta_2(i(g)) = \beta_1(g)$.

Definition 5.4.33 Let (Γ, β) be a Hölder Complex. A set $X \subset \mathbb{R}^n$ is called a (semialgebraic) **Geometric Hölder Complex** corresponding to (Γ, β) with the principal vertex $a \in X$ if

1. There exists a homeomorphism $\psi : C\Gamma \rightarrow X$, $C\Gamma$ is the topological cone over Γ,

2. If $\alpha \subset C\Gamma$ is the vertex of $C\Gamma$ then $a = \psi(\alpha)$,
3. For each $g \in E_\Gamma$, the set $\psi(Cg)$ is a (semialgebraic) $\beta(g)$-Hölder triangle with principal vertex a, where $Cg \subset C\Gamma$ is the subcone over g.

Theorem 5.4.34 (Theorem 6.1, [Bir99]) *Let $X \subset \mathbb{R}^n$ be a two-dimensional closed semialgebraic set and let $a \in X$. Then there exist a number $\delta > 0$ and a Hölder Complex (Γ, β) such that $B(a, \delta) \cap X$ is a semialgebraic Geometric Hölder Complex corresponding to (Γ, β) with the principal vertex a, where $B(a, \delta)$ is the closed ball centered at a of radius δ.*

Definition 5.4.35 We say that b is a *non-critical vertex* of Γ if it is incident with exactly two different edges g_1 and g_2 and these edges connect two different vertices b_1 and b_2 with b. If this vertex b is connected by g_1 and g_2 with only one vertex $b', b \neq b'$, we say that b is a *loop vertex*. The other vertices of Γ (which are neither non-critical nor loop) are called *critical vertices* of Γ.

Given a Geometric Hölder Complex, there is a simplification process described by Lev Birbrair (Theorem 7.3, [Bir99]), allowing to assume that every vertex in V_Γ is a either a critical vertex or a loop vertex. The resulting Geometric Hölder Complex is called a **Canonical Hölder Complex** of X at a. Two simplifications of the same Hölder Complex are combinatorially equivalent (Fig. 5.9).

Theorem 5.4.36 (Birbrair Classification Theorem, Theorem 8.1 [Bir99]) *Let $X_1, X_2 \subset \mathbb{R}^n$ be two dimensional semialgebraic subsets with $a_1 \in X_1$ and $a_2 \in X_2$. The germs (X_1, a_1) and (X_2, a_2) are bi-Lipschitz inner equivalent if and only if the Canonical Hölder Complexes of X_1 at a_1 and X_2 at a_2 are combinatorially equivalent.*

The computation of the Hölder complex of weighted homogeneous surfaces was discussed in [BF08b]. We state two theorems of their work.

Theorem 5.4.37 (Theorem 4.1, [BF08b]) *Let $X \subset \mathbb{R}^n$ be a semialgebraic surface. Let $x_0 \in X$ be a point such that X is a β−Hölder Triangle at x_0 or a β−Horn at x_0. Then*

$$\beta(X, x_0) =$$

$$\inf\{t_{\mathrm{ord}}(\gamma_1, \gamma_2) | \gamma_1 \text{ and } \gamma_2 \text{ are semialgebraic arcs in } X \text{ with } \gamma_1(0) = \gamma_2(0)\},$$

Fig. 5.9 Simplification process

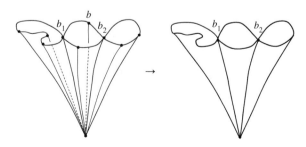

where $t_{ord}(\gamma_1, \gamma_2)$ *denote the contact between* γ_1 *and* γ_2.

Theorem 5.4.38 (Theorem 2.6, [BF08b]) *Let* $X \subset \mathbb{R}^3$ *be a semialgebraic* (a_1, a_2, a_3)-*weighted homogeneous surface,* $a_1 \geq a_2 \geq a_3 > 0$. *If* 0 *is an isolated singular point and the local link of* X *at* 0 *is connected, then the germ of* X *at* 0 *is bi-Lipschitz equivalent with respect to the inner metric, to a germ at* 0 *of a* β−*Horn where* β *is equal to* 1 *or* $\frac{a_2}{a_1}$.

In the next example we apply these results to compute the β−exponent of the E_8−singularity.

Example 5.4.39 Let $f : \mathbb{R}^3 \to \mathbb{R}$, $f(x, y, z) = x^2 - y^3 + z^5$ and $X = \{(x, y, z) \in \mathbb{R}^3 |\ f(x, y, z) = 0\}$ as in Example 5.3.9. The polynomial f is weighted homogeneous of type $(a_1, a_2, a_3) = (15, 10, 6)$, with $a_1 \geq a_2 \geq a_3$ as required by Theorem 5.4.38.

Then from Theorem 5.4.37 for each pair of arcs γ_1 and γ_2 at 0 in X one has to compute $t_{ord}(\gamma_1, \gamma_2)$. As $\beta = \frac{p}{q} \geq 1$, to prove that $\beta = 1$ is sufficient to find arcs γ_1 and γ_2 in X with this property.

In this case $X \cap \{z = 0\}$ is the curve $x^2 - y^3 = 0$. Let $\gamma^1 = (+t^{\frac{3}{2}}, t, 0)$ and $\gamma^2 = (-t^{\frac{3}{2}}, t, 0)$ be the parametrizations of its branches. On the other hand $X \cap \{x = 0\}$ is the curve $y^3 - z^5 = 0$ parametrized by $\alpha(s) = (0, s^{\frac{5}{3}}, s)$. Then their tangent cones are not equal and therefore their contact is 1.

5.5 Bi-Lipschitz Invariance of the Tangent Cone

The notion of the tangent cone at a singular point is a generalization of the notion of tangent space at a smooth point of a real or complex variety. Whitney gave different algebraic and geometric definitions of the tangent cone in [Whi65c] and [Whi65b] and proved some equivalences between these definitions in case of complex varieties.

Zariski in his famous paper *Some open questions in the theory of singularieties* [Zar71] asks whether the fact that two germs of complex hypersurfaces $(X, 0)$ and $(Y, 0)$ in $\mathbb{C}^n$ having the same topological type would imply that their tangent cones are homeomorphic. In [FdB05] Fernandez de Bobadilla gave a counterexample for this conjecture.

We define in this section the tangent cone and prove some of its properties. The main result is that ambient bi-Lipschitz equivalent semialgebraic sets have bi-Lipschitz equivalent tangent cones. In the last section we discuss applications of this result.

The results of this section are due to Edson Sampaio [Sam16].

5.5.1 Properties of the Tangent Cone

Definition 5.5.1 Let X be a semialgebraic set such that $x_0 \in \bar{X}$. A vector $v \in \mathbb{R}^n$ is a **tangent vector** of X at $x_0 \in \mathbb{R}^n$ if there is a sequence of points $\{x_i\} \subset X \setminus x_0$, $x_i \to x_0$ and a sequence of positive real numbers $\{t_i\}$ such that

$$\lim_{i \to \infty} \frac{1}{t_i}(x_i - x_0) = v$$

The **tangent cone** of X at x_0 is the set

$$C(X, x_0) = \{v \in \mathbb{R}^n \,|\, v \text{ is a tangent vector of X at } x_o \in \bar{X}.\}$$

Remark 5.5.2 $C(X, x_0)$ coincides with the cone $C_3(X, x_0)$ defined by Whitney in [Whi65c] and [Whi65b].

Remark 5.5.3 It follows from the Curve Selection Lemma for semialgebraic sets that if $X \subset \mathbb{R}^n$ is a semialgebraic set and $x_0 \in \bar{X}$, then the following holds:

$$C(X, x_0) = \{v \in \mathbb{R}^n \,|\, \exists\, a\ C^1 - \text{semialgebraic arc } \alpha : [0, \epsilon) \to \mathbb{R}^n,$$

$$\alpha(0) = x_0,\ \alpha((0, \epsilon)) \subset X,\ \text{and } \alpha(t) - x_0 = tv + o(t).\}$$

Exercise 5.5.4 Let $(X, 0)$ be the germ of a semialgebraic set in $\mathbb{R}^n$. Define the **tangent link** of X at 0 as

$$L_0 X = \lim_{\epsilon \to 0} \frac{1}{\epsilon}(X \cap S(0, \epsilon)).$$

Prove that $\mathrm{Cone}(L_0 X) = C(X, 0)$.

Definition 5.5.5 Let $X = V(I)$ be an algebraic set in $\mathbb{C}^n$, where I is the ideal generated by the complex polynomials $g_1, \dots, g_s$. For each $f \in I$, denote by H_f the lowest degree homogeneous polynomial of f and by I_0 the ideal generated by H_f for all $f \in I$. The **algebraic tangent** cone is the set

$$C_a(X, 0) = \{x \,|\, H_f(x) = 0, \forall H_f \in I_0\}.$$

Notice that $C_a(X, 0)$ is a union of complex lines passing through 0.

For complex algebraic varieties, it follows from [Whi72, Theorem 4D] that $C_a(X, 0) = C(X, 0)$.

Example 5.5.6 Let $f(x) = f_m(x) + f_{m+1}(x) + \cdots + f_{m+k}(x)$ be a complex polynomial of degree $d = m + k$, f_i homogeneous of degree i, $x \in \mathbb{C}^n$. The tangent cone of $X = f^{-1}(0)$ at $x = 0$ is the algebraic set $C(X, 0) = \{x \in \mathbb{C}^n \,|\, f_m(x) = 0\}$.

Exercise 5.5.7 Let $\Phi : \mathbb{R}^2 \to \mathbb{R}^3$, $\Phi(u, v) = (u, v^2, uv)$ be the parametrization of the Whitney umbrella and let $X = \Phi(\mathbb{R}^2)$. Find $C(X, 0)$.

Lemma 5.5.8 (Lemma 3.1, [Val05]) *Let $X, Y \subset \mathbb{R}^n$ be semialgebraic sets. If X and Y are bi-Lipschitz equivalent with respect to the outer metric, then there is a bi-Lipschitz homeomorphism $\Phi : \mathbb{R}^{2n} \to \mathbb{R}^{2n}$ such that $\Phi(X \times \{0\}) = \{0\} \times Y$.*

Proof Let $\phi : X \to Y$ be a bi-Lipschitz homeomorphism. By Mcshane-Whitney-Kirszbraun's Theorem (see [McS34, Theorem 1.16]) there exist Lipschitz maps $\tilde{\phi} : \mathbb{R}^n \to \mathbb{R}^n$ and $\tilde{\psi} : \mathbb{R}^n \to \mathbb{R}^n$ such that $\tilde{\phi}|_X = \phi$ and $\tilde{\psi}|_Y = \phi^{-1}$. We define $\Phi, \Psi : \mathbb{R}^n \times \mathbb{R}^n \to \mathbb{R}^n \times \mathbb{R}^n$, as follows

$$\Phi(x, y) = (x - \tilde{\psi}(y) + \tilde{\phi}(x)), y + \tilde{\phi}(x))$$

and

$$\Psi(z, w) = (z + \tilde{\psi}(w), w - \tilde{\phi}(z) + \tilde{\psi}(x))).$$

Then Φ and Ψ are Lipschitz maps and to show that $\Psi = \Phi^{-1}$ choose $(x, y) \in \mathbb{R}^n \times \mathbb{R}^n$. Then

$$\Psi(\Phi(x, y)) = \Psi(x - \tilde{\psi}(y + \tilde{\phi}(x)), y + \tilde{\phi}(x)) =$$

$$= (x - \tilde{\psi}(y + \tilde{\phi}(x)) + \tilde{\psi}(y + \tilde{\phi}(x)), y + \tilde{\phi}(x) - \phi(x - \tilde{\psi}(y + \tilde{\phi}(x)) + \tilde{\psi}(y + \tilde{\phi}(x))$$

$$= (x, y).$$

Similarly $\Phi(\Psi(z, w)) = (z, w)$. Therefore $\Psi = \Phi^{-1}$. It is clear that $\Phi(X \times \{0\}) = \{0\} \times Y$. □

The main theorem in [Sam16] is the following

Theorem 5.5.9 *Let $X \subset \mathbb{R}^n$, $Y \subset \mathbb{R}^n$ be semialgebraic sets. If the germs (X, x_0) and (Y, y_0) are bi-Lipschitz equivalent, then $C(X, x_0)$ and $C(Y, y_0)$ are also bi-Lipschitz equivalent.*

Remark 5.5.10 This result was proved by Bernig and Lytchak [BL07] under the additional assumption that the bi-Lipschitz homeomorphism is also semialgebraic. A simpler proof in the complex case was given by Birbrair, Fernandes and Neumann in [BFN10].

Proof We assume that $x_0 = y_0 = 0$. By Lemma 5.5.8 we can suppose that $X, Y \subset \mathbb{R}^N$ and there exists a bi-Lipschitz map $\phi : \mathbb{R}^N \to \mathbb{R}^N$ such that $\phi(X) = Y$. Let $c > 0$ a constant such that

$$\frac{1}{c}\|x - y\| \leq \|\phi(x) - \phi(y)\| \leq c\|x - y\|, \ \forall x, y \in \mathbb{R}^N.$$

Let $\psi = \phi^{-1}$. We define two sequences of mappings: for each $m \in \mathbb{N}$, let $\phi_m : \overline{B}_1 \to \mathbb{R}^N$ and $\psi_m : \overline{B}_c \to \mathbb{R}^N$, where $\overline{B}_r = \{x \in \mathbb{R}^N \mid \|x\| \leq r\}$ be defined by $\phi_m(v) = m\phi(\frac{v}{m})$ and $\psi_m(v) = m\psi(\frac{v}{m})$.

Notice that for any $m \in \mathbb{N}$, we have

$$\frac{1}{c}\|u - v\| \leq \|\phi_m(u) - \phi_m(v)\| \leq c\|u - v\|, \ \forall u, v \in \overline{B}_1$$

and

$$\frac{1}{c}\|z - w\| \leq \|\psi_m(z) - \psi_m(w)\| \leq c\|z - w\|, \ \forall z, w \in \overline{B}_c$$

Then by Arzela-Ascoli Theorem there exist a subsequence $\{m_j\} \subset \mathbb{N}$ and mappings $d\phi : \overline{B}_1 \to \mathbb{R}^N$ and $d\psi : \overline{B}_c \to \mathbb{R}^N$ such that $\phi_{m_j} \to d\phi$ and $\psi_{m_j} \to d\psi$ uniformly as $j \to \infty$. Clearly,

$$\frac{1}{c}\|u - v\| \leq \|d\phi(u) - d\phi(v)\| \leq c\|u - v\|, \ \forall u, v \in \overline{B}_1$$

and

$$\frac{1}{c}\|z - w\| \leq \|d\psi(z) - d\psi(w)\| \leq c\|z - w\|, \ \forall z, w \in \overline{B}_c$$

Let $U = d\phi(B_1)$. Since $d\phi$ is continuous and injective, U is an open set. We claim that

$$d\psi \circ d\phi = \mathrm{id}_{B_1} \text{ and } d\phi \circ d\psi = \mathrm{id}_U.$$

Let $v \in B_1$ and $w = d\phi(v) = \lim_{j \to \infty} \frac{\phi(t_j v)}{t_j}$ with $t_j = \frac{1}{m_j}$.
Then

$$\|d\psi(w) - v\| = \| \lim_{j \to \infty} \frac{\psi(t_j v)}{t_j} - v\| = \lim_{j \to \infty} \| \frac{\psi(t_j v)}{t_j} - \frac{t_j v}{t_j}\|$$

$$= \lim_{j \to \infty} \frac{1}{t_j}\|\psi(t_j v) - t_j v\| = \lim_{j \to \infty} \frac{1}{t_j}\|\phi^{-1}(t_j w) - \phi^{-1}(\phi(t_j v))\|$$

$$\leq \lim_{j \to \infty} \frac{c}{t_j}\|t_j w - \phi(t_j v)\| \leq \lim_{j \to \infty} c\|w - \frac{\phi(t_j v)}{t_j}\|$$

$$= 0$$

Hence $d\psi(w) = d\psi(d\phi(v)) = v$ for all $v \in \bar{B}_1$. Similarly, we obtain $d\phi \circ d\psi|_U = \text{id}|_U$.

We now claim that $d\phi(C(X, 0) \cap B_1) \subset C(Y, 0)$ and $d\psi(C(Y, 0)) \subset C(X, 0)$.

If $v \in C(X, 0) \cap B_1$, then there exists $\alpha : [0, \epsilon) \to X$ such that $\alpha(t) = tv + o(t)$. Thus, $\phi(\alpha(t)) \in Y$ for all $t \in [0, \epsilon)$, and since ϕ is Lipschitz, we have $\phi(\alpha(t)) = \phi(tv) + o(t)$. On the other hand, by definition of the map $d\phi$, we get $\phi(t_j v) = t_j d\phi(v) + o(t_j)$. Hence

$$
d\phi(v) = \lim_{j \to \infty} \phi_{m_j}(v) = \lim_{j \to \infty} \frac{\phi(t_j v)}{t_j} =
$$

$$
= \lim_{j \to \infty} \frac{\phi(\alpha(t_j))}{t_j} \in C(Y, 0)
$$

Then $d\phi(C(X, 0) \cap B_1) \subset C(Y, 0)$.   □

5.5.2 Applications

The main reference here is [BFN10].

We say that a semialgebraic set $X \subset \mathbb{R}^n$ is metrically conical at a point x_0 if there exists an Euclidean ball $B \subset \mathbb{R}^n$ centered at x_0 such that $X \cap B$ is bi-Lipschitz homeomorphic, with respect to the inner metric, to the metric cone over its link at x_0. When such a bi-Lipschitz homeomorphism is semialgebraic we say that X is semialgebraically metrically conical at x_0.

Since the germ of a semialgebraic subset is locally (semialgebraically) homeomorphic to the cone over its link, we can ask whether the germ can be bi-Lipschitz inner equivalent to the cone over its link (or to its tangent cone).

This is not always so, as we can see in the following example [BF08a].

Example 5.5.11 The β−Horn in $\mathbb{R}^3$ defined by

$$
X = \{(x_1, x_2, x_3) \in \mathbb{R}^3 \,\|\, (x_1^2 + x_2^2)^q = x_3^{2p}, \frac{p}{q} \geq 1\}
$$

are metrically conical if and only if $\beta = \frac{p}{q} = 1$.

In fact, if $\beta > 1$ the tangent cone of X has dimension 1 and hence X cannot be bi-Lipschitz equivalent to a cone.

The tangent cone of a complex algebraic set is also an algebraic set of the same dimension [Whi65a]. In [BF08a] L. Birbrair and A. Fernandes show there exists a big class of complex algebraic surfaces, with isolated singularities that do admit a metrically conical structure. This was pioneer result that motivated the developments of this topic since then.

Example 5.5.12 It follows from Theorem 6.1 in [BF08a] that the complex A_{2k-1} singularity given by $x^2 + y^2 = z^{2k}$, $k > 2$ is not metrically conical.

The following is a Corollary of Theorem 5.5.9.

Corollary 5.5.13 (Corollary 2.5 in [BFN10]) *Let $X \subset \mathbb{R}^n$ be a normally embedded semialgebraic set. If X is semialgebraically conical at a point $x \in X$, then the germ (X, x) is semialgebraically bi-Lipschitz homeomorphic to the germ $(C_x X, 0)$.*

Proof The result follows from Theorem 5.5.9 since the tangent cone of a metric cone at the vertex is the cone itself. □

Definition 5.5.14 A semialgebraic subset $X \subset \mathbb{R}^n$ is **Lipschitz regular** (resp. semialgebraically Lipschitz regular) at $x_0 \in X$ if there is and open neighborhood of $x_0 \in X$ which is bi-Lipschitz homeomorphic (resp. semialgebraically bi-Lipschitz homeomorphic) to an Euclidean ball.

Theorem 5.5.15 (Theorem 3.2 in [BFLS16], Theorem 4.2 in [Sam16]) *Let $X \subset \mathbb{C}^n$ be a complex analytic set. If X is Lipschitz regular at $x_0 \in X$, then x_0 is smooth point of X.*

The proof of this theorem follows from Theorem 5.5.17 and Lemma 5.5.16 bellow.

Lemma 5.5.16 *If X is Lipschitz regular at $x_0 \in X$, there exists a neighborhood U of x_0 in X which is normally embedded in $\mathbb{C}^n$.*

Theorem 5.5.17 (Prill [Pri67]) *Let $V \subset \mathbb{C}^n$ be a complex cone. If $0 \in V$ has a neighborhood homeomorphic to a Euclidean ball, then V is linear subspace of $\mathbb{C}^n$.*

5.6 Lipschitz Theory of Singularities

Here we consider the problem of bi-Lipschitz $\mathcal{G}$-classification of analytic function-germs, where $\mathcal{G} = \mathcal{R}, \mathcal{C}, \mathcal{K}$. The group $\mathcal{R}$ is the group of changes of coordinates in the source, and $\mathcal{K}$-equivalence is the contact equivalence defined by Mather [Mat69]. The aim is to introduce the bi-Lipschitz singularity theory of germs of smooth functions, including some invariants of bi-Lipschitz equivalence and discussion on finiteness theorems for the classification with respect to these equivalences.

Finiteness theorems for analytic map-germs, in the real and in the complex case, with respect to the topological equivalence were the subject of investigation of various authors (see, for example, [Sab83, Nak84, BS91, Cos98]) and many interesting results were obtained in this direction. Mostowski [Mos85] and Parusiński [Par88] proved that the set of equivalence classes of semialgebraic sets with bounded degree is finite. A finiteness result holds for bi-Lipschitz $\mathcal{K}$-classification of function and map germs [ABCF10] and [RV11]. However Henry and Parusiński [HP03] show that the bi-Lipschitz $\mathcal{R}$- classification of function germs has moduli.

The main references for this section are [NRT20] and [HP03].

5.6.1   Basic Definitions and Results

We denote by ε_n the set of all smooth (when $\mathbb{K} = \mathbb{R}$) or holomorphic (when $\mathbb{K} = \mathbb{C}$) function germs at $0 \in \mathbb{K}^n$, $\mathbb{K} = \mathbb{R}$ or $\mathbb{C}$ and by $\mathcal{M}_n$ the set of germs in ε_n vanishing at 0. Notice that ε_n is a local ring and $\mathcal{M}_n$ is the maximal ideal of ε_n. We define the groups $\mathcal{G} = \mathcal{R}, \mathcal{C}$ and $\mathcal{K}$, acting on ε_n.

Definition 5.6.1 Denote by $\mathcal{R}$ the group of smooth diffeomorphism germs $H : (\mathbb{K}^n, 0) \to (\mathbb{K}^n, 0)$. Then $\mathcal{R}$ acts on ε_n by composition $H.f = f \circ H^{-1}$. Two germs $f, g \in \varepsilon_n$ are called $\mathcal{R}$-**equivalent** (or right equivalent) if they lie in the same orbit of this action. When H is a bi-Lipschitz homeomorphism and $g = f \circ H^{-1}$ we say that f and g are bi-Lipschitz $\mathcal{R}$-equivalent.

Given a germ $f \in \varepsilon_n$, denote by Jf the **Jacobian ideal** of f, i.e., the ideal in ε_n generated by the partial derivatives of f. The **codimension** (also called the **Milnor number** of f) of a germ $f \in \varepsilon_n$ is defined to be $\dim_{\mathbb{K}} \varepsilon_n / Jf$. It is well-known that in the complex case the Milnor number is a topological invariant (Milnor [Mil68]), and is therefore a bi-Lipschitz invariant, i.e. if $f, g \in \varepsilon_n$ are bi-Lipschitz $\mathcal{R}$-equivalent then their codimensions are equal. For families of reduced complex plane curves, Zariski [Zar65a, Zar65b] proved that the converse holds, that is, a μ-constant family of germs is bi-Lipschitz trivial. We would like to remark that the Milnor number is not a bi-Lipschitz $\mathcal{R}$-invariant in the real case. Consider the family $f_t(x, y) = x^4 + y^4 + tx^2y^2 + y^6, t \geq 0$. This family is bi-Lipschitz $\mathcal{R}$-trivial, however, $\mu(f_t) = 9$, $t \neq 2$ and $\mu(f_2) = 13$ (see [NRT20] for more details).

Denote by $J_0^k(n, 1)$ the k-jet space of $\mathcal{M}_n$, by $\mathcal{R}^k$ the set of k-jets of elements in $\mathcal{R}$. A germ $f \in \mathcal{M}_n$ is called $\mathcal{R}$- k-**determined** if for any $g \in \mathcal{M}_n$ such that $j^k g(0) = j^k f(0)$ then $f \sim_{\mathcal{R}} g$; f is called **finitely determined** if f is k-determined for some $k \in \mathbb{N}$. If a germ $f : (\mathbb{K}^n, 0) \to (\mathbb{K}, 0)$ is finitely determined then it has an isolated singularity at 0. The converse holds for $\mathbb{K} = \mathbb{C}$ (see Wall [Wal81] or [AGZV12]).

The action of $\mathcal{R}^k$ on $J_0^k(n, 1)$ is defined by taking composition and then truncating the Taylor expansion. A germ $f \in \mathcal{M}_n$ is said to be $\mathcal{R}$-**simple** if there exists $k \in \mathbb{N}$ sufficiently large and a neighborhood U of $j^k f(0)$ in $J_0^k(n, 1)$ that meets only finitely many $\mathcal{R}^k$-orbits in $J_0^k(n, 1)$. The germs that are not $\mathcal{R}$-simple are called $\mathcal{R}$-**modal**.

Two germs $z, w \in J_0^k(n, 1)$ are **bi-Lipschitz** $\mathcal{R}$-**equivalent** if there exists a bi-Lipschitz homeomorphism germ $\phi : (\mathbb{K}^n, 0) \to (\mathbb{K}^n, 0)$ such that $z \circ \phi = w$. The Lipschitz $\mathcal{R}$-orbits are the equivalence classes of the bi-Lipschitz $\mathcal{R}$-equivalences. A germ $f \in \mathcal{M}_n$ is **Lipschitz** $\mathcal{R}$- **simple** if for $k \in \mathbb{N}$ sufficiently large f, there is a neighborhood of $j^k f(0)$ in $J^k(n, 1)$ that meets only finitely many Lipschitz $\mathcal{R}$-orbits. The germs that are not Lipschitz simple are called **Lipschitz modal**. In other words, these germs have Lipschitz modality. We remark that if a germ is smoothly simple then it is also Lipschitz simple. The converse is not true in general.

Given $f \in \varepsilon_n$, the **corank** of f at 0 is defined to be the nullity (dimension of the kernel) of the Hessian $\left(\frac{\partial^2 f}{\partial x_i \partial x_j}(0) \right)$. The following result is known as the Splitting lemma; see Ebeling [Ebe07] or Gibson [Gib79] for example.

Lemma 5.6.2 *Let $f \in M_n^2$ be a finitely determined germ of corank c. Then there exists $g \in M_c^3$ such that*

$$f(x_1, \ldots, x_n) \sim_{\mathcal{R}} g(x_1, \ldots, x_c) \pm x_{c+1}^2 \pm \ldots \pm x_n^2.$$

The codimension of f and g are equal. Moreover, g is uniquely determined up to smooth equivalence, i.e. if $f \sim_{\mathcal{R}} g + Q \sim_{\mathcal{R}} h + Q$ then $g \sim_{\mathcal{R}} h$, where Q is the non-degenerate quadratic form.

Definition 5.6.3 Two function-germs $f, g : (\mathbb{K}^n, 0) \to (\mathbb{K}, 0)$ are called **bi-Lipschitz $\mathcal{K}$-equivalent** (or **contact bi-Lipschitz equivalent**) if there exist two germs of bi-Lipschitz homeomorphisms $h : (\mathbb{K}^n, 0) \to (\mathbb{K}^n, 0)$ and $H : (\mathbb{K}^n \times \mathbb{K}, 0) \to (\mathbb{K}^n \times \mathbb{K}, 0)$ such that $H(\mathbb{K}^n \times \{0\}) = \mathbb{K}^n \times \{0\}$ and the following diagram is commutative:

$$
\begin{array}{ccccc}
(\mathbb{K}^n, 0) & \xrightarrow{(id, f)} & (\mathbb{K}^n \times \mathbb{K}, 0) & \xrightarrow{\pi_n} & (\mathbb{K}^n, 0) \\
h \downarrow & & H \downarrow & & h \downarrow \\
(\mathbb{K}^n, 0) & \xrightarrow{(id, g)} & (\mathbb{K}^n \times \mathbb{K}, 0) & \xrightarrow{\pi_n} & (\mathbb{K}^n, 0)
\end{array}
$$

where $id : \mathbb{K}^n \to \mathbb{K}^n$ is the identity map and $\pi_n : \mathbb{K}^n \times \mathbb{K} \to \mathbb{K}^n$ is the canonical projection.

The function-germs f and g are called **bi-Lipschitz C-equivalent** if $h = id$.

In other words, two function-germs f and g are bi-Lipschitz $\mathcal{K}$-equivalent if there exists a germ of a bi-Lipschitz map $H : (\mathbb{K}^n \times \mathbb{K}, 0) \longrightarrow (\mathbb{K}^n \times \mathbb{K}, 0)$ such that $H(x, y)$ can be written in the form $H(x, y) = (h(x), \tilde{H}(x, y))$, where h is a bi-Lipschitz map-germ, $\tilde{H}(x, 0) = 0$ and H maps the germ of the graph (f) onto the graph (g).

When H and h are C^∞ diffeomorphisms we get Mather's contact group $\mathcal{K}$ (see [Mat69].)

Exercise 5.6.4 If the functions f and g are bi-Lipschitz $\mathcal{K}$-equivalent then the set germs $f^{-1}(0)$ and $g^{-1}(0)$ are ambient bi-Lipschitz equivalent in $\mathbb{K}^n$.

Definition 5.6.5 Two functions $f, g : \mathbb{R}^n \to \mathbb{R}$ are called **of the same contact** at a point $x_0 \in \mathbb{R}^n$ if there exist a neighborhood U_{x_0} of x_0 in $\mathbb{R}^n$ and two positive numbers c_1 and c_2 such that, for all $x \in U_{x_0}$, we have

$$c_1 |f(x)| \leq |g(x)| \leq c_2 |f(x)|$$

and $f(x)g(x) \geq 0$. We use the notation: $f \approx g$.

Remark 5.6.6 It is clear that if two function-germs f and g are of the same contact then the germs of their zero-sets are equal.

5.6.2   Finiteness of Bi-Lipschitz $\mathcal{K}$-Classification

The following theorem appears in [BCFR07]. An analogous result holds for map-germs $f : (\mathbb{R}^n, 0) \to (\mathbb{R}^p, 0)$, see [RV11].

Theorem 5.6.7 *Let $f, g : (\mathbb{R}^n, 0) \to (\mathbb{R}, 0)$ be two germs of Lipschitz functions. Then, f and g are bi-Lipschitz C-equivalent if and only if one of the two conditions is true:*

1. $f \approx g$,
2. $f \approx -g$.

As a consequence of Theorem 5.6.7, one can prove the following finiteness theorem.

Theorem 5.6.8 ([BCFR07], Theorem 2.7) *Let $\mathcal{P}_k(\mathbb{R}^n)$ be the set of all polynomials in n variables with degree less than or equal to k. Then the set of equivalence classes of germs at 0 of polynomials in $\mathcal{P}_k(\mathbb{R}^n)$, with respect to bi-Lipschitz $\mathcal{K}$-equivalence, is finite.*

In particular there is only a finite number of non-equivalent bi-Lipschitz real algebraic hypersurfaces $f = 0$, where $f \in \mathcal{P}_k(\mathbb{R}^n)$. For a proof of this Theorem see [BCFR07]. This result extends to polynomial mappings $f : \mathbb{R}^n \to \mathbb{R}^p$ with degree less than or equal to k, [RV11].

Example 5.6.9 Let $f_t : (\mathbb{K}^2, 0) \to (\mathbb{K}, 0)$ be the one parameter family of germs in the plane given by $f_t(x, y) = f(x, y, t) = x^3 - 3t^2 x y^4 + y^6$. Notice that f_t has isolated singularity for all $t \neq \pm \frac{1}{\sqrt[3]{2}}$. It follows from Theorem 5.6.8 that the number of bi-Lipschitz $\mathcal{K}$-equivalence classes of f_t is finite. When $\mathbb{K} = \mathbb{C}$ the complement of $t = \pm \frac{1}{\sqrt[3]{2}}$ in the parameter space is a connected subset of $\mathbb{C} \simeq \mathbb{R}^2$ and the Milnor number $\mu(f_t)$ is constant and equal to 10, for all $t \in \mathbb{C} \setminus \{\pm \frac{1}{\sqrt[3]{2}}\}$.

Let $X_t = f_t^{-1}(0)$ be the germ of the reduced plane curve in $\mathbb{C}^2$. Since $\mu(f_t)$ is constant in $\mathbb{C} \setminus \{\pm \frac{1}{\sqrt[3]{2}}\}$ it follows that X_t and $X_{t'}$ are bi-Lipschitz equivalent for all t, t' in this set, see Zariski [Zar65a, Zar65b]. As we shall see in Sect. 5.6.4, this family is modal with respect to bi-Lipschitz $\mathcal{R}$-equivalence.

5.6.2.1   Proof of Theorem 5.6.7

Suppose that the germs of Lipschitz functions f and g are bi-Lipschitz C-equivalent. Let $H : (\mathbb{R}^n \times \mathbb{R}, 0) \longrightarrow (\mathbb{R}^n \times \mathbb{R}, 0)$ be the germ of a bi-Lipschitz homeomorphism satisfying the conditions of Definition 5.6.3. Let V_+ be the subset of $\mathbb{R}^n \times \mathbb{R}$ of points (x, y) where $y > 0$, and V_- be the subset of $\mathbb{R}^n \times \mathbb{R}$ where $y < 0$. We have one of the following possibilities:

(1) $H(V_+) = V_+$ and $H(V_-) = V_-,$ or

(2) $H(V_+) = V_-$ and $H(V_-) = V_+$.

Let us consider the first possibility. In this case, the functions f and g have the same sign on each connected component of the set $f(x) \neq 0$. Moreover,

$$|g(x)| = \|(x, 0) - (x, g(x))\| = \|H(x, 0) - H(x, f(x))\|$$
$$\leq c_2 \|(x, 0) - (x, f(x))\| = c_2 |f(x)|,$$

where c_2 is a positive real number. Using the same argument we can show

$$c_1 |f(x)| \leq |g(x)|, \qquad c_1 > 0.$$

Hence, $f \approx g$.

Let us consider the second possibility. Let $\xi : (\mathbb{R}^n \times \mathbb{R}, 0) \to (\mathbb{R}^n \times \mathbb{R}, 0)$ be a map-germ defined as follows:

$$\xi(x, y) = (x, -y).$$

Applying the same arguments to a map $\xi \circ H$, we will conclude that $f \approx -g$. Reciprocally, suppose that $f \approx g$. Let us construct a map-germ

$$H : (\mathbb{R}^n \times \mathbb{R}, 0) \to (\mathbb{R}^n \times \mathbb{R}, 0)$$

in the following way:

$$H(x, y) = \begin{cases} (x, 0) & \text{if } y = 0, \\ (x, \dfrac{g(x)}{f(x)} y) & \text{if } 0 \leq |y| \leq |f(x)|, \\ (x, y - f(x) + g(x)) & \text{if } y \geq f(x) \geq 0 \text{ or } y \leq f(x) \leq 0, \\ (x, y + f(x) - g(x)) & \text{if } y \geq -f(x) \geq 0 \text{ or } y \leq -f(x) \leq 0. \end{cases}$$

$$(5.5)$$

The map $H(x, y) = (x, \tilde{H}(x, y))$ above defined is bi-Lipschitz. In fact, H is injective because, for any fixed $x^\star$, we can show that $\tilde{H}(x^\star, y)$ is a continuous and monotone function. Moreover, H is Lipschitz if $0 \leq |f(x)| \leq |y|$. Let us show that H is Lipschitz if $0 \leq |y| \leq |f(x)|$. By Rademacher's theorem, in almost every x near $0 \in \mathbb{R}^n$, all the partial derivatives $\dfrac{\partial f}{\partial x_i}, \dfrac{\partial g}{\partial x_i}$ exist, hence the derivative $\dfrac{\partial \tilde{H}}{\partial x_i}$ exist in almost every x near $0 \in \mathbb{R}^n$. By the Mean Value Theorem and continuity of $\tilde{H}$, it is enough to show that the derivatives $\dfrac{\partial \tilde{H}}{\partial x_i}$ are bounded on the domain $0 \leq |y| \leq |f(x)|$, for all $i = 1, \ldots, n$. We have,

$$\frac{\partial \tilde{H}}{\partial x_i} = \frac{(\frac{\partial g}{\partial x_i} f(x) - \frac{\partial f}{\partial x_i} g(x)) y}{(f(x))^2} = \frac{\partial g}{\partial x_i} \frac{y}{f(x)} - \frac{\partial f}{\partial x_i} \frac{g(x)}{f(x)} \frac{y}{f(x)}.$$

Since $|y| \leq |f(x)|$, then $\dfrac{y}{f(x)}$ is bounded. The expression $\dfrac{g(x)}{f(x)}$ is bounded since $f \approx g$. Moreover, $\dfrac{\partial g}{\partial x_i}$ and $\dfrac{\partial f}{\partial x_i}$ are bounded because f and g are Lipschitz functions.

Since H^{-1} can be constructed in the same form as (5.5), we conclude that H^{-1} is also Lipschitz and, thus, H is a bi-Lipschitz map.

Remark 5.6.10 A complete invariant for contact equivalence of semialgebraic function germs was recently defined by Birbrair, Fernandes, Gabrielov and Grandjean in [BFGG17]. They describe a partition of the neighborhood of the origin in $\mathbb{R}^2$ into zones where the function has explicit asymptotic behavior. This partition is called a **pizza**. Each function germ admits a "minimal" pizza, unique up to combinatorial equivalence. It follows from their main result that two semialgebraic function germs are semialgebraically bi-Lipschitz $\mathcal{K}$–equivalent if and only if their corresponding minimal pizzas are equivalent.

Other recent references on bi-Lipschitz equivalences of function germs in the plane are Koike and Parusiński[KP13] and Birbrair, Fernandes, Grandjean and Gaffney in [BFGG18].

5.6.3 Invariants of Bi-Lipschitz $\mathcal{R}$-Equivalence

We first prove the bi-Lipschitz invariance of the rank.

Theorem 5.6.11 (Theorem 4.1, [NRT20]) *Let* $f, g : (\mathbb{K}^n, 0) \to (\mathbb{K}^p, 0)$ *be two smooth map germs. If* f *and* g *are bi-Lipschitz* $\mathcal{R}$-*equivalent then the rank of* f *is equal to the rank of* g *at* 0.

Proof Since f and g are bi-Lipschitz equivalent, there exists a germ of a bi-Lipschitz homeomorphism $\varphi : (\mathbb{K}^n, 0) \to (\mathbb{K}^n, 0)$ such that $f \circ \varphi = g$. Then, there exists a sequence $\{m_i\}$ in $\mathbb{N}$ such that the sequence of maps $\varphi_{m_i} : (\mathbb{K}^n, 0) \to (\mathbb{K}^n, 0)$ defined by $\varphi_{m_i} = m_i \varphi(\frac{x}{m_i})$ converges uniformly to a bi-Lipschitz germ $d\varphi$ on a neighborhood of 0 as i tends to ∞. Write $f(x) = Df(0)(x) + o(\|x\|)$ and $g(y) = Dg(0)(y) + o(\|y\|)$ on a neighborhood of 0. Since $f \circ \varphi(y) = g(y)$

$$m_i f \circ \varphi(\frac{y}{m_i}) = m_i g(\frac{y}{m_i})$$

$$\implies \quad m_i(Df(0)(\varphi(\frac{y}{m_i})) + m_i o \|\varphi(\frac{y}{m_i})\| = m_i Dg(0)(\frac{y}{m_i}) + m_i o \|\frac{y}{m_i}\|$$

$$\implies \quad \lim_{i \to \infty} Df(0)(m_i \varphi(\frac{y}{m_i})) + \lim_{i \to \infty} m_i o \|\varphi(\frac{y}{m_i})\| = Dg(0)(y) + \lim_{i \to \infty} m_i o \|\frac{y}{m_i}\|.$$

Hence,

$$Df(0)(d\phi(y)) = Dg(0)(y)$$

for y in a small neighborhood of 0. This implies that the rank of f and g are equal.

□

It has been known that the multiplicity is a bi-Lipschitz invariant (see [RT97, FR04]). A new proof was recently given by Nguyen et al. [NRT20, Lemma 4.2]. We present here a shorter argument suggested by the referee to whom we are very grateful.

Lemma 5.6.12 *If f and g are bi-Lipschitz $\mathcal{R}$-equivalent, then the multiplicities m_f and m_g are equal.*

Proof If the multiplicity m_f of f is m the following holds:

(1) There is a $c > 0$ such that $\|f(x)\| \leq c\|x\|^m$ in a neighborhood of the origin (by Taylor's formula).
(2) There is a sequence $(x_k)_k$ converging to 0 and a constant $A > 0$ such that $\|f(x_k)\| \geq A\|x_k\|^m$. In fact any sequence whose limit of secants is not a zero of the initial form of f at the origin satisfies this property. This follows again from Taylor's formula.
Since $g = f \circ \phi$ for a ϕ bi-Lipschitz, g satisfies 1 and 2.   □

The Thom–Levine criterion for bi-Lipschitz triviality is as follows:

Theorem 5.6.13 *Let $\mathbb{K} = \mathbb{C}$ or $\mathbb{R}$ and $F : (\mathbb{K}^n \times \mathbb{K}, 0) \rightarrow (\mathbb{K}, 0)$ be a one-parameter deformation of a germ $f : (\mathbb{K}^n, 0) \rightarrow (\mathbb{K}, 0)$. If there exists a germ of continuous vector field of the form*

$$X(x, t) = \frac{\partial}{\partial t} + \sum_{i=1}^{n} X_i(x, t)\frac{\partial}{\partial x_i}$$

Lipschitz in x, (i.e. there exists a number $C > 0$ with

$$\|X(x_1, t) - X(x_2, t)\| \leq C\|x_1 - x_2\|$$

for all t), such that $X.F = 0$, then F is a bi-Lipschitz trivial deformation of f.

Proof The existence of the flow of the vector field $X(x, t)$, denoted $\Phi(x, t)$, follows from the classical Picard-Lindelof theorem. Applying Gronwall's Lemma, we can show that $\Phi_t : \mathbb{K}^n, 0 \rightarrow \mathbb{K}^n, 0$ is a bi-Lipschitz homeomorphism. Therefore, this flow induces the bi-Lipschitz triviality of F.   □

It seems to be unknown whether the converse of the above statement holds. For this reason we call the deformation F in Theorem 5.6.13 strongly bi-Lipschitz trivial. Fernandes and Ruas [FR04] gave sufficient conditions for a deformation of a quasihomogeneous polynomial (in the real case) to be strongly bi-Lipschitz trivial.

5.6.4 Henry-Parusiński's Example

In this subsection our main object of study are the holomorphic function germs $f : (\mathbb{C}^2, 0) \to (\mathbb{C}, 0)$. We first introduce the notion of relative polar curve of f.

Definition 5.6.14 The relative polar curve of f is the curve determined by an equation $\Gamma(f) : af_x + bf_y = 0$ where $(a : b)$ ia a general point in the projective line $\mathbb{P}^1_{\mathbb{C}}$.

Up to a linear change of coordinates in $\mathbb{C}^2$ we can assume that $(a : b) = (1 : 0)$ and the polar curve of f is given by $\Gamma(f) : f_x = 0$ (see Section 8.1 in Anne Pichon's lecture notes for more details on polar curves and generic projections).

In [HP03], Henry and Parusiński showed that the bi-Lipschitz $\mathcal{R}$-equivalence of analytic function germs $f : (\mathbb{C}^2, 0) \to (\mathbb{C}, 0)$ admits continuous moduli.

They consider the one parameter family of germs

$$f_t(x, y) = f(x, y, t) = x^3 - 3t^2 xy^4 + y^6$$

and show that if t, t' are sufficiently generic then f_t and $f_{t'}$ are not strongly bi-Lipschitz $\mathcal{R}$-equivalent function germs. We review here Theorem 1.1 of Henry and Parusiński [HP03]:

Theorem 5.6.15 (Theorem 1.1 [HP03]) *There is no Lipschitz vector field*

$$v(x, y, t) = \frac{\partial}{\partial t} + v_1(x, y, t)\frac{\partial}{\partial x} + v_2(x, y, t)\frac{\partial}{\partial y},$$

$v_1(0, 0, t) = v_2(0, 0, t) = 0$ *defined in a neighborhood of $(0, 0, t_0)$ and tangent to the levels of*

$$f(x, y, t) = x^3 - 3t^2 xy^4 + y^6.$$

Proof Let us suppose, by contradiction, that such v exists. Then $\frac{\partial f}{\partial v} \equiv 0$.

Let Γ be the family of polar curves

$$\Gamma = \{(x, y, t)| \frac{\partial f}{\partial x} = 3(x^2 - t^2 y^4) = 0\}.$$

The branches of Γ are $x = \pm ty^2$. We now develop $\frac{\partial f}{\partial v} \equiv 0$ along each branch of Γ and substitute $x = \pm ty^2$ to get

$$0 = \frac{\partial f}{\partial v} = \frac{\partial f}{\partial t} + v_2 \frac{\partial f}{\partial y}$$
$$= -6txy^4 + v_2(-12t^2xy^3 + 6y^5)$$
$$= \mp 6t^2y^6 + v_2(\mp 12t^3y^5 + 6y^5)$$

Hence

$$v_2(\pm ty^2, y, t) = \frac{\pm t^2 y}{1 \mp 2t^3}.$$

Comparing v_2 between the two branches of Γ, we get

$$v_2(ty^2, y, t) - v_2(-ty^2, y, t) = \frac{t^2 y}{1 - 2t^3} - \frac{-t^2 y}{1 + 2t^3} \sim y \qquad (5.6)$$

On the other hand if v_2 is Lipschitz, with Lipschitz constant L then

$$|v_2(ty^2, y, t) - v_2(-ty^2, y, t)| \le 2Lt|ty^2|,$$

which contradicts (5.6).   □

Henry and Parusiński introduce a new invariant based on the observation that a bi-Lipschitz homeomorphism does not move much the regions around the relative polar curves. For a single germ f defined at $(\mathbb{C}^2, 0)$, the invariant is given in terms of the leading coefficients of the asymptotic expansions of f along the branches of its generic polar curve.

Fernandes and Ruas in [FR13] study the **strong bi-Lipschitz triviality**: two function germs f and g are **strongly bi-Lipschitz equivalent** if they can be embedded in a bi-Lipschitz trivial family, whose triviality is given by integrating a Lipschitz vector field. They show that if two weighted homogeneous (but not homogeneous) function-germs $(\mathbb{C}^2, 0) \to (\mathbb{C}, 0)$ are strongly bi-Lipschitz equivalent, then they are analytically equivalent. This result does not hold for families of homogeneous germs with isolated singularities and same degree since they are Lipschitz equivalent [FR04].

In some sense, the problem of bi-Lipschitz classification for weighted homogeneous (and not homogeneous) real function-germs in two variables is quite close to the problem of analytic classification (see [CR18]).

The moduli space of bi-Lipschitz equivalence is not completely understood yet, not even in the case of weighted homogeneous function germs.

Motivated by [HP03], in [FR13] Fernandes and Ruas prove the following result

Theorem 5.6.16 (Theorem 3.4, [FR13]) *Let $F(x, y, t)$ be a polynomial function such that for all $t \in U$ the function $f_t(x, y) = F(x, y, t)$ is a w–homogeneous ($w_1 > w_2$) with an isolated singularity at $(0, 0) \in \mathbb{C}^2$, where $U \subset \mathbb{C}^2$ is a domain. If $\{f_t : t \in U\}$, as a family of function-germs at $(0, 0) \in \mathbb{C}^2$, is strongly bi-Lipschitz trivial then f_{t_1} is analytically equivalent to f_{t_2} for any $t_1, t_2 \in U$.*

Let $f : (\mathbb{C}^2, 0) \to (\mathbb{C}, 0)$ be a germ of reduced analytic function with Taylor expansion

$$f(x, y) = H_k(x, y) + H_{k+1}(x, y) + \dots \tag{5.7}$$

where H_i is homogeneous polynomial of degree i.

Let us suppose that f is mini-regular in x of order k, that is, $H_k(1, 0) \neq 0$. A polar arc $x = \gamma(y)$ is a branch of the polar curve $\Gamma : \frac{\partial f}{\partial x} = 0$. As f is mini-regular in x, it follows that $x = \gamma(y)$ is not tangent to the x–axis. As in [HP03] we distinguish two classes of polar arcs: polar arcs that are tangent to the tangent cone $C_0(X) = \{H_k = 0\}$ to $X = f^{-1}(0)$ at the origin and polar arcs that are not tangent to the tangent cone $C_0(X)$.

The polar arcs of the first type are called **tangential**. The non-tangential polar arcs are the moving ones as their tangents at zero moves when we make change of coordinates.

Let γ be a polar arc. Let $h_0 = h_0(\gamma) \in Q_+$ and $c_0 = c_0(\gamma) \in \mathbb{C}^*$ be given by the expansion

$$f(\gamma(y), y) = c_0 y^{h_0} + \dots, \quad c_0 \neq 0.$$

Let $l \in C_0(X)$ be a fixed tangent line to X at 0. We denote by $\Gamma(l)$ the set of all polar arcs tangent to l at 0. $\Gamma(l)$ is nonempty if and only if

$$l \subset Sing(C_0(X)) = \{\frac{\partial H_k}{\partial x} = \frac{\partial H_k}{\partial y} = 0\}.$$

We define $I(l)$ as the set of formal expressions

$$I(l) = \{c_0(\gamma) y^{h_0(\gamma)} | \gamma \in \Gamma(l)\}/\mathbb{C}^*$$

divided by the action of $\mathbb{C}^*$ where $c \in \mathbb{C}^*$, acts by multiplication on γ :

$$\{c_{01} y^{h_{01}}, \dots, c_{0k} y^{h_{0k}}\} \sim \{c_{01} c^{h_{01}} y^{h_{01}}, \dots, c_{0k} c^{h_{0k}} y^{h_{0k}}\}.$$

Definition 5.6.17 (Definition 4.2, [HP03]) Let $f(x, y)$ be an analytic function germ. Henry-Parusiński invariant $Inv(f)$ of f is the set of all $I(l)$, where l runs over all lines in $Sing(C_0(X))$.

Theorem 5.6.18 (Theorem 4.3, [HP03]) *Let f_1, f_2 be two analytic functions germs $(\mathbb{C}^2, 0) \to (\mathbb{C}, 0)$ mini-regular in x. If f_1 and f_2 are bi-Lipschitz equivalent then $Inv(f_1) = Inv(f_2)$.*

Example 5.6.19 (Proposition 3.5[FR04]) Let

$$f_t(x, y) = \frac{1}{3}x^3 - t^2 xy^{3n-2} + y^{3n}, \ n \geq 3.$$

The function germ $f_0(x, y) = \frac{1}{3}x^3 + y^{3n}$ is weighted homogeneous of type $(n, 1; 3n)$. The weighted degree of xy^{3n-2} with respect to these weights is $4n - 2 > 3n$ when $n \geq 3$. In [FR04, Proposition 3.5] the authors prove that the family f_t is not strongly bi-Lipschitz trivial.

The polar curve $\frac{\partial f_t}{\partial x} = 0$ of f_t given by $x^2 - t^2 y^{3n-2} = 0$, has branches $x = \pm t y^{\frac{3n-2}{2}}$.

Then $f_t(\pm t y^{\frac{3n-2}{2}}, y) = y^6$ and it follows that $Inv(f_t)$ does not depend on the parameter t. Hence this invariant does not distinguish the elements of the family f_t.

An open question is how to define bi-Lipschitz $\mathcal{R}-$invariants associated to higher order terms of the Taylor expansion of the function germ f.

5.7 o-Minimal Structures (by Nhan Nguyen)

We overview basic properties of definable sets in o-minimal structures. For more details, we refer the reader to [Cos98, vdDM96, vdD98, Loi].

A **structure** on the ordered field of the real numbers $(\mathbb{R}, +, .)$ is a family $\mathcal{D} = (\mathcal{D}_n)_{n \in \mathbb{N}}$ such that for each n the following properties hold: (1) $\mathcal{D}_n$ is a Boolean algebra of subsets of $\mathbb{R}^n$; (2) if $A \in \mathcal{D}_n$, then $\mathbb{R} \times A \in \mathcal{D}_{n+1}$ and $A \times \mathbb{R} \in \mathcal{D}_{n+1}$; (3) $\mathcal{D}_n$ contains the zero sets of all polynomials in n variables; (4) if $A \in \mathcal{D}_n$, $\pi(A) \in \mathcal{D}_{n-1}$ where $\pi : \mathbb{R}^n \to \mathbb{R}^{n-1}$ denotes the orthogonal projection onto the first $(n - 1)$ coordinates. The structure $\mathcal{D}$ is said to be **o-minimal** if (5) every set in $\mathcal{D}_1$ is a finite union of intervals and points.

A set in a structure $\mathcal{D}$ is called a $\mathcal{D}$-**set** (or a definable set) and a map whose graph is in the structure $\mathcal{D}$ is called a $\mathcal{D}$-**map** (or a definable map).

Let $f_i : \mathbb{R}^{n_i} \to \mathbb{R}, n_i \in \mathbb{N}, i \in I$ be functions. We denote by $(\mathbb{R}, +, .., (f_i)_{I \in i})$ the smallest structure on $(\mathbb{R}, +, .)$ containing the graphs of all functions $(f_i)_{i \in I}$ and call it the **structure generated by the** f_i's. The following are some examples of o-minimal structures on $(\mathbb{R}, +, .)$.

Example 5.7.1

1. Class of semialgebraic sets: $\mathbb{R}_{alg} = (\mathbb{R}, +, .., \mathcal{P})$ where $\mathcal{P}$ is the class of all polynomial functions (by Tarski–Seidenberg).

2. Class of **globally subanalytic** sets: $\mathbb{R}_{an} = (\mathbb{R}, +, ., \mathcal{A})$ where $\mathcal{A}$ is the class of functions $f : \mathbb{R}^n \to \mathbb{R}$ such that $f = 0$ outside $[-1, 1]^n$ and the restriction of f to $[-1, 1]^n$ is analytic (by Gabrielov).
3. $\mathbb{R}_{an}^{\mathbb{R}} = (\mathbb{R}, +, ., \mathcal{A}, \mathcal{R})$ where $\mathcal{A}$ is defined as above and $\mathcal{R}$ is the class of all functions of the following form

$$f(x) = \begin{cases} x^r, & x > 0 \\ 0, & x \le 0 \end{cases}, \qquad \text{where } r \in \mathbb{R}.$$

(by Miller [Mil94a]).
4. $\mathbb{R}_{an,exp} = (\mathbb{R}, +, ., \mathcal{A}, (\exp))$, where $\mathcal{A}$ is defined as above and exp is the exponential function (by van den Dries and Miller [vdDM94]).

Notice that

$$\mathbb{R}_{alg} \subset \mathbb{R}_{an} \subset \mathbb{R}_{an}^{\mathbb{R}} \subset \mathbb{R}_{an,exp}.$$

A structure $\mathcal{D}$ is said to be **polynomially bounded** if for any $\mathcal{D}$-function $f : \mathbb{R} \to \mathbb{R}$, there exist $a > 0$ and $m \in \mathbb{N}$ such that $|f(x)| < x^m$ for all $x > a$.

Theorem 5.7.2 (Miller's Dichotomy [Mil94b]) *Let $\mathcal{D}$ be an o-minimal structure. Then, either $\mathcal{D}$ is polynomially bounded or it is exponential i.e., $\mathcal{D}$ contains the graph of the exponential function $\exp(x)$.*

From Miller's Dichotomy, the structures $\mathbb{R}_{alg}, \mathbb{R}_{an}$ and $\mathbb{R}_{an}^{\mathbb{R}}$ are polynomially bounded o-minimal, while $\mathbb{R}_{an,exp}$ is not.

Let $\mathcal{D}$ be a structure. Then, the closure, interior and boundary of a $\mathcal{D}$-set are $\mathcal{D}$-sets; the image and inverse images of a $\mathcal{D}$-set under a $\mathcal{D}$-map are $\mathcal{D}$-sets; composition of $\mathcal{D}$-maps is a $\mathcal{D}$-map. In the sequel, we assume that $\mathcal{D}$ is an o-minimal structure.

Theorem 5.7.3 (Monotonicity) *Let $f : (a, b) \to \mathbb{R}$ be a $\mathcal{D}$-function and $p \in \mathbb{N}$. Then, there are $a_0, \ldots, a_k$ with $a = a_0 < a_1 < \ldots < a_k = b$ such that $f|_{(a_i, a_{i+1})}$ is C^p, and either constant or strictly monotone, for $i = 0, \ldots, k - 1$.*

Given $p \in \mathbb{N}$, we say that a subset C of $\mathbb{R}^n$ is a C^p $\mathcal{D}$-**cell** if, for:

$n = 1$: C is either a point or an open interval.
$n > 1$: C has one of the following forms

$$\Gamma(\xi) := \{(x, y) \in B \times \mathbb{R} : y = \xi(x)\},$$

$$(\xi_1, \xi_2) := \{(x, y) \in B \times \mathbb{R} : \xi_1(x) < y < \xi_2(x)\},$$

$$(-\infty, \xi) := \{(x, y) \in B \times \mathbb{R} : y < \xi(x)\},$$

$$(\xi, +\infty) := \{(x, y) \in B \times \mathbb{R} : \xi(x) < y\},$$

where B is a C^p cell in $\mathbb{R}^{n-1}$, ξ, ξ_1, ξ_2 are $\mathcal{D}$-functions of class C^p on B and $\xi_1(x) < \xi_2(x)$, $\forall x \in B$. The cell B is called the **base** of C.

A C^p **cylindrical $\mathcal{D}$-cell decomposition** (C^p **cdcd**) of $\mathbb{R}^n$ is defined by induction as follows

1. A C^p cdcd of $\mathbb{R}$ is a finite collection of points and intervals

$$(-\infty, a_1), \ldots, (a_k, +\infty), \{a_1\}, \ldots \{a_k\}, \quad \text{where } a_1 < a_2 < \ldots < a_k.$$

2. A C^p cdcd of $\mathbb{R}^n$ is a partition of $\mathbb{R}^n$ into C^p cells such that the collection of all images of these cells under the natural projection onto the first $(n-1)$ coordinates $\pi : \mathbb{R}^n \to \mathbb{R}^{n-1}$ forms a C^p cdcd of $\mathbb{R}^{n-1}$.

We say that a C^p cdcd of $\mathbb{R}^n$ is **compatible** with $X = \{X_1, \ldots X_k\}$, a family of $\mathcal{D}$-subsets of $\mathbb{R}^n$, if each X_i is the union of some C^p $\mathcal{D}$-cells of the decomposition.

Theorem 5.7.4 (Cell Decomposition)

(1) Let $X = \{X_1, \ldots, X_k\}$ be a family of $\mathcal{D}$-subsets of $\mathbb{R}^n$. Then, there exists a C^p cdcd of $\mathbb{R}^n$ compatible with X.
(2) Let $f : X \to \mathbb{R}$ be a $\mathcal{D}$-function. Then, there exists a C^p cdcd of $\mathbb{R}^n$ compatible with X such that the restriction of f to each cell of the cdcd is of class C^p.

Theorem 5.7.5 (On Components) *Every $\mathcal{D}$-set has finitely many connected components and each of these components is also a $\mathcal{D}$-set.*

Theorem 5.7.6 (Uniform Bound) *Let A be a $\mathcal{D}$-subset of $\mathbb{R}^{n+m}$. Then there exists $N \in \mathbb{N}$ such that for all $x \in \mathbb{R}^m$ the set $A_x = \{y \in \mathbb{R}^n : (x, y) \in A\}$ has at most N connected components.*

Theorem 5.7.7 (Definable Choice) *Let A be a $\mathcal{D}$-subset of $\mathbb{R}^{n+m}$ and let $\pi : \mathbb{R}^{n+m} \to \mathbb{R}^n$ be the orthogonal projection onto the first n coordinates. Then, there exists a $\mathcal{D}$-map $\rho : \pi(A) \to \mathbb{R}^{n+m}$ such that $\pi(\rho(x)) = x$ for all $x \in \pi(A)$.*

Theorem 5.7.8 (Curve Selection) *Let A be a $\mathcal{D}$-subset of $\mathbb{R}^n$ and let $a \in \overline{A} \setminus A$. Let $p \in \mathbb{N}$. Then, there is a C^p $\mathcal{D}$-curve $\gamma : (0, 1) \to A \setminus \{a\}$ such that $\lim_{t \to 0} \gamma(t) = a$.*

Theorem 5.7.9 (Local Conical Structure) *Let A be a $\mathcal{D}$-subset of $\mathbb{R}^n$, and p be a point in A. Then there are $r > 0$ and definable homeomorphism h from the cone with vertex p and base $A \cap \mathbb{S}(p, r)$ onto $A \cap \overline{B}(p, r)$ such that $h|_{A \cap \mathbb{S}(p,r)} = Id$ and $\|h(x) - p\| = \|x - p\|$ for every x in the cone.*

Given a $\mathcal{D}$-set $A \subset \mathbb{R}^n$, the dimension of A is defined as follows:

$$\dim A = \sup\{\dim C : C \text{ is a } C^1 \text{ submanifold contained in } A\}.$$

Theorem 5.7.10

1. If $A \subset B$ are $\mathcal{D}$-sets, then $\dim A \leq \dim B$.
2. If A is a $\mathcal{D}$-set, then $\dim(\overline{A} \setminus A) < \dim A$.

3. *Let $f : A \to \mathbb{R}^m$ be a $\mathcal{D}$-map. If $\dim f^{-1}(x) \le k$ for every $x \in f(A)$, then*

$$\dim f(A) \le \dim A \le \dim f(A) + k.$$

For $p \in \mathbb{N}$, we denote by Φ^p the set of all odd, strictly increasing C^p definable bijection from $\mathbb{R}$ onto $\mathbb{R}$ and p-flat at 0.

Theorem 5.7.11 (Łojasiewicz Inequality) *Let $\mathcal{D}$ be a o-minimal structure. Let $A \subset \mathbb{R}^n$ be compact and let $f, g : A \to \mathbb{R}$ be continuous $\mathcal{D}$-functions with $f^{-1}(0) \subseteq g^{-1}(0)$. Then, there exists $\phi \in \Phi^p$ such that $\phi(g(x)) \le |f(x)|$ for all $x \in A$. In particular, if $\mathcal{D}$ is a polynomially bounded o-minimal structure then there exist $N > 0$ and $C > 0$ such that $|g(x)|^N \le C|f(x)|$ for all $x \in A$.*

Let $X \subset \mathbb{R}^m \times \mathbb{R}^n$ be a $\mathcal{D}$-set. Let A be a $\mathcal{D}$-subset of $\mathbb{R}^m$. Consider X as a family of $\mathcal{D}$-sets parameterized by $\mathbb{R}^m$. We denote by

$$X|_A = \{(x, y) \in X : x \in A\}.$$

We say that X is **definably trivial** over A if there exist a $\mathcal{D}$-set Z and a $\mathcal{D}$-homeomorphism $h : A \times Z \to X|_A$ such that the following diagram commutes

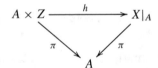

where π is the orthogonal projection onto $\mathbb{R}^m$. The map h is called a definable trivialization over A. Furthermore, if there exists a definable continuous function $C : A \to \mathbb{R}$ such that for $x, x' \in Z$ and $t \in A$: $|h_t(x) - h_t(x')| \le C(t)|x - x'|$ and for any $(x, x') \in X_t \times X_t$, $|h_t^{-1}(x) - h_t^{-1}(x')| \le C(t)|x - x'|$ then we call X definably bi-Lipschitz trivial and h a definable bi-Lipschitz trivialization.

The map h as above is called a **definable trivialization** of X over A. The trivialization of h is said to be **compatible** with Y if there is a $\mathcal{D}$-subset K of Z such that $h(A \times K) = Y|_A$.

Theorem 5.7.12 (Hardt's Triviality Theorem) *Let $X \subset \mathbb{R}^m \times \mathbb{R}^n$ be a $\mathcal{D}$-set. Let $X_1, \ldots, X_k$ be $\mathcal{D}$-subsets of X. Then, there exists a finite partition of $\mathbb{R}^m$ into $\mathcal{D}$-sets $C_1, \ldots, C_l$ such that X is definably trivial over each C_i and the trivializations over each C_i are compatible with $X_1, \ldots, X_k$.*

Valette in [Val05] proved that if $\mathcal{D}$ is a polynomially bounded o-minimal structure, then the trivializations in Theorem 5.7.12 can be chosen to be definable bi-Lipschitz trivializations.

References

[ABCF10] S. Alvarez, L. Birbrair, J.C.F. Costa, A. Fernandes, Topological $\mathcal{K}$-equivalence of analytic function-germs. Cent. Eur. J. Math. **8**(2), 338–345 (2010) 139

[ACCM19] H. Aguilar Cabrera, J.L. Cisneros Molina, Geometric viewpoint of Milnor's fibration theorem, in *The International School on Singularities and Lipschitz Geometry*, present volume, 2019. 112, 117

[AGZV12] V.I. Arnold, S.M. Gusein-Zade, A.N. Varchenko, *Singularities of Differentiable Maps. Volume 1: Classification of Critical Points, Caustics and Wave Fronts.* Modern Birkhäuser Classics (Birkhäuser/Springer, New York, 2012); Translated from the Russian by Ian Porteous based on a previous translation by M. Reynolds, Reprint of the 1985 edition 140

[BC57] F. Bruhat, H. Cartan, Sur la structure des sous-ensembles analytiques réels. C. R. Acad. Sci. Paris **244**, 988–990 (1957) 116

[BCFR07] L. Birbrair, J.C.F. Costa, A. Fernandes, M.A.S. Ruas, $\mathcal{K}$-bi-Lipschitz equivalence of real function-germs. Proc. Amer. Math. Soc. **135**(4), 1089–1095 (2007) 142

[BCR98] J. Bochnak, M. Coste, M.-F. Roy, *Real Algebraic Geometry*. Ergebnisse der Mathematik und ihrer Grenzgebiete (3) [Results in Mathematics and Related Areas (3)], vol. 36 (Springer, Berlin, 1998); Translated from the 1987 French original, Revised by the authors 115, 116, 117

[BF00] L. Birbrair, A.C.G. Fernandes, Metric theory of semialgebraic curves. Rev. Mat. Complut. **13**(2), 369–382 (2000) 122, 125

[BF08a] L. Birbrair, A. Fernandes, Inner metric geometry of complex algebraic surfaces with isolated singularities. Comm. Pure Appl. Math. **61**(11), 1483–1494 (2008) 112, 138, 139

[BF08b] L. Birbrair, A. Fernandes, Local Lipschitz geometry of real weighted homogeneous surfaces. Geom. Dedicata **135**, 211–217 (2008) 133, 134

[BFGG17] L. Birbrair, A. Fernandes, V. Grandjean, A. Gabrielov, Lipschitz contact equivalence of function germs in $\mathbb{R}^2$. Ann. Sc. Norm. Super. Pisa Cl. Sci. **17**(1), 81–92 (2017) 144

[BFGG18] L. Birbrair, A. Fernandes, V. Grandjean, T. Gaffney, Blow-analytic equivalence versus contact bi-Lipschitz equivalence. J. Math. Soc. Japan **70**(3), 989–1006 (2018) 144

[BFLS16] L. Birbrair, A. Fernandes, D.T. Lê, J.E. Sampaio, Lipschitz regular complex algebraic sets are smooth. Proc. Amer. Math. Soc. **144**(3), 983–987 (2016) 139

[BFN10] L. Birbrair, A. Fernandes, W.D. Neumann, On normal embedding of complex algebraic surfaces, in *Real and Complex Singularities*. London Mathematical Society Lecture Note series, vol. 380 (Cambridge University Press, Cambridge, 2010), pp. 17–22 136, 138, 139

[Bir99] L. Birbrair, Local bi-Lipschitz classification of 2-dimensional semialgebraic sets. Houston J. Math. **25**(3), 453–472 (1999) 122, 129, 132, 133

[BL07] A. Bernig, A. Lytchak, Tangent spaces and Gromov-Hausdorff limits of subanalytic spaces. J. Reine Angew. Math. **608**, 1–15 (2007) 136

[BM00] L. Birbrair, T. Mostowski, Normal embeddings of semialgebraic sets. Michigan Math. J. **47**(1), 125–132 (2000) 120

[BR90] R. Benedetti, J.-J. Risler, *Real Algebraic and Semi-algebraic Sets*. Actualités Mathématiques. [Current Mathematical Topics] (Hermann, Paris, 1990) 116, 117

[BS91] R. Benedetti, M. Shiota, Finiteness of semialgebraic types of polynomial functions. Math. Z. **208**(4), 589–596 (1991) 139

[Cos98] M. Coste, Topological types of fewnomials, in *Singularities Symposium—Łojasiewicz 70 (Kraków, 1996; Warsaw, 1996)*. Banach Center Publications, vol. 44 (Polish Academy of Sciences Institute of Mathematics, Warsaw, 1998), pp. 81–92 139, 149

[CR18] L. Câmara, M.A.S. Ruas, On the moduli space of quasi-homogeneous functions (2018, preprint), arXiv:2004.03778 147

[Ebe07] W. Ebeling, *Functions of Several Complex Variables and Their Singularities*. Graduate Studies in Mathematics, vol. 83 (American Mathematical Society, Providence, 2007) 140

[FdB05] J. Fernández de Bobadilla, Answers to some equisingularity questions. Invent. Math. **161**(3), 657–675 (2005) 134

[Fer03] A. Fernandes, Topological equivalence of complex curves and bi-Lipschitz homeomorphisms. Michigan Math. J. **51**(3), 593–606 (2003) 122

[FKK98] T. Fukui, S. Koike, T.-C. Kuo, Blow-analytic equisingularities, properties, problems and progress, in *Real Analytic and Algebraic Singularities (Nagoya/Sapporo/Hachioji, 1996)*. Pitman Research Notes in Mathematics Series, vol. 381 (Longman, Harlow, 1998), pp. 8–29 122

[FR04] A. Fernandes, M.A.S. Ruas, Bi-Lipschitz determinacy of quasihomogeneous germs. Glasg. Math. J. **46**(1), 77–82 (2004) 145, 146, 147, 149

[FR13] A. Fernandes, M. Ruas, Rigidity of bi-Lipschitz equivalence of weighted homogeneous function-germs in the plane. Proc. Amer. Math. Soc. **141**(4), 1125–1133 (2013) 147, 148

[GBGPPP17] E.R. García Barroso, P.D. González Pérez, P. Popescu-Pampu, Variations on inversion theorems for Newton-Puiseux series. Math. Ann. **368**(3–4), 1359–1397 (2017) 116

[Gib79] C.G. Gibson, *Singular Points of Smooth Mappings*. Research Notes in Mathematics, vol. 25 (Pitman (Advanced Publishing Program), Boston, 1979) 140

[HP03] J-P. Henry, A. Parusiński, Existence of moduli for bi-Lipschitz equivalence of analytic functions. Compositio Math. **136**(2), 217–235 (2003) 112, 139, 146, 147, 148, 149

[Kin78] H.C. King, Topological type of isolated critical points. Ann. Math. **107**, 385–397 (1978) 112

[KP13] S. Koike, A. Parusiński, Equivalence relations for two variable real analytic function germs. J. Math. Soc. Japan **65**(1), 237–276 (2013) 144

[KPR18] D. Kerner, H.M. Pedersen, M.A.S. Ruas, Lipschitz normal embeddings in the space of matrices. Math. Z. **290**, 485–507 (2018) 119

[Loi] T.L. Loi, Lecture 1: o-minimal structures, in *The Japanese-Australian Workshop on Real and Complex Singularities-JARCS III*. Proceedings of the Centre for Mathematics and its Applications, vol. 43 (The Australian National University, Canberra, 2010), pp. 19–30 149

[Mat69] J.N. Mather, Stability of C^∞ mappings. II. Infinitesimal stability implies stability. Ann. Math. **89**, 254–291 (1969) 139, 141

[McS34] E.J. McShane, Extension of range of functions. Bull. Amer. Math. Soc. **40**(12), 837–842 (1934) 121, 136

[Mil68] J. Milnor, *Singular Points of Complex Hypersurfaces*. Annals of Mathematics Studies, no. 61 (Princeton University Press, Princeton, 1968) 112, 116, 117, 140

[Mil94a] C. Miller, Expansions of the real field with power functions. Ann. Pure Appl. Logic **68**(1), 79–94 (1994) 150

[Mil94b] C. Miller, Exponentiation is hard to avoid. Proc. Amer. Math. Soc. **122**(1), 257–259 (1994) 150

[Mos85] T. Mostowski, Lipschitz equisingularity. Diss. Math. (Rozprawy Mat.) **243**, 46 (1985) 112, 139

[MR09] C.G. Moreira, M.A.S. Ruas, The curve selection lemma and the Morse-Sard theorem. *Manuscripta Math.* **129**(3), 401–408 (2009) 116

[Nak84] I. Nakai, On topological types of polynomial mappings. Topology **23**(1), 45–66 (1984) 139

[NRT20] N. Nguyen, M. Ruas, S. Trivedi, Classification of lipschitz simple function germs. Proc. London Math. Soc. **121**(1), 51–82 (2020) 139, 140, 144, 145

[Par88] A. Parusiński, Lipschitz properties of semi-analytic sets. Ann. Inst. Fourier (Grenoble) **38**(4), 189–213 (1988) 139

[Per85] B. Perron, Conjugaison topologique des germes de fonctions holomorphes à singularité isolée en dimension trois. Invent. Math. **82**, 27–35 (1985) 112

[Pic19] A. Pichon, An introduction to lipschitz geometry of complex singularities, in *The International School on Singularities and Lipschitz Geometry*. Lecture Notes in Mathematics (Springer, Cham, 2020) 112, 118

[Pri67] D. Prill, Cones in complex affine space are topologically singular. Proc. Amer. Math. Soc. **18**, 178–182 (1967) 139

[PT69] F. Pham, B. Teissier, Fractions lipschitziennes d'une algèbre analytique complexe et saturation de Zariski. Preprint No. M17.0669 (1969). *Lipschitz Fractions of a Complex Analytic Algebra and Zariski Saturation*. English translation In *The International School on Singularities and Lipschitz Geometry*. Lecture Notes in Mathematics (Springer, Cham, 2020) 112

[RT97] J.J. Risler, D.J.A Trotman, Bi-Lipschitz invariance of the multiplicity. Bull. London Math. Soc. **29**(2), 200–204 (1997) 145

[RV11] M.A. Soares Ruas, G. Valette, C^0 and bi-Lipschitz $\mathcal{K}$-equivalence of mappings. Math. Z. **269**(1–2), 293–308 (2011) 139, 142

[Sab83] C. Sabbah, Le type topologique éclaté d'une application analytique, in *Singularities, Part 2 (Arcata, Calif., 1981)*. Proceedings of Symposia in Pure Mathematics, vol. 40 (American Mathematical Society, Providence, 1983), pp. 433–440 139

[Sam15] J.E. Sampaio, Regularidade Lipschitz, invariância da multiplicidade e a geometria dos cones tangentes de conjuntos analíticos. PhD Thesis, Universidade Federal do Ceará, 2015 119

[Sam16] J.E. Sampaio, Bi-Lipschitz homeomorphic subanalytic sets have bi-Lipschitz homeomorphic tangent cones. Selecta Math. (N.S.) **22**(2), 553–559 (2016) 134, 136, 139

[Sam20] J.E. Sampaio, Some classes of homeomorphisms that preserve multiplicity and tangent cones, in *A Panorama of Singularities*. Contemporary Mathematics, vol. 742 (American Mathematical Society, Providence, 2020), pp. 189–200 119

[Val05] G. Valette, Lipschitz triangulations. Illinois J. Math. **49**(3), 953–979 (2005) 129, 130, 136, 152

[vdD98] L. van den Dries, *Tame Topology and o-Minimal Structures*. London Mathematical Society Lecture Note Series, vol. 248 (Cambridge University Press, Cambridge, 1998) 149

[vdDM94] L. van den Dries, C. Miller, On the real exponential field with restricted analytic functions. Israel J. Math. **85**(1–3), 19–56 (1994) 150

[vdDM96] L. van den Dries, C. Miller, Geometric categories and o-minimal structures. Duke Math. J. **84**(2), 497–540 (1996) 149

[Wal58] A.H. Wallace, Algebraic approximation of curves. Canad. J. Math. **10**, 242–278 (1958) 116

[Wal81] C.T.C. Wall, Finite determinacy of smooth map-germs. Bull. London Math. Soc. **13**(6), 481–539 (1981) 140

[Wal04] C.T.C. Wall, *Singular Points of Plane Curves*. London Mathematical Society Student Texts, vol. 63 (Cambridge University Press, Cambridge, 2004) 115

[Whi65a] H. Whitney, Tangents to an analytic variety. Ann. Math. **81**, 496–549 (1965) 138

[Whi65b] H. Whitney, Local properties of analytic varieties, in *Differential and Combinatorial Topology (A Symposium in Honor of Marston Morse)* (Princeton University Press, Princeton, 1965), pp. 205–244 134, 135

[Whi65c] H. Whitney, Tangents to an analytic variety. Ann. Math. **81**, 496–549 (1965) 134, 135

[Whi72] H. Whitney, *Complex Analytic Varieties* (Addison-Wesley, Reading, 1972) 135

[Zar65a] O. Zariski, Studies in equisingularity. I. Equivalent singularities of plane algebroid curves. Amer. J. Math. **87**, 507–536 (1965) 140, 142

[Zar65b] O. Zariski, Studies in equisingularity. II. Equisingularity in codimension 1 (and characteristic zero). Amer. J. Math. **87**, 972–1006 (1965) 140, 142

[Zar71] O. Zariski, Some open questions in the theory of singularities. Bull. Amer. Math. Soc. **77**, 481–491 (1971) 134

Chapter 6
Surface Singularities in $\mathbb{R}^4$: First Steps Towards Lipschitz Knot Theory

Lev Birbrair and Andrei Gabrielov

Abstract A link of an isolated singularity of a two-dimensional semialgebraic surface in $\mathbb{R}^4$ is a knot (or a link) in S^3. Thus the ambient Lipschitz classification of surface singularities in $\mathbb{R}^4$ can be interpreted as a metric refinement of the topological classification of knots (or links) in S^3. We show that, given a knot K in S^3, there are infinitely many distinct ambient Lipschitz equivalence classes of outer Lipschitz equivalent singularities in $\mathbb{R}^4$ with the links topologically ambient equivalent to K.

6.1 Introduction

There are three kinds of equivalence relations in Lipschitz Geometry of Singularities. One equivalence relation, **inner Lipschitz equivalence**, is bi-Lipschitz homeomorphism (of the germs at the origin) of singular sets with respect to the inner metric, where the distance between two points of a set X is defined as infimum of the lengths of paths inside X connecting the two points. The second equivalence relation, **outer Lipschitz equivalence**, is bi-Lipschitz homeomorphism with respect to the outer metric, where the distance is defined as the distance between the points in the ambient space. A set X is called **Lipschitz normally embedded** if its inner and outer metrics are equivalent.

In [BG19], we considered the third equivalence relation, ambient Lipschitz equivalence. Two germs X and Y of semialgebraic sets at the origin of $\mathbb{R}^n$ are called **ambient Lipschitz equivalent** if there exists a germ of a bi-Lipschitz homeomorphism h of $(\mathbb{R}^n, 0)$ such that $Y = h(X)$. In particular, such sets X and Y are outer Lipschitz equivalent. Two outer Lipschitz equivalent sets are always

L. Birbrair (✉)
Department of Matemática, Universidade Federal do Ceará (UFC), Fortaleza-Ce, Brasil
e-mail: birb@ufc.br

A. Gabrielov
Department of Mathematics, Purdue University, West Lafayette, IN, USA
e-mail: gabriea@purdue.edu

© The Author(s), under exclusive license to Springer Nature Switzerland AG 2020
W. Neumann, A. Pichon (eds.), *Introduction to Lipschitz Geometry of Singularities*,
Lecture Notes in Mathematics 2280, https://doi.org/10.1007/978-3-030-61807-0_6

inner Lipschitz equivalent, but can be ambient topologically non-equivalent (see Neumann–Pichon [NP17]).

Let X and Y be two semialgebraic surface singularities (two-dimensional germs at the origin) in $\mathbb{R}^n$ which are outer Lipschitz equivalent. Suppose also that X and Y are topologically ambient equivalent. Does it imply that the sets X and Y are ambient Lipschitz equivalent? It seems plausible that the answer is "yes" when $n \geq 5$, or when X and Y are Lipschitz normally embedded. However, examples in [BG19] show that the answer may be "no" when $n = 3$ or $n = 4$.

One class of examples in $\mathbb{R}^3$ and $\mathbb{R}^4$ is based on the theorem of Sampaio [Sam16]: ambient Lipschitz equivalence of two sets implies ambient Lipschitz equivalence of their tangent cones. Thus any two sets with topologically ambient non-equivalent tangent cones cannot be ambient Lipschitz equivalent.

The case $n = 4$ is especially interesting, as in that case the link of a two-dimensional germ X in $\mathbb{R}^4$ is a knot (or a link) in S^3, and the arguments are based on the knot theory. For a given surface $X \subset \mathbb{R}^3$ there are finitely many distinct ambient Lipschitz equivalence classes of the surfaces which are topologically ambient equivalent and outer Lipschitz equivalent to X. However, there may be infinitely many such ambient Lipschitz equivalence classes for a surface in $\mathbb{R}^4$. Moreover, a more delicate argument, based on the "bridge construction" below, provides infinitely many distinct Lipschitz ambient equivalence classes of surfaces which are topologically ambient equivalent to a given surface $X \subset \mathbb{R}^4$ and all belong to the same outer Lipschitz equivalence class, even when each of these surfaces has a tangent cone consisting of a single ray.

In this paper we use the **topological ambient equivalence** relation for the germs at the origin of singular sets, and for their links. It means that there exists a homeomorphism of a small ball (or a small sphere) centered at the origin, mapping one singular set (or its link) to another. This definition corresponds to the classical topological equivalence in Singularity Theory. For the links of surfaces in $\mathbb{R}^4$ this equivalence is closely related to isotopy of knots. There is, however, a minor difference between the topological ambient equivalence and isotopy: in Knot Theory the homeomorphism is required to be isotopic to identity. Two knots may be topologically ambient equivalent but not isotopic. This difference between two equivalence relations is non-essential, as the number of ambient Lipschitz equivalence classes is infinite in both cases.

The authors thank the anonymous referee whose thoughtful suggestions helped us to improve the exposition.

6.2   Examples in $\mathbb{R}^3$ and $\mathbb{R}^4$ Based on Sampaio's Theorem

In this paper, an **arc** in $\mathbb{R}^n$ is a germ at the origin of a semialgebraic mapping $\gamma : [0, \epsilon) \rightarrow \mathbb{R}_n$ such that $\|\gamma(t)\| = t$. The **tangency order** or **contact order** $\mathrm{tord}(\gamma_1, \gamma_2)$ of two arcs is the smallest Puiseux exponent at zero of the function

Fig. 6.1 The links of the surfaces X_1 and X_2 in Example 6.1, and of their tangent cones

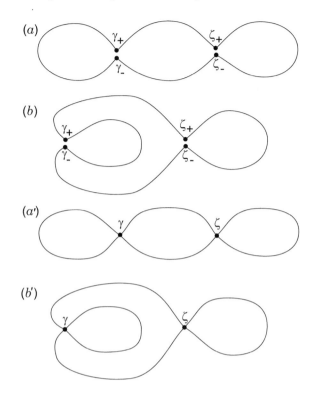

$\|\gamma_2(t) - \gamma_1(t)\|$. Sometimes in the literature the tangency order is called the tangency exponent or the order of contact (see [BFGG17, BM18, Par88]).

Example 6.1 (See [BG19, Example 2]) Let X_1 and X_2 be two surfaces in $\mathbb{R}^3$ with the links at the origin shown in Fig. 6.1(a), (b).

There are two pairs of "pinched" arcs, $\gamma_\pm$ and $\zeta_\pm$, with $\mathrm{tord}(\gamma_+, \gamma_-) > 1$ and $\mathrm{tord}(\zeta_+, \zeta_-) > 1$, while both surfaces are straight-line cones over their links outside small conical neighborhoods of $\gamma_\pm$ and $\zeta_\pm$. The arcs γ_+ and γ_- correspond to a single ray γ of the tangent cone, and the arcs ζ_+ and ζ_- correspond to a single ray ζ. One can define X_1 and X_2 by explicit semialgebraic formulas. For small conical neighborhoods of the pinched arcs $\gamma_\pm$ and $\zeta_\pm$ it can be done as in [BG19, Example 3], and outside those neighborhoods the links of X_1 and X_2 can be approximated by piecewise linear curves. Both surfaces X_1 and X_2 are topologically ambient equivalent to a cone over a circle in S^2 and outer Lipschitz equivalent, but not ambient Lipschitz equivalent by Sampaio's theorem, since their tangent cones are not topologically ambient equivalent: there is a connected component of the complement in S^2 of the link of X_1 (see Fig. 6.1(a')) with the whole link as its boundary, while there is no such component of the complement of the link of X_2 (see Fig. 6.1(b')).

Fig. 6.2 The links of the
surfaces X_1 and X_2 in
Example 6.2, and of their
tangent cones

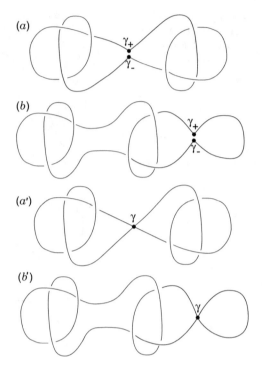

Example 6.2 (See [BG19]) Let X_1 and X_2 be two surfaces in $\mathbb{R}^4$ with the links
at the origin shown in Fig. 6.2(a) and (b), and the links of their tangent cones at the
origin shown in Fig. 6.2(a') and in (b'). The tangency exponent of the arcs γ_+ and γ_-
is $\alpha > 1$, thus the arcs γ_+ and γ_- correspond to a single ray γ of the tangent cone.
One can define X_1 and X_2 by explicit semialgebraic formulas. Both surfaces X_1
and X_2 are topologically ambient equivalent to a cone over a circle in S^3 embedded
as the connected sum of two trefoil knots, and outer Lipschitz equivalent but not
ambient Lipschitz equivalent by Sampaio's theorem, since their tangent cones at
the origin are not topologically ambient equivalent: the link of the tangent cone of
X_1 (see Fig. 6.1(a')) is a bouquet of two knotted circles, while the link of X_2 (see
Fig. 6.1(b')) is a bouquet of a knotted circle and an unknotted one.

6.3   Bridge Construction

A (q, β)-**bridge** is the set $A_{q,\beta} = T_+ \cup T_- \subset \mathbb{R}^4$ where $1 < \beta < q$ and

$$T_\pm = \left\{0 \leq t \leq 1, \ -t^\beta \leq x \leq t^\beta, \ y = \pm t^q, \ z = 0\right\}.$$

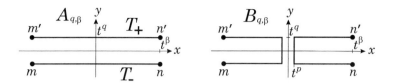

Fig. 6.3 The links of a (q, β)-bridge $A_{q,\beta}$ and a broken (q, β)-bridge $B_{q,\beta}$

Its link is shown in Fig. 6.3 (left). A **broken** (q, β)-**bridge** $B_{q,\beta}$ is obtained from $A_{q,\beta}$ by the **saddle operation**, removing from $T_\pm$ two p-Hölder triangles

$$\left\{ t \geq 0, \; |x| \leq t^p, \; y = \pm t^q, \; z = 0 \right\}$$

where $p > q$, and replacing them by two q-Hölder triangles

$$\left\{ 0 \leq t \leq 1, \; x = \pm t^p, \; |y| \leq t^q, \; z = 0 \right\}.$$

Its link is shown in Fig. 6.3 (right). We call (q, β)-bridge any surface outer Lipschitz equivalent to $A_{q,\beta}$. It was shown in [BG19] that ambient Lipschitz equivalence $h :$ $X \to Y$ of two surfaces in $\mathbb{R}^4$ maps a (q, β)-bridge in X to a (q, β)-bridge in Y, and that the two surfaces remain ambient Lipschitz equivalent when their (q, β)-bridges are replaced by the broken (q, β)-bridges.

Remark 6.3.1 Our definition of a broken bridge is slightly different from the definition in Example 4 of [BG19], where it was defined with $p < q$. Condition $p > q$ makes the "broken bridge" operation invertible: two surface germs with the same (q, β)-bridge are ambient Lipschitz equivalent if and only if they are ambient Lipschitz equivalent after the bridge is broken (with the same $p > q$). Note that this invertibility is never used here or in [BG19].

Example 6.3 (See [BG19]) The common boundary of $A_{q,\beta}$ and $B_{q,\beta}$ consists of the four arcs $\left\{ 0 \leq t \leq 1, \; x = \pm t^\beta, \; y = \pm t^q, \; z = 0 \right\}$ shown as m, n, m', n' in Fig. 6.3. Let $G \subset \mathbb{R}^4$ be a semialgebraic surface containing $A_{q,\beta}$ and bounded by the four straight line segments $\{0 \leq t \leq 1, \; \pm x = \pm y = t, \; z = 0\}$ (see Fig. 6.4 where the boundary arcs of G are shown as M, N, M', N'). Let H be the surface obtained from G by replacing the bridge $A_{\beta,q}$ by the broken bridge $B_{\beta,q}$ (see Fig. 6.5).

Consider two topologically trivial knots K and L in the hyperplane $\{t = 1\} \subset$ $\mathbb{R}^4_{x,y,z,t}$ as shown in Fig. 6.6a, b. Each of these two knots contains the curve $g = G \cap \{t = 1\}$, We define the surface X as the union of G and a straight cone over $K \setminus g$, and the surface Y as the union of G and a straight cone over $L \setminus g$.

Theorem 6.3.2 (See [BG19, Theorem 3.2]) *The germs of the surfaces X and Y at the origin are outer Lipschitz equivalent, topologically ambient equivalent, but not ambient Lipschitz equivalent.*

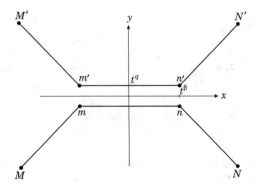

Fig. 6.4 The link of the surface G in Example 6.3

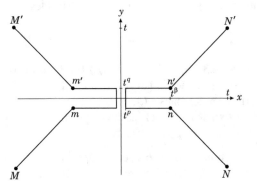

Fig. 6.5 The link of the surface H in Example 6.3

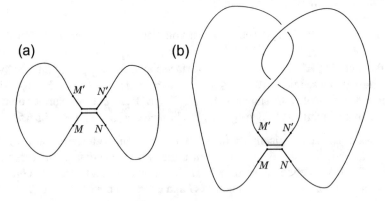

Fig. 6.6 The links of the surfaces X and Y in Example 6.3

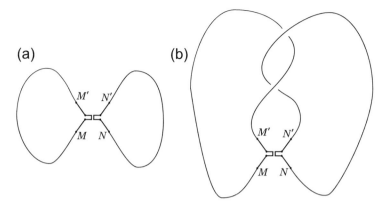

Fig. 6.7 The links of the surfaces X' and Y' in Example 6.3

This is proved by replacing the (q, β)-bridges in X and Y by the broken (q, β)-bridges, resulting in the new surfaces X' and Y', shown in Fig. 6.7. The link of X' consists of two unlinked circles while the two circles in the link of Y' are linked. Thus X' and Y' are not topologically ambient equivalent, which implies that X and Y are not ambient Lipschitz equivalent.

Remark 6.3.3 Notice that the tangent cones of both surfaces X and Y in Example 6.3 are topologically ambient equivalent to a cone over two unknotted circles, pinched at one point. Thus Sampaio's theorem does not apply, and we need the bridge construction in this example. Notice also that the bridge construction employed in this example allows one to construct examples of outer Lipschitz equivalent, topologically ambient equivalent but ambient Lipschitz non-equivalent surface singularities in $\mathbb{R}^4$ with the tangent cones as small as a single ray.

The surfaces X and Y in Example 6.3 differ by a "twist" of the (q, β)-bridge, which can be extended to a homeomorphism of the ambient space, but not to a bi-Lipschitz homeomorphism. One can iterate such a twist to obtain infinitely many ambient Lipschitz non-equivalent surfaces. On can also attach an additional knot to the links of both surfaces X and Y (see Fig. 6.8). This yields the following "universality" result (see [BG19] Theorem 4.1).

Theorem 6.3.4 *For any semialgebraic surface germ $S \subset \mathbb{R}^4$ there exist infinitely many semialgebraic surface germs $X_i \subset \mathbb{R}^4$ such that*

(1) *For all i, the germs X_i are topologically ambient equivalent to S;*
(2) *All germs X_i are outer Lipschitz equivalent;*
(3) *The tangent cones of all germs X_i at the origin are topologically ambient equivalent;*
(4) *For $i \neq j$ the germs X_i and X_j are not ambient Lipschitz equivalent.*

Other versions of universality can be formulated:

Fig. 6.8 The links of the
surfaces X and Y with an
extra knot attached

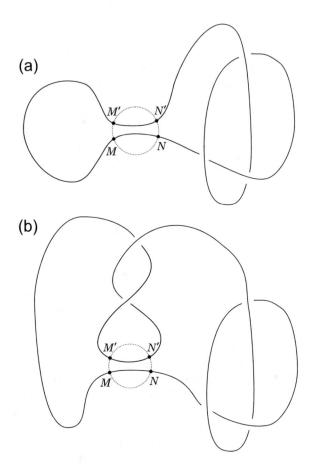

Theorem 6.3.5 (Universality Theorem) *For each knot $K \subset S^3$, there exists a germ at the origin of a semialgebraic surface $X_K \subset \mathbb{R}^4$ such that*

(1) *The germs X_K are outer Lipschitz equivalent for all knots K.*
(2) *The links of all germs X_K are topologically trivial knots in $S^3 \subset \mathbb{R}^4$.*
(3) *The germs X_{K_1} and X_{K_2} are ambient Lipschitz equivalent only if the knots K_1 and K_2 are topologically ambient equivalent.*

Proof Idea of the proof is illustrated in Fig. 6.9. The knot K is realized as a smooth semialgebraic circle embedded in S^3. Let L be a semialgebraic strip homeomorphic to $K \times [0, \epsilon]$ embedded in S^3 so that K is one of the two boundary curves of L. Let K' be the other boundary curve of L, so that the knots K and K' are isotopic. Let $C \subset \mathbb{R}^4$ be a cone over $K \cup K'$ with the vertex at the origin. Let y be a point such that the intersection of a small ball $B \subset S^3$ centered at y with $K \cup K'$ consists of two segments, $[a, b] \subset K$ and $[a', b'] \subset K'$. We can remove from C the cone C_B

Fig. 6.9 The link of the surface X_K in Theorem 6.3.5

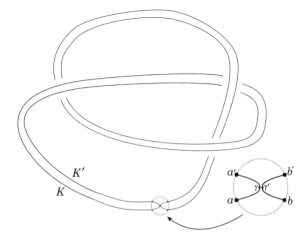

over $B \cap (K \cup K')$ and replace it by a semialgebraic surface G inside the cone over B, with the link shown in the zoomed part of Fig. 6.9, such that

(a) the link of G consists of two segments $[a, a']$ and $[b, b']$,
(b) the boundary of G is the same as the boundary of C_B,
(c) two arcs γ and γ' in G have the tangency order $\beta > 1$.

This results in the surface X_K with the link shown in Fig. 6.9. The link of X_K is a trivial knot (it is contractible to a point inside L). The link of the tangent cone of the surface X_K consists of two knots, each of them isotopic to K, pinched at a point corresponding to the common tangent line to γ and γ'. Thus the tangent cones of the surfaces X_K are not topologically ambient equivalent for the knots which are not topologically ambient equivalent. It follows from Sampaio's theorem that the same is true for the surfaces X_K themselves. $\square$

Remark 6.3.6 Theorem 6.3.5 is called "Universality Theorem" because it implies that the ambient Lipschitz classification problem for the surface germs in $\mathbb{R}^4$ in a single outer Lipschitz equivalence class, with the topologically ambient trivial links, contains all of the Knot Theory.

Acknowledgments Lev Birbrair research supported under CNPq 302655/2014-0 grant and by Capes-Cofecub. Andrei Gabrielov research supported by the NSF grant DMS-1665115.

References

[BFGG17] L. Birbrair, A. Fernandes, V. Grandjean, A. Gabrielov, Lipschitz contact equivalence of function germs in $\mathbb{R}^2$. Ann. Sc. Norm. Super. Pisa Cl. Sci. **17**(1), 81–92 (2017) 159
[BG19] L. Birbrair, A. Gabrielov, Ambient Lipschitz equivalence of real surface singularities. Int. Math. Res. Not. **20**, 6347–6361 (2019) 157, 158, 159, 160, 161, 163

[BM18] L. Birbrair, R. Mendes, Arc criterion of normal embedding, in *Singularities and Foliations. Geometry, Topology and Applications*. Springer Proceedings in Mathematics & Statistics, vol. 222 (Springer, Cham, 2018), pp. 549–553 159

[NP17] W.D. Neumann, A. Pichon, Lipschitz geometry does not determine embedded topological type, in *Singularities in Geometry, Topology, Foliations and Dynamics*, Trends in Mathematics (Birkhäuser/Springer, Cham, 2017), pp. 183–195 158

[Par88] A. Parusiński, Lipschitz properties of semi-analytic sets. Ann. Inst. Fourier **38**(4), 189–213 (1988) 159

[Sam16] J.E. Sampaio, Bi-Lipschitz homeomorphic subanalytic sets have bi-Lipschitz homeomorphic tangent cones. Sel. Math. **22**(2), 553–559 (2016) 158

Chapter 7
An Introduction to Lipschitz Geometry of Complex Singularities

Anne Pichon

Abstract The aim of this paper to introduce the reader to a recent point of view on the Lipschitz classifications of complex singularities. It presents the complete classification of Lipschitz geometry of complex plane curves singularities and in particular, it introduces the so-called bubble trick, which is a key tool to study Lipschitz geometry of germs. It describes also the thick-thin decomposition of a normal complex surface singularity and built two geometric decompositions of a normal surface germ into standard pieces which are invariant by respectively inner and outer bilipschitz homeomorphisms. This leads in particular to the complete classification of Lipschitz geometry for the inner metric.

7.1 Introduction

The aim of this paper is to introduce the reader to a recent point of view on the Lipschitz classifications of complex singularities.

The notes are structured as follows. Section 7.2 explains what is Lipschitz geometry for the inner and outer metrics of singularities and why it is interesting for the classification of space singularities. Section 7.3 gives the complete classification of Lipschitz geometry of complex curves and covers the results of [NP14]. In particular, it introduces the so-called bubble trick, which is a key tool to study Lipschitz geometry of germs. Section 7.4 describes the thick-thin decomposition of a normal complex surface germ following [BNP14]. Section 7.5 describes two geometric decompositions of a normal surface germ into standard pieces which are invariant by respectively inner and outer bilipschitz homeomorphism, following the results of [BNP14] and [NP12]. This leads in particular to the complete classification of Lipschitz geometry for the inner metric.

A. Pichon (✉)
Aix Marseille Univ, CNRS, Centrale Marseille, I2M, Marseille, France
e-mail: anne.pichon@univ-amu.fr

© The Author(s), under exclusive license to Springer Nature Switzerland AG 2020
W. Neumann, A. Pichon (eds.), *Introduction to Lipschitz Geometry of Singularities*,
Lecture Notes in Mathematics 2280, https://doi.org/10.1007/978-3-030-61807-0_7

The paper contains a lot of detailed examples which were presented and discussed during the afternoon exercise sessions of the school and also an appendix (Part 7.6) which gives the computation of the resolution graph of a surface singularity with equation $x^2 + f(y, z) = 0$ following Hirzebruch–Jung and Laufer's method. This enables the readers to produce a lot of examples by themself.

In these notes, I do not give the detailed proofs of the invariance of the inner and outer Lipschitz decompositions (Theorems 7.5.30 and 7.5.36). We refer to [BNP14] and [NP12] respectively. However, it has to be noted that even if the two statements look similar, the techniques used in the proofs are radically different. The invariance of the inner decomposition uses the Lipschitz invariance of fast loops (introduced in Sect. 7.4) of minimal length in their homology class ([BNP14, Section 14]) while that of the outer invariance uses sophisticated bubble trick arguments [NP12].

Notice that the pioneering paper [BNP14] is written for a normal complex surface, as well as the initial version of [NP12]. However, the extensions of the inner and outer geometric decompositions to the general case of a reduced singularity (not necessarily isolated) are fairly easy. A version of [NP12] in this general setting will appear soon.

Finally, notice that the inner and outer geometric decompositions are the analogs of the pizza decompositions of a real surface germ presented in the lecture notes of Maria Aparecida Ruas (for inner metric) and Lev Birbrair (for the outer metric) in the present volume. In the real surface case, these decompositions give complete classifications for the inner and outer metrics. As already mentioned, the inner decomposition in the complex case also gives a complete classification after adding a few more invariants (Theorem 7.5.30). In contrast, a complete classification for the outer metric of complex surface singularities would need more work and is still an open question.

7.2 Preliminaries

7.2.1 What is Lipschitz Geometry of Singular Spaces?

In the sequel, $\mathbb{K}$ will denote either $\mathbb{R}$ or $\mathbb{C}$.

Let $(X, 0)$ be a germ of analytic space in $\mathbb{K}^n$ which contains the origin. So X is defined by

$$X = \{(x_1, \ldots, x_n) \in \mathbb{K}^n \mid f_j(x_1, \ldots, x_n) = 0, j = 1, \ldots, r\},$$

where the f_j's are convergent power series, $f_j \in \mathbb{K}\{x_1, \ldots, x_n\}$ and $f_j(0) = 0$.

Question 7.1 How does X look in a small neighbourhood of the origin?

There are multiple answers to this vague question depending on the category we work in, i.e., on the chosen equivalence relation between germs.

First, we can consider the topological equivalence relation:

Definition 7.2.1 Two analytic germs $(X, 0)$ and $(X', 0)$ are **topologically equivalent** if there exists a germ of homeomorphism $\psi \colon (X, 0) \to (X', 0)$. The **topological type** of $(X, 0)$ is the equivalence class of $(X, 0)$ for this equivalence relation.

Two analytic germs $(X, 0) \subset (\mathbb{K}^n, 0)$ and $(X', 0) \subset (\mathbb{K}^n, 0)$ are **topologically equisingular** if there exists a germ of homeomorphism $\psi \colon (\mathbb{K}^n, 0) \to (\mathbb{K}^n, 0)$ such that $\psi(X) = X'$. We call **embedded topological type** of $(X, 0)$ the equivalence class of $(X, 0)$ for this equivalence relation.

The embedded topological type of $(X, 0) \subset (\mathbb{R}^n, 0)$ is completely determined by the embedded topology of its link as stated in the following famous Conical Structure Theorem, presented in Sect. 7.4 of the course of José Luis Cisneros Molina in the present volume. Let us recall its statement.

Theorem 7.2.2 (Conical Structure Theorem) *Let B_ϵ^n be the ball with radius $\epsilon > 0$ centered at the origin of $\mathbb{R}^n$ and let S_ϵ^{n-1} be its boundary.*

Let $(X, 0) \subset (\mathbb{R}^n, 0)$ be an analytic germ. For $\epsilon > 0$, set $X^{(\epsilon)} = S_\epsilon^{n-1} \cap X$. There exists $\epsilon_0 > 0$ such that for every $\epsilon > 0$ with $0 < \epsilon \leq \epsilon_0$, the pair $(B_\epsilon^n, X \cap B_\epsilon^n)$ is homeomorphic to the pair $(B_{\epsilon_0}^n, Cone(X^{(\epsilon_0)}))$, where $Cone(X^{(\epsilon_0)})$ denotes the cone over $X^{(\epsilon_0)}$, i.e., the union of the segments $[0, x]$ joining the origin to a point $x \in X^{(\epsilon_0)}$.

In other words, the homeomorphism class of the pair $(S_\epsilon^{n-1}, X^{(\epsilon)})$ does not depend on ϵ when $\epsilon > 0$ is sufficiently small and it determines completely the embedded topological type of $(X, 0)$.

Definition 7.2.3 When $0 < \epsilon \leq \epsilon_0$, the intersection $X^{(\epsilon)}$ is called the **link** of $(X, 0)$.

Example 7.2.4

1. Assume that X is the real cusp in $\mathbb{R}^2$ with equation $x^3 - y^2 = 0$. Then its link at 0 consists of two points in the circle S_ϵ^1.
2. If X is the complex cusp in $\mathbb{C}^2$ with equation $x^3 - y^2 = 0$, its link at 0 is the trefoil knot in the 3-sphere S_ϵ^3.
3. If X is the complex surface E_8 in $\mathbb{C}^3$ with equation $x^2 + y^3 + z^5 = 0$, its (non embedded) link at 0 is a Seifert manifold whose homeomorphism class is completely described through plumbing theory by its minimal resolution graph. See the course of Walter Neumann in the present volume for more details. The resolution graph is explicitely computed in appendix section 7.6 of the present notes.

The Conical Structure Theorem gives a complete answer to Question 7.1 in the topological category, but it completely ignores the geometric properties of the set $(X, 0)$. In particular, a very interesting question is:

Question 7.2 How does the link $X^{(\epsilon)}$ evolve metrically as ϵ tends to 0?

In other words, is $X \cap B_\epsilon$ bilipschitz homeomorphic to the straight cone $Cone(X^{(\epsilon)})$? Or are there some parts of $X^{(\epsilon)}$ which shrink faster than linearly when ϵ tends to 0?

This question can be studied from different points of view depending on the choice of the metric. If $(X, 0) \subset (\mathbb{R}^n, 0)$ is the germ of a real analytic space, there are two natural metrics on $(X, 0)$ which are defined from the Euclidean metric of the ambient space $\mathbb{R}^n$:

Definition 7.2.5 The **outer metric** d_o on X is the metric induced by the ambient Euclidean metric, i.e., for all $x, y \in X$, $d_o(x, y) = \|x - y\|_{\mathbb{R}^n}$.

The **inner metric** d_i on X is the length metric defined for all $x, y \in X$ by: $d_i(x, y) = \inf length(\gamma)$, where $\gamma : [0, 1] \to X$ varies among rectifyable arcs on X such that $\gamma(0) = x$ and $\gamma(1) = y$.

Definition 7.2.6 Let (M, d) and (M', d') be two metric spaces. A map $f : M \to M'$ is a **bilipschitz homeomorphism** if f is a bijection and there exists a real constant $K \geq 1$ such that for all $x, y \in M$,

$$\frac{1}{K}d(x, y) \leq d'(f(x), f(y)) \leq Kd(x, y).$$

Definition 7.2.7 Two real analytic germs $(X, 0) \subset (\mathbb{R}^n, 0)$ and $(X', 0) \subset (\mathbb{R}^m, 0)$ are **inner Lipschitz equivalent** (resp. **outer Lipschitz equivalent**) if there exists a germ of bilipschitz homeomorphism $\psi : (X, 0) \to (X', 0)$ with respect to the inner (resp. outer) metrics.

The equivalence classes of the germ $(X, 0) \subset (\mathbb{R}^n, 0)$ for these equivalence relations are called respectively the **inner Lipschitz geometry** and the **outer Lipschitz geometry** of $(X, 0)$.

Throughout these notes, we will use the "big-Theta" asymptotic notations of Bachmann–Landau in the following form:

Definition 7.2.8 Given two function germs $f, g : ([0, \infty), 0) \to ([0, \infty), 0)$, we say that f is **big-Theta** of g and we write $f(t) = \Theta(g(t))$ if there exist real numbers $\eta > 0$ and $K > 0$ such that for all t such that for all $t \in [0, \eta)$, $\frac{1}{K}g(t) \leq f(t) \leq Kg(t)$.

Example 7.2.9 Consider the real cusp C with equation $y^2 - x^3 = 0$ in $\mathbb{R}^2$ (see Fig. 7.1). For a real number $t > 0$, consider the two points $p_1(t) = (t^2, t^3)$ and $p_2(t) = (t^2, -t^3)$ on C. Then $d_o(p_1(t), p_2(t)) = \Theta(t^{3/2})$ while the inner distance is obtained by taking infimum of lengths of paths on C between the two points $p_1(t)$ and $p_2(t)$. The shortest length is obtained by taking a path going through the origin, and we have $d_i(p_1(t), p_2(t)) = \Theta(t)$. Therefore, taking the limit of the quotient as t tends to 0, we obtain:

$$\frac{d_o(p_1(t), p_2(t))}{d_i(p_1(t), p_2(t))} = \Theta(t^{1/2}) \to 0.$$

Fig. 7.1 The real cusp
$y^2 - x^3 = 0$

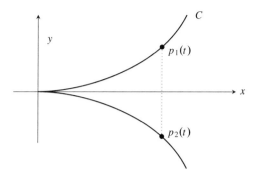

Using this, you are ready to make the following:

Exercise 7.2.10

1. Prove that there is no bilipschitz homeomorphism between the outer and inner metrics on the real cusp C with equation $y^2 - x^3 = 0$ in $\mathbb{R}^2$.
2. Prove that $(C, 0)$ equipped with the inner metric is metrically conical, i.e. bilipschitz equivalent to the cone over its link.

Example 7.2.11 Consider the real surface S in $\mathbb{R}^3$ with equation $x^2 + y^2 - z^3 = 0$ in $\mathbb{R}^2$. For a real number $t > 0$, consider the two points $p_1(t) = (t^3, 0, t^2)$ and $p_2(t) = (t^3, 0, -t^2)$ on S. Then $d_o(p_1(t), p_2(t)) = \Theta(t^{3/2})$. We also have $d_i(p_1(t), p_2(t)) = \Theta(t^{3/2})$ since $d_i(p_1(t), p_2(t))$ is the length of a half-circle joining $p_1(t)$ and $p_2(t)$ on the circle $\{x = t^3\} \cap S$.

Exercise 7.2.12 Consider the real surface S of Example 7.2.11.

1. Prove that the identity map is a bilipschitz homeomorphism between the outer and inner metrics on $(S, 0)$.
2. Prove that $(S, 0)$ equipped with the inner metric is not metrically conical.

7.2.2 Independence of the Embedding and Motivations

If $(X, 0)$ is a germ of a real analytic space, the two metrics d_o and d_i defined above obviously depend on the choice of an embedding $(X, 0) \subset (\mathbb{R}^n, 0)$ since they are defined by using the Euclidean metric of the ambient $\mathbb{R}^n$. The aim of this section is to give a proof of one of the main results which motivates the study of Lipschitz geometry of singularities:

Proposition 7.2.13 *The Lipschitz geometries of $(X, 0)$ for the outer and inner metrics are independent of the embedding $(X, 0) \subset (\mathbb{R}^n, 0)$.*

In other words, bilipschitz classes of $(X, 0)$ just depend on the analytic type of $(X, 0)$. Before proving this result, let us give some consequences which motivate the study of Lipschitz geometry of germs of singular spaces.

The outer Lipschitz geometry determines the inner Lipschitz geometry since the inner metric is determined by the outer one through integration along paths. Moreover, the inner Lipschitz geometry obviously determines the topological type of $(X, 0)$. Therefore, an important consequence of Proposition 7.2.13 is that the Lipschitz geometries give two intermediate classifications between the analytical type and the topological type.

A very small amount of analytic invariants are determined by the topological type of an analytic germ (even if one considers the embedded topological type). In particular, a natural question is to ask whether the Lipschitz classification is sufficiently rigid to catch analytic invariants:

Question 7.3 Which analytical invariants are in fact Lipschitz invariants?

Recent results show that in the case of a complex surface singularity, a large amount of analytic invariants are determined by the outer Lipschitz geometry. For example, the multiplicity of a complex surface singularity is an outer Lipschitz invariant ([NP12] for a normal surface, [Sam17] for a hypersurface in $\mathbb{C}^3$ and [FdBFS18] for the general case). However it is now known that the multiplicity is not a Lipschitz invariant in higher dimensions [BFSV18]. In [NP12] it is shown that many other data are in fact Lipschitz invariants in the case of surface singularities, such as the geometry of hyperplane sections and the geometry of polar curves and discriminant curves of generic projections (Theorem 7.5.38); higher dimensions remain almost unexplored. This shows that the outer Lipschitz class contains potentially a lot of information on the singularity and that outer Lipschitz geometry of singularities is a very promising area to explore.

Here is another motivation. Analytic types of singular space germs contain continuous moduli, and this is why it is difficult to describe a complete analytic classification. For example, consider the family of curves germs $(X_t, 0)_{t \in \mathbb{C}}$ where X_t is the union of four transversal lines with equation $xy(x - y)(x - ty) = 0$. For every pair (t, t') with $t \neq t'$, $(X_t, 0)$ is not analytically equivalent to $(X_{t'}, 0)$. On the contrary, it is known since the works of T. Mostowki in the complex case [Mos85], and Parusiński in the real case [Par88] and [Par94], that the outer Lipschitz classification of germs of singular spaces is **tame**, which means that it admits a discrete complete invariant. Then a complete classification of Lipschitz geometry of singular spaces seems to be a more reachable goal.

Proof of Proposition 7.2.13 Let $(f_1, \ldots, f_n)$ and $(g_1, \ldots, g_m)$ be two systems of generators of the maximal ideal $\mathcal{M}$ of $(X, 0)$. We will first prove that the outer metrics d_I and d_J for the embeddings

$$I = (f_1, \ldots, f_n): (X, 0) \to (\mathbb{R}^n, 0) \quad \text{and} \quad J = (g_1, \ldots, g_m): (X, 0) \to (\mathbb{R}^m, 0)$$

are bilipschitz equivalent. It suffices to prove that the outer metric for the embedding $(f_1, \ldots, f_n, g_1, \ldots, g_m)$ is bilipschitz equivalent to the metric d_I. By induction, we just have to prove that for any $g \in \mathcal{M}$, the metric $d_{I'}$ associated with the embedding $I' = (f_1, \ldots, f_n, g): (X, 0) \to (\mathbb{R}^{n+1}, 0)$ is bilipschitz equivalent to d_I.

Since g is in the ideal $\mathcal{M}$, it may be expressed as $G(f_1, \ldots, f_n)$ where $G: (\mathbb{R}^n, 0) \to (\mathbb{R}, 0)$ is real analytic. Let Γ be the graph of the function $G(x_1, \ldots, x_n)$ in $(\mathbb{R}^n, 0) \times \mathbb{R}$. It is defined over a neighbourhood of 0 in $\mathbb{R}^n$. The projection $\pi: \Gamma \to (\mathbb{R}^n, 0)$ is bilipschitz over any compact neighbourhood of 0 in $\mathbb{R}^n$ on which it is defined. We have $I'(X, 0) \subset \Gamma \subset (\mathbb{R}^n, 0) \times \mathbb{R}$, so $\pi|_{I'(X,0)}: I'(X, 0) \to I(X, 0)$ is bilipschitz for the outer metrics $d_{I'}$ and d_I.   □

7.3 The Lipschitz Geometry of a Complex Curve Singularity

7.3.1 Complex Curves Have Trivial Inner Lipschitz Geometry

Let $X \subset \mathbb{C}^2$ be the complex cusp with equation $y^2 - x^3 = 0$. Let $t \in \mathbb{R}$ and consider the two points $p_1(t) = (t^2, t^3)$ and $p_2(t) = (t^2, -t^3)$ on X. Since these two points are on two distinct strands of the braid $X \cap (S^1_{|t|} \times \mathbb{C})$, it is easy to see that the shortest path in X from $p_1(t)$ to $p_2(t)$ passes through the origin and that $d_i(p_1(t), p_2(t)) = \Theta(t)$. This suggests that $(X, 0)$ is locally inner bilipschitz homeomorphic to the cone over its link. This means that the inner Lipschitz geometry tells one no more than the topological type, i.e., the number of connected components of the link (which are circles), and is therefore uninteresting. The aim of this section is to prove this for any complex curve.

Definition 7.3.1 An analytic germ $(X, 0)$ is **metrically conical** if it is inner Lipschitz homeomorphic to the straight cone over its link.

In this paper, a **complex curve germ** or **complex curve singularity** will mean a germ of reduced complex analytic space of dimension 1.

Proposition 7.3.2 *Any complex space curve germ $(C, 0) \subset (\mathbb{C}^N, 0)$ is metrically conical.*

Proof Take a linear projection $p: \mathbb{C}^N \to \mathbb{C}$ which is generic for the curve $(C, 0)$ (i.e., its kernel contains no tangent line of C at 0) and let $\pi := p|_C$, which is a branched cover of germs. Let $D_\epsilon = \{z \in \mathbb{C} : |z| \le \epsilon\}$ with ϵ small, and let E_ϵ be the part of C which branched covers D_ϵ. Since π is holomorphic away from 0 we have a local Lipschitz constant $K(x)$ at each point $x \in C \backslash \{0\}$ given by the absolute value of the derivative map of π at x. On each branch γ of C this $K(x)$ extends continuously over 0 by taking for $K(0)$ the absolute value of the restriction $p|_{T_0\gamma}: T_0\gamma \to \mathbb{C}$ where $T_0\gamma$ denotes the tangent cone to γ at 0. So the infimum and supremum K^- and K^+ of $K(x)$ on $E_\epsilon \backslash \{0\}$ are defined and positive. For any arc γ in E_ϵ which is smooth except where it passes through 0 we have $K^-\ell(\gamma) \le \ell'(\gamma) \le K^+\ell(\gamma)$, where ℓ respectively ℓ' represent arc length using inner metric on e_ϵ respectively the metric lifted from B_ϵ. Since E_ϵ with the latter metric is strictly conical, we are done.   □

7.3.2   The Outer Lipschitz Geometry of a Complex Curve

Let $\mathbf{G}(n-2, \mathbb{C}^n)$ be the Grassmanian of $(n-2)$-planes in $\mathbb{C}^n$.

Let $\mathcal{D} \in \mathbf{G}(n-2, \mathbb{C}^n)$ and let $\ell_\mathcal{D} : \mathbb{C}^n \to \mathbb{C}^2$ be the linear projection with kernel $\mathcal{D}$. Suppose $(C, 0) \subset (\mathbb{C}^n, 0)$ is a complex curve germ. There exists an open dense subset Ω_C of $\mathbf{G}(n-2, \mathbb{C}^n)$ such that for $\mathcal{D} \in \Omega_C$, $\mathcal{D}$ contains no limit of secant lines to the curve C ([Tei82, pp. 354]).

Definition 7.3.3 The projection $\ell_\mathcal{D}$ is said to be **generic for** C if $\mathcal{D} \in \Omega_C$.

In the sequel, we will use extensively the following result

Theorem 7.3.4 ([Tei82, pp. 352–354]) *If $\ell_\mathcal{D}$ is a generic projection for C, then the restriction $\ell_\mathcal{D}|_C : C \to \ell_\mathcal{D}(C)$ is a bilipschitz homeomorphism for the outer metric.*

As a consequence of Theorem 7.3.4, in order to understand Lipschitz geometry of curve germs, it suffices to understand Lipschitz geometry of plane curve germs.

Let us start with an example.

Example 7.3.5 Consider the plane curve germ $(C, 0)$ with two branches having Puiseux expansions

$$y = x^{3/2} + x^{13/6}, \quad y = x^{5/2}.$$

Its topological type is completely described by the sets of characteristic exponents of the branches: $\{3/2, 13/6\}$ and $\{5/2\}$ and by the contact exponents between the two branches: $3/2$. Those data are summarized in the Eggers-Wall tree of the curve germ (see [Wal04, GBGPPP19]), or equivalently, in what we will call the **carrousel tree** (see the proof of Lemma 7.3.8 and Fig. 7.2), which is exactly the Kuo-Lu tree defined in [KL77] but with the horizontal bars contracted to points.

Now, for small $t \in \mathbb{R}^+$, consider the intersection $C \cap \{x = t\}$. This gives 8 points $p_i(t), i = 1 \ldots, 8$ and then, varying t, this gives 8 real semi-analytic arcs $p_i : [0, 1) \to X$ such that $p_i(0) = 0$ and $\|p_i(t)\| = \Theta(t)$.

Figure 7.3 gives pictures of sections of C with complex lines $x = 0.1, 0.05, 0.025$ and 0. The central two-points set corresponds to the branch $y = x^{5/2}$ while the two lateral three-points sets correspond to the other branch.

It is easy to see on this example that for each pair (i, j) with $i \neq j$, we have $d_o(p_i(t), p_j(t)) = \Theta(t^{q(i,j)})$ where $q(i, j) \in \mathbb{Q}^+$ and that the set of such $q(i, j)$'s is exactly the set of essential exponents $\{3/2, 13/6, 5/2\}$. This shows that one can recover the essential exponents by measuring the outer distance between points of C.

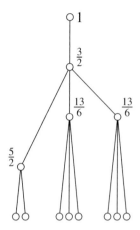

Fig. 7.2 The carrousel tree

0.1

0.05

0.025

0

Fig. 7.3 Sections of C

More generally, we will show that we can actually recover the carrousel tree by measuring outer distances on X even after a bilipschitz change of the metric. Conversely, the outer Lipschitz geometry of a plane curve is determined by its embedded topological type. This gives the complete classification of the outer geometry of complex plane curve germs:

Theorem 7.3.6 *Let $(E_1, 0) \subset (\mathbb{C}^2, 0)$ and $(E_2, 0) \subset (\mathbb{C}^2, 0)$ be two germs of complex curves. The following are equivalent:*

1. *$(E_1, 0)$ and $(E_2, 0)$ have same outer Lipschitz geometry.*
2. *there is a meromorphic germ $\phi \colon (E_1, 0) \rightarrow (E_2, 0)$ which is a bilipschitz homeomorphism for the outer metric;*
3. *$(E_1, 0)$ and $(E_2, 0)$ have the same embedded topological type;*
4. *there is a bilipschitz homeomorphism of germs $h \colon (\mathbb{C}^2, 0) \rightarrow (\mathbb{C}^2, 0)$ with $h(E_1) = E_2$.*

As a corollary of Theorems 7.3.4 and 7.3.6, we obtain:

Corollary 7.3.7 *The outer Lipschitz geometry of a curve germ* $(C, 0) \subset (\mathbb{C}^N, 0)$ *determines and is determined by the embedded topological type of any generic linear projection* $(\ell(C), 0) \subset (\mathbb{C}^2, 0)$.

The equivalence of (1), (3), and (4) of Theorem 7.3.6 is proved in [NP14]. The equivalence of (2) and (3) was first proved by Pham and Teissier [PT69] (published in the present volume) by developing the theory of Lipschitz saturation and revisited by Fernandes in [Fer03]. In the present lecture notes, we will give the proof of (1) $\Rightarrow$ (3), since it is based on the so-called *bubble trick* argument which can be considered as a prototype for exploring Lipschitz geometry of singular spaces in various settings. Another more sophisticated bubble trick argument is developed in [NP12] to study Lipschitz geometry of surface germs (namely in the proof of Theorem 7.5.36).

Proof of (1) $\Rightarrow$ (3) ***of Theorem 7.3.6*** We want to prove that the embedded topological type of a plane curve germ $(C, 0) \subset (\mathbb{C}^2, 0)$ is determined by the outer Lipschitz geometry of $(C, 0)$.

We first prove this using the analytic structure and the outer metric on $(C, 0)$. The proof is close to Fernandes' approach in [Fer03]. We then modify the proof to make it purely topological and to allow a bilipschitz change of the metric.

The tangent cone to C at 0 is a union of lines $L^{(j)}$, $j = 1, \ldots, m$, and by choosing our coordinates we can assume they are all transverse to the y-axis.

There is $\epsilon_0 > 0$ such that for every $\epsilon \in (0, \epsilon_0]$, the curve C meets transversely the set

$$T_\epsilon := \{(x, y) \in \mathbb{C}^2 : |x| = \epsilon\}.$$

Let M be the multiplicity of C. The hypothesis of transversality to the y-axis means that the lines $x = t$ for $t \in (0, \epsilon_0]$ intersect C in M points $p_1(t), \ldots, p_M(t)$. Those points depend continuously on t. Denote by $[M]$ the set $\{1, 2, \ldots, M\}$. For each $j, k \in [M]$ with $j < k$, the distance $d(p_j(t), p_k(t))$ has the form $O(t^{q(j,k)})$, where $q(j, k) = q(k, j) \in \mathbb{Q} \cap [1, +\infty)$ is either a characteristic Puiseux exponent for a branch of the plane curve C or a coincidence exponent between two branches of C in the sense of e.g., [TMW89]. We call such exponents **essential**.

For $j \in [M]$, define $q(j, j) = \infty$. $\qquad\qquad\qquad\qquad\qquad\qquad\qquad\qquad$ $\square$

Lemma 7.3.8 *The map* $q : [M] \times [M] \to \mathbb{Q} \cup \{\infty\}$, $(j, k) \mapsto q(j, k)$, *determines the embedded topology of* C.

Proof To prove the lemma we will construct from q the so-called *carrousel tree*. Then, we will show that it encodes the same data as the Eggers tree. This implies that it determines the embedded topology of C.

The $q(j, k)$ have the property that $q(j, l) \geq min(q(j, k), q(k, l))$ for any triple j, k, l. So for any $q \in \mathbb{Q} \cup \{\infty\}$, $q > 0$, the binary relation on the set $[M]$ defined by $j \sim_q k \Leftrightarrow q(j, k) \geq q$ is an equivalence relation.

Name the elements of the set $q([M] \times [M]) \cup \{1\}$ in decreasing order of size: $\infty = q_0 > q_1 > q_2 > \cdots > q_s = 1$. For each $i = 0, \ldots, s$ let $G_{i,1}, \ldots, G_{i,M_i}$ be the equivalence classes for the relation $\sim_{q_i}$. So $M_0 = M$ and the sets $G_{0,j}$ are singletons while $M_s = 1$ and $G_{s,1} = [M]$. We form a tree with these equivalence classes $G_{i,j}$ as vertices, and edges given by inclusion relations: the singleton sets $G_{0,j}$ are the leaves and there is an edge between $G_{i,j}$ and $G_{i+1,k}$ if $G_{i,j} \subseteq G_{i+1,k}$. The vertex $G_{s,1}$ is the root of this tree. We weight each vertex with its corresponding q_i.

The **carrousel tree** is the tree obtained from this tree by suppressing valence 2 vertices (i.e., vertices with exactly two incident edges): we remove each such vertex and amalgamate its two adjacent edges into one edge. We follow the computer science convention of drawing the tree with its root vertex at the top, descending to its leaves at the bottom (see Fig. 7.2).

At any non-leaf vertex v of the carrousel tree we have a weight q_v, $1 \leq q_v \leq q_1$, which is one of the q_i's. We write it as m_v/n_v, where n_v is the lcm of the denominators of the q-weights at the vertices on the path from v up to the root vertex. If v' is the adjacent vertex above v along this path, we put $r_v = n_v/n_{v'}$ and $s_v = n_v(q_v - q_{v'})$. At each vertex v the subtrees cut off below v consist of groups of r_v isomorphic trees, with possibly one additional tree. We label the top of the edge connecting to this additional tree at v, if it exists, with the number r_v, and then delete all but one from each group of r_v isomorphic trees below v. We do this for each non-leaf vertex of the carrousel tree. The resulting tree, with the q_v labels at vertices and the extra label on a downward edge at some vertices is easily recognized as a mild modification of the Eggers tree: there is a natural action of the Galois group whose quotient is the Eggers tree. □

As already noted, this reconstruction of the embedded topology involved the complex structure and the outer metric. We must show that we can reconstruct it without using the complex structure, even after applying a bilipschitz change to the outer metric. We will use what we call a **bubble trick**.

Recall that the tangent cone of C is a union of lines $L^{(j)}$. We denote by $C^{(j)}$ the part of C tangent to the line $L^{(j)}$. It suffices to recover the topology of each $C^{(j)}$ independently, since the $C^{(j)}$'s are distinguished by the fact that the distance between any two of them outside a ball of radius ϵ around 0 is $\Theta(\epsilon)$, even after bilipschitz change of the metric. We therefore assume from now on that the tangent cone of C is a single complex line.

We now arrive at a crucial moment of the proof and of the paper.

The Bubble Trick The points $p_1(t), \ldots, p_M(t)$ which we used in order to find the numbers $q(j, k)$ were obtained by intersecting C with the line $x = t$. The arc $p_1(t)$, $t \in [0, \epsilon_0]$ satisfies $d(0, p_1(t)) = \Theta(t)$. Moreover, the other points $p_2(t), \ldots, p_M(t)$ are in the transverse disk of radius rt centered at $p_1(t)$ in the plane $x = t$. Here r can be as small as we like, so long as ϵ_0 is then chosen sufficiently small.

Instead of a transverse disk of radius rt, we can use a ball $B(p_1(t), rt)$ of radius rt centered at $p_1(t)$. This ball $B(p_1(t), rt)$ intersects C in M disks

$D_1(t), \ldots, D_M(t)$, and we have $d(D_j(t), D_k(t)) = \Theta(t^{q(j,k)})$, so we still recover the numbers $q(j, k)$. In fact, the ball in the outer metric on C of radius rt around $p_1(t)$ is $B_C(p_1(t), rt) := C \cap B(p_1(t), rt)$, which consists of these M disks $D_1(t), \ldots, D_M(t)$.

We now replace the arc $p_1(t)$ by any continuous arc $p_1'(t)$ on C with the property that $d(0, p_1'(t)) = \Theta(t)$. If r is sufficiently small it is still true that $B_C(p_1'(t), rt)$ consists of M disks $D_1'(t), \ldots, D_M'(t)$ with $d(D_j'(t), D_k'(t)) = \Theta(t^{q(j,k)})$. So at this point, we have gotten rid of the dependence on analytic structure in discovering the topology, but not yet of the dependence on the outer geometry.

Let now d' be a metric on C such that the identity map is a K-bilipschitz homeomorphism in a neighbourhood of the origin. We work inside this neighbourhood, taking t, ϵ_0 and r sufficiently small. $B'(p, \eta)$ will denote the ball in C for the metric d' centered at $p \in C$ with radius $\eta \geq 0$.

The bilipschitz change of the metric may disintegrate the balls in many connected components, as sketched on Fig. 7.4, where the round ball $B_C(p_1'(t), rt)$ has 3 components (3 is the mulitplicity of C), while $B'(p_1'(t), rt)$ has 6 components (for clarity of the picture, we draw the ball $B'(p_1'(t), rt)$ as if the distance d' were induced by an ambient metric, but this is not the case in general).

If we try to perform the same argument as before using the balls $B'(p_1'(t), rt)$ instead of $B_C(p_1'(t), rt)$, we get a problem since $B'(p_1'(t), rt)$ may have many irrelevant components and we can no longer simply use distance between connected components. To resolve this, we consider the two balls $B_1'(t) = B'(p_1'(t), \frac{rt}{K^3})$ and $B_2'(t) = B'(p_1'(t), \frac{rt}{K})$, we have the inclusions:

$$B_C\left(p_1'(t), \frac{rt}{K^4}\right) \subset B_1'(t) \subset B_C\left(p_1'(t), \frac{rt}{K^2}\right) \subset B_2'(t) \subset B_C\left(p_1'(t), rt\right)$$

Using these inclusions, we obtain that only M components of $B_2'(t)$ intersect $B_1'(t)$ and that naming these components $D_1'(t), \ldots, D_M'(t)$ again, we still have $d(D_j'(t), D_k'(t)) = \Theta(t^{q(j,k)})$ so the $q(j, k)$ are determined as before (prove this as an exercise). See Fig. 7.5 for a schematic picture of the situation (again, for clarity

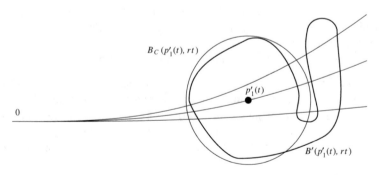

Fig. 7.4 Change of the metric

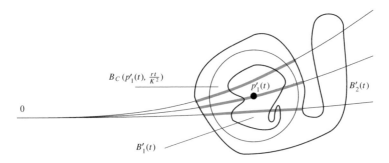

$B_C\,(p'_1(t),\,\frac{r_1}{K^2})$

$p'_1(t)$

$B'_2(t)$

0

$B'_1(t)$

Fig. 7.5 The bubble trick

of the picture, we draw the balls $B'_1(t)$ and $B'_2(t)$ as if the distance d' were induced by an ambient metric, but this is not the case in general).

7.3.3   The Bubble Trick with Jumps

The "bubble trick" introduced in the proof of Theorem 7.3.6 is a powerful tool to capture invariants of Lipschitz geometry of a complex curve germ. However, this first version of the bubble trick is not well adapted to explore the outer Lipschitz geometry of a singular space of dimension ≥ 2 for the following reason. In the case of a plane curve germ $(C, 0)$, the bubble trick is based on the fact that the distance orders between points of $\ell^{-1}(t) \cap C$ with respect to $t \in \mathbb{R}$ are Lipschitz invariants, where $\ell \colon (C, 0) \to (\mathbb{C}, 0)$ denotes a generic projection of the curve germ. Now, assume that $(X, 0)$ is a complex surface germ with multiplicity $m \geq 2$ (so it has a singularity at 0), and consider a generic projection $\ell \colon (X, 0) \to (\mathbb{C}^2, 0)$. Then the critical locus of ℓ is a curve germ $(\Pi_\ell, 0) \subset (X, 0)$ called the **polar curve**, and its image $\Delta_\ell = \ell(\Pi_\ell)$ is a curve germ $(\Delta_\ell, 0) \subset (\mathbb{C}^2, 0)$ called the **discriminant curve** of ℓ. Let $x \in \mathbb{C}^2 \setminus \{0\}$. The number of points in $\ell^{-1}(x) \cap C$ depends on x: it equals $m - 1$ if $x \in \Delta_\ell$, where m denotes the multiplicity of $(X, 0)$, and m otherwise. Moreover, consider a semialgebraic real arc germ $p \colon t \in [0, \eta) \mapsto p(t) \in \mathbb{C}^2$ such that $\|p(t)\| = |t|$ and $\forall t \neq 0$, $p(t) \notin \Delta_\ell$; then the distance orders between the m points $p_1(t), \ldots, p_m(t)$ of $\ell^{-1}(p(t))$ will depend on the position of the arc $p(t)$ with respect to the curve Δ_ℓ. So the situation is much more complicated, even in dimension 2.

In [NP14], we use an adapted version of the bubble trick which enables us to explore the outer Lipschitz geometry of a complex surface $(X, 0)$. We call it the **bubble trick with jumps**. Roughly speaking, it consists in using horns

$$\mathcal{H}(p(t), r|t|^q) = \bigcup_{t \in [0,1)} B((p(t), r|t|^q),$$

where $B(x, a)$ denotes the ball in X with center x and radius a and where $p(t)$ is a real arc on $(X, 0)$ such that $||p(t)|| = \Theta(t)$ and $r \in]0, +\infty[$, and in exploring "jumps" in the topology of $\mathcal{H}(p(t), a|t|^q)$ when q varies from $+\infty$ to 1, for example, jumps of the number of connected components of $\mathcal{H}(p(t), r|t|^q) \setminus \{0\}$.

In order to give a flavour of this bubble trick with jumps, we will perform it on a plane curve germ, giving an alternative proof of the implication $(1) \Rightarrow (3)$ of Theorem 7.3.6.

The Bubble Trick with Jumps

We use again the notations of the version 1 of the bubble trick from the proof of Theorem 7.3.6. Let $(C, 0)$ be a plane curve germ with multiplicity M and with s branches $C_1, \ldots, C_s$. Let $p'_1(t)$ be a continuous arc on C_1 with the property that $d(0, p'_1(t)) = \Theta(t)$. Let us order the numbers $q(1, k), k = 2, \ldots, M$ in decreasing order:

$$1 \leq q(1, M) < q(1, M - 1) < \cdots < q(1, 2) < q(1, 1) = \infty.$$

Let us consider the horns $\mathcal{H}_{q,r} = \mathcal{H}(p'_1(t), r|t|^q)$ with $q \in [1, +\infty[$.

For $q \gg 1$ and small $\epsilon > 0$, the number of connected components of $B(0, \epsilon) \cap (\mathcal{H}_{q,r} \setminus \{0\})$ equals 1. Now, let us decrease q. For every $\eta > 0$ small enough, the number of connected components of $\mathcal{H}_{q(1,2+\eta),r} \setminus \{0\}$ equals 1, while the number of connected components of $\mathcal{H}_{q(1,2-\eta),r} \setminus \{0\}$ is > 2. Decreasing q, we have a jump in the number of connected components exactly when passing one of the rational numbers $q(1, k)$. So this enables one to recover all the characteristic exponents of C_1 and its contact exponents with the other branches of C. We can do the same for a real arc $p'_i(t)$ in each branch C_i of $(C, 0)$ and this will recover the integers $q(i, k)$ for $k = 1, \ldots, M$. We then reconstruct the function $q : [M] \times [M] \to \mathbb{Q}_{\geq 1}$ which characterizes the embedded topology of $(C, 0)$, or equivalently the carrousel tree of $(C, 0)$.

Moreover, the same jumps appear when one uses instead horns

$$\mathcal{H}'(p'(t), r|t|^q) = \bigcup_{t \in [0,1)} B'((p'(t), r|t|^q),$$

where B' denotes balls with respect to a metric d' which is bilipschitz equivalent to the initial outer metric. Indeed, if K is the bilipschitz constant of such a bilipschitz change, then we have the inclusions

$$\mathcal{H}\left(p'(t), \frac{rt}{K^4}\right) \subset \mathcal{H}'\left(p'(t), \frac{rt}{K^3}\right) \subset \mathcal{H}\left(p'(t), \frac{rt}{K^2}\right)$$

$$\subset \mathcal{H}'\left(p'(t), \frac{rt}{K^3}\right) \subset \mathcal{H}\left(p'(t), rt\right).$$

Fig. 7.6 Sections of C associated to the arc $p_1'(t)$

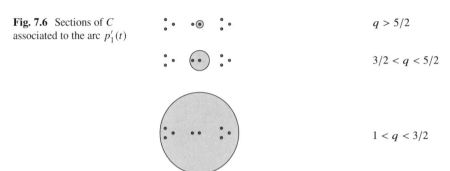

$q > 5/2$

$3/2 < q < 5/2$

$1 < q < 3/2$

Then the same argument as in the version 1 of the bubble trick shows that for q fixed and different from $q(1, k)$, $k = 2, \ldots, M$, the numbers of connected components of $B(0, \epsilon) \cap \left(\mathcal{H}_{q,r} \setminus \{0\} \right)$ and $B(0, \epsilon) \cap \left(\mathcal{H}_{q,r}' \setminus \{0\} \right)$ are equal.

Example 7.3.9 Consider again the plane curve singularity with two branches of Example 7.3.5 given by the Puiseux series:

$$C_1 : \quad y = x^{3/2} + x^{13/6}, \quad C_2 : \quad y = x^{5/2}.$$

Consider first an arc $p_1'(t)$ inside C_1 parametrized by $x = t \in [0, 1)$. Then $p_1'(t)$ corresponds to one of the two extremities of the carrousel tree of Fig. 7.2 whose neighbour vertex is weighted by $5/2$. Figure 7.6 represents the intersection of the horn $\mathcal{H}_{q,r}$ with the line $\{x = t\}$ for different values of $q \in [1, +\infty[$ and for $t \in \mathbb{C}^*$ of sufficiently small absolute value. This shows two jumps: a first jump at $q = 5/2$, which says that $5/2$ is a characteristic exponent of a branch since $p_1'(t)$ and the new point appearing in the intersection belong to the same connected component C_1 of $C \setminus \{0\}$, while the second jump at $3/2$ says that $3/2$ is the contact exponent of C_1 with the other component since the new points appearing at $q = 3/2 - \eta$ belong to C_2.

This first exploration enables one to construct the left part of the carrousel tree of C shown on Fig. 7.7, i.e., the one corresponding to the carrousel tree of C_1.

To complete the picture, we now consider an arc $p_2'(t)$ inside C_2 corresponding to a component of $C_2 \cap \{x = t\}$. This means that $p_2'(t)$ corresponds to one of the 6 extremities of the carrousel tree of Fig. 7.2 whose neighbour vertex is weighted by $13/6$. Figure 7.8 represents the jumps for the horns $\mathcal{H}_{q,r}$, centered on $p_2'(t)$. This shows two jumps: a first jump at $q = 13/6$, which says that $13/6$ is a characteristic exponent of C_2, then a second jump at $3/2$ corresponding to the contact exponent of C_1 and C_2.

This exploration of C_2 enables one to construct the right part of the carrousel tree of C shown on Fig. 7.9, i.e., the one corresponding to the carrousel tree of C_2.

Merging the two partial carrousel trees above, we obtain the carrousel tree of Fig. 7.2, recovering the embedded topology of $(C, 0)$.

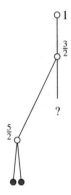

Fig. 7.7 Partial carrousel tree associated to the arc $p'_1(t)$

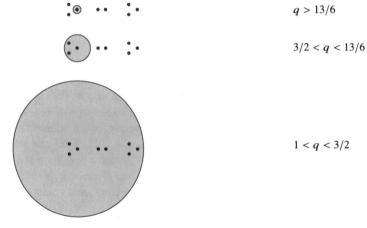

Fig. 7.8 Sections of C associated to the arc $p'_2(t)$

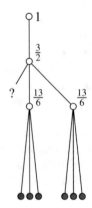

Fig. 7.9 Partial carrousel tree associated to the arc $p'_2(t)$

7.4 The Thick-Thin Decomposition of a Surface Singularity

7.4.1 Fast Loops as Obstructions to Metric Conicalness

We know that every complex curve germ $(C, 0) \subset (\mathbb{C}^N, 0)$ is metrically conical for the inner geometry (Proposition 7.3.2). This is no longer true in higher dimensions. The first example of a non-metrically-conical complex analytic germ $(X,0)$ appeared in [BF08]: for $k \geq 2$, the surface singularity $A_k \colon x^2 + y^2 + z^{k+1} = 0$ is not metrically conical for the inner metric.[1] The examples in [BFN08, BFN09, BFN10] then suggested that failure of metric conicalness is common. For example, among ADE singularities of surfaces, only A_1 and D_4 are metrically conical (Exercise 7.4.22). In [BFN10] it is also shown that the inner Lipschitz geometry of a singularity may not be determined by its topological type.

A complete classification of the Lipschitz inner geometry of normal complex surfaces is presented in [BNP14]. It is based on the existence of the so-called thick-thin decomposition of the surface into two semi-algebraic sets. The aim of Sects. 7.4.1–7.4.3 is to describe this decomposition.

The simplest obstruction to the metric conicalness of a germ $(X, 0)$ is the existence of *fast loops* (see Definition 7.4.2 below).

Let p and q be two pairwise coprime positive integers such that $p \geq q$. Set $\beta = \frac{p}{q}$. The prototype of a fast loop is the β-**horn**, which is the following semi-algebraic surface in $\mathbb{R}^3$ (Fig.7.10):

$$\mathcal{H}_\beta = \{(x, y, z) \in \mathbb{R}^2 \times \mathbb{R}^+ : (x^2 + y^2)^q = (z^2)^p\}.$$

Exercise 7.4.1 Show that $\mathcal{H}_\beta$ is inner bilipschitz homeomorphic to $\mathcal{H}_{\beta'}$ if and only if $\beta = \beta'$.[2]

$\mathcal{H}_1$ is a straight cone, so it is metrically conical. As a consequence of Exercise 7.4.1, we obtain that for $\beta > 1$, $\mathcal{H}_\beta$ is not metrically conical. For $t > 0$, set $\gamma_t = \mathcal{H}_\beta \cap \{z = t\}$. When $\beta > 1$, the family of curves $(\gamma_t)_{t>0}$ is a *fast loop* inside $\mathcal{H}_\beta$. More generally:

---

[1] Notice that in the real algebraic setting, it is easy to construct germs with dimension ≥ 3 which are not metrically conical for the inner geometry. For example a three-dimensional horn-shaped germ $(X, 0)$ whose link $X^{(\epsilon)}$ is a 2-torus with diameter $\Theta(\epsilon^2)$.

[2] *Hint:* the length of the family of curves $C_t = \mathcal{H}_\beta \cap \{z = t\}$ is a $\Theta(t^\beta)$ and this is invariant by a bilipschitz change of the metric. Show that such a family of curves cannot exist in $\mathcal{H}_{\beta'}$ if $\beta' \neq \beta$.

Fig. 7.10 The β-horns $\mathcal{H}_\beta$

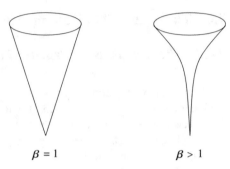

$\beta = 1$ $\beta > 1$

Definition 7.4.2 Let $(X, 0) \subset (\mathbb{R}^n, 0)$ be a semianalytic germ. A **fast loop** in $(X, 0)$ is a continuous family of loops $\{\gamma_\epsilon : S^1 \to X^{(\epsilon)}\}_{0 < \epsilon \leq \epsilon_0}$ such that:

1. γ_ϵ is essential (i.e., homotopically non trivial) in the link $X^{(\epsilon)} = X \cap S_\epsilon$;
2. there exists $q > 1$ such that

$$\lim_{\epsilon \to 0} \frac{\text{length}(\gamma_\epsilon)}{\epsilon^q} = 0.$$

In the next section, we will define what we call **the thick-thin decomposition of a normal surface germ** $(X, 0)$. It consists in decomposing $(X, 0)$ as a union of two semi-algebraic sets $(X, 0) = (Y, 0) \bigcup (Z, 0)$ where $(Z, 0)$ is *thin* (Definition 7.4.5) and where $(Y, 0)$ is *thick* (Definition 7.4.10). The thin part $(Z, 0)$ will contain all the fast loops of $(X, 0)$ inside a Milnor ball with radius ϵ_0. The thick part $(Y, 0)$ is the closure of the complement of the thin part and has the property that it contains a maximal metrically conical set. This enables one to characterize the germs $(X, 0)$ which are metrically conical:

Theorem 7.4.3 ([BNP14, Theorem 7.5, Corollary 1.8]) *Let $(X, 0)$ be a normal complex surface and let*

$$(X, 0) = (X_{thick}, 0) \bigcup (X_{thin}, 0)$$

be its thick-thin decomposition.

Then $(X, 0)$ is metrically conical if and only if $X_{thin} = \emptyset$, so $(X, 0) = (X_{thick}, 0)$.

7.4.2 Thick-Thin Decomposition

Definition 7.4.4 Let $(Z, 0) \subset (\mathbb{R}^n, 0)$ be a semi-algebraic germ. The *tangent cone* of $(Z, 0)$ is the set $T_0 Z$ of vectors $v \in \mathbb{R}^n$ such that there exists a sequence of points

(x_k) in $Z \setminus \{0\}$ tending to 0 and a sequence of positive real numbers (t_k) such that

$$\lim_{k \to \infty} \frac{1}{t_k} x_k = v.$$

Definition 7.4.5 A semi-algebraic germ $(Z, 0) \subset (\mathbb{R}^n, 0)$ of pure dimension is **thin** if the dimension of its tangent cone T_0X at 0 satisfies $\dim(T_0Z) < \dim(Z)$.

Example 7.4.6 For every $\beta > 1$, the β-horn $\mathcal{H}_\beta$ is thin since $\dim(\mathcal{H}_\beta) = 2$ while $T_0\mathcal{H}_\beta$ is a half-line. On the other hand, $\mathcal{H}_1$ is not thin.

Example 7.4.7 Let $\lambda \in \mathbb{C}^*$ and denote by C_λ the plane curve with Puiseux parametrization $y = \lambda x^{5/3}$. Let $a, b \in \mathbb{R}$ such that $0 < a < b$. Consider the semi-algebraic germ $(Z, 0) \subset (\mathbb{C}^2, 0)$ defined by $Z = \bigcup_{a \leq |\lambda| \leq b} C_\lambda$. The tangent cone T_0Z is the complex line $y = 0$, while Z is 4-dimensional, so $(Z, 0)$ is thin.

Let $Z^{(\epsilon)}$ be the intersection of Z with the boundary of the bidisc $\{|x| \leq \epsilon\} \times \{|y| \leq \epsilon\}$. By [Dur83], one obtains, up to homeomorphism (or diffeomorphism in a stratified sense), the same link $Z^{(\epsilon)}$ as when intersecting with a round sphere. When $\epsilon > 0$ is small enough, $Z^{(\epsilon)} \subset \{|x| = \epsilon\} \times \{|y| \leq \epsilon\}$ and the projection $Z^{(\epsilon)} \to S_\epsilon^1$ defined by $(x, y) \to x$ is a locally trivial fibration whose fibers are the flat annuli $A_t = Z \cap \{x = t\}$, $|t| = \epsilon$, and the lengths of the boundary components of A_t are $\Theta(\epsilon^{5/3})$.

Notice that Z can be described through a resolution as follows. Let $\sigma : Y \to \mathbb{C}^2$ be the minimal embedded resolution of the curve $E_1: y = x^{5/3}$. It decomposes into four successive blow-ups of points. Denote $E_1, \ldots, E_4$ the corresponding components of the exceptional divisor $\sigma^{-1}(0)$ indexed by their order of creation. Then σ is a simultaneous resolution of the curves C_λ. Therefore, the strict transform of Z by σ is a neighbourhood of E_4 minus neighbourhoods of the intersecting points $E_4 \cap E_2$ and $E_4 \cap E_3$ as pictured in Fig. 7.11. The tree T on the left is the dual tree of σ. Its vertices are weighted by the self-intersections E_i^2 and the arrow represents the strict transform of C_1.

Definition 7.4.8 Let $1 < q \in \mathbb{Q}$. A **q-horn neighbourhood** of a semi-algebraic germ $(A, 0) \subset (\mathbb{R}^N, 0)$ is a set of the form $\{x \in \mathbb{R}^n \cap B_\epsilon : d(x, A) \leq c|x|^q\}$ for some $c > 0$, where d denotes the Euclidean metric.

The following proposition helps picture "thinness"

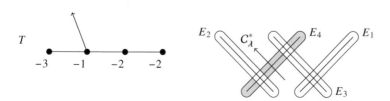

Fig. 7.11 The strict transform of Z by σ

Fig. 7.12 Trying to glue a
thin germ with a metrically
conical germ

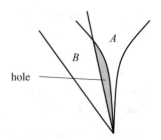

Proposition 7.4.9 ([BNP14, Proposition 1.3]) *Any thin semi-algebraic germ* $(Z, 0) \subset (\mathbb{R}^N, 0)$ *is contained in some q-horn neighbourhood of its tangent cone* $T_0 Z$.

We will now define thick semi-algebraic sets. The definition is built on the following observation. Let $(X, 0) \subset (\mathbb{R}^n, 0)$ be a real algebraic germ; we would like to decompose $(X, 0)$ into two semialgebraic sets $(A, 0)$ and $(B, 0)$ glued along their boundary germs, where $(A, 0)$ is thin and $(B, 0)$ is metrically conical. But try to glue a thin germ $(A, 0)$ with a metrically conical germ $(B, 0)$ so that they intersect only along their boundary germs.... It is not possible! There would be a hole between them (see Fig. 7.12). So we have to replace $(B, 0)$ by something else than conical.

"Thick" is a generalization of "metrically conical." Roughly speaking, a thick algebraic set is obtained by slightly inflating a metrically conical set in order that it can interface along its boundary with thin parts. The precise definition is as follows:

Definition 7.4.10 Let $B_\epsilon \subset \mathbb{R}^N$ denote the ball of radius ϵ centered at the origin, and S_ϵ its boundary. A semi-algebraic germ $(Y, 0) \subset (\mathbb{R}^N, 0)$ is **thick** if there exists $\epsilon_0 > 0$ and $K \geq 1$ such that $Y \cap B_{\epsilon_0}$ is the union of subsets Y_ϵ, $\epsilon \leq \epsilon_0$ which are metrically conical with bilipschitz constant K and satisfy the following properties (see Fig. 7.1):

1. $Y_\epsilon \subset B_\epsilon$, $Y_\epsilon \cap S_\epsilon = Y \cap S_\epsilon$ and Y_ϵ is metrically conical;
2. For $\epsilon_1 < \epsilon_2$ we have $Y_{\epsilon_2} \cap B_{\epsilon_1} \subset Y_{\epsilon_1}$ and this embedding respects the conical structures. Moreover, the difference $(Y_{\epsilon_1} \cap S_{\epsilon_1}) \setminus (Y_{\epsilon_2} \cap S_{\epsilon_1})$ of the links of these cones is homeomorphic to $\partial(Y_{\epsilon_1} \cap S_{\epsilon_1}) \times [0, 1)$.

Clearly, a semi-algebraic germ cannot be both thick and thin (Fig. 7.13).

Example 7.4.11 The set $Z = \{(x, y, z) \in \mathbb{R}^3 : x^2 + y^2 \leq z^3\}$ gives a thin germ at 0 since it is a three-dimensional germ whose tangent cone is half the z-axis. The intersection $Z \cap B_\epsilon$ is contained in a closed $3/2$-horn neighbourhood of the z-axis. The complement in $\mathbb{R}^3$ of this thin set is thick.

Example 7.4.12 Consider again the thin germ $(Z, 0) \subset (\mathbb{C}^2, 0)$ of Example 7.4.7. Then the germ $(Y, 0)$ defined by $Y = \overline{\mathbb{C}^2 \setminus Z}$ is a thick germ. To give an imaged picture of it, fix $\eta > 0$ and consider the conical set $W \subset \mathbb{C}^2$ defined as the union of the complex lines $y = \alpha x$ for $|\alpha| \geq \eta$; then $(Y, 0)$ is obtained by "slightly

Fig. 7.13 Thick germ

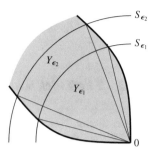

inflating" W. Notice that the strict transform of Y by the resolution σ introduced in Example 7.4.7 is a neighbourhood of the union of curves $E_1 \cup E_3$.

For any semi-algebraic germ $(A, 0)$ of $(\mathbb{R}^N, 0)$, we write $A^{(\epsilon)} := A \cap S_\epsilon \subset S_\epsilon$. When ϵ is sufficiently small, $A^{(\epsilon)}$ is the ϵ-link of $(A, 0)$.

Definition 7.4.13 (Thick-Thin Decomposition) A **thick-thin decomposition** of the normal complex surface germ $(X, 0)$ is a decomposition of it as a union of semi-algebraic germs of pure dimension 4 called **pieces**:

$$(X, 0) = \bigcup_{i=1}^{r}(Y_i, 0) \cup \bigcup_{j=1}^{s}(Z_j, 0), \tag{7.1}$$

such that the $Y_i \setminus \{0\}$ and $Z_j \setminus \{0\}$ are connected and:

1. Each Y_i is thick and each Z_j is thin;
2. The $Y_i \setminus \{0\}$ are pairwise disjoint and the $Z_j \setminus \{0\}$ are pairwise disjoint;
3. If ϵ_0 is chosen small enough such that S_ϵ is transverse to each of the germs $(Y_i, 0)$ and $(Z_j, 0)$ for $\epsilon \leq \epsilon_0$, then $X^{(\epsilon)} = \bigcup_{i=1}^{r} Y_i^{(\epsilon)} \cup \bigcup_{j=1}^{s} Z_j^{(\epsilon)}$ decomposes the 3-manifold $X^{(\epsilon)} \subset S_\epsilon$ into connected submanifolds with boundary, glued along their boundary components.

Definition 7.4.14 A thick-thin decomposition is **minimal** if

1. the tangent cone of its thin part $\bigcup_{j=1}^{s} Z_j$ is contained in the tangent cone of the thin part of any other thick-thin decomposition and
2. the number s of its thin pieces is minimal among thick-thin decompositions satisfying (1).

The following theorem expresses the existence and uniqueness of a minimal thick-thin decomposition of a normal complex surface singularity.

Theorem 7.4.15 ([BNP14, Section 8]) *Let $(X, 0)$ be a normal complex surface germ. Then a minimal thick-thin decomposition of $(X, 0)$ exists. For any two minimal thick-thin decompositions of $(X, 0)$, there exists $q > 1$ and a homeomorphism of the* *germ* $(X, 0)$ *to itself which takes one decomposition to the other and moves each* $x \in X$ *by a distance at most* $\|x\|^q$.

7.4.3   The Thick-Thin Decomposition in a Resolution

In this section, we describe explicitly the minimal thick-thin decomposition of a normal complex surface germ $(X, 0) \subset (\mathbb{C}^n, 0)$ in terms of a suitable resolution of $(X, 0)$ as presented in [BNP14, Section 2]. The uniqueness of the minimal thick-thin decomposition is proved in [BNP14, Section 8]. We refer to Sect. 7.5 of the lecture notes of Jawad Snoussi and to Sect. 7.6 of those of Walter Neumann in the present volume for classical background about resolution of complex surfaces.

Let $\pi : (\widetilde{X}, E) \to (X, 0)$ be the minimal resolution with the following properties:

1. It is a good resolution, i.e., the irreducible components of the exceptional divisor are smooth and meet transversely, at most two at a time.
2. It factors through the blow-up $e_0 : X_0 \to X$ of the origin. An irreducible component of the exceptional divisor $\pi^{-1}(0)$ which projects surjectively on an irreducible component of $e_0^{-1}(0)$ will be called an $\mathcal{L}$-**curve**.
3. No two $\mathcal{L}$-curves intersect.

This is achieved by starting with a minimal good resolution, then blowing up to resolve any base points of a general system of hyperplane sections, and finally blowing up any intersection point between $\mathcal{L}$-curves.

Definition 7.4.16 Let Γ be the resolution graph of the above resolution. A vertex of Γ is called a **node** if it has valence ≥ 3 or represents a curve of genus >0 or represents an $\mathcal{L}$-curve. If a node represents an $\mathcal{L}$-curve it is called an $\mathcal{L}$-*node* . By the previous paragraph, $\mathcal{L}$-nodes cannot be adjacent to each other. Other types of nodes will be introduced in Definitions 7.5.23 and 7.5.31.

A **string** is a connected subgraph of Γ containing no nodes. A **bamboo** is a string ending in a vertex of valence 1.

For each irreducible curve E_v in E, let $N(E_v)$ be a small closed tubular neighborhood of E_v in $\widetilde{X}$. For any subgraph Γ' of Γ define (see Fig. 7.14):

$$N(\Gamma') := \bigcup_{v \in \Gamma'} N(E_v) \quad \text{and} \quad \mathcal{N}(\Gamma') := \overline{N(\Gamma) \setminus \bigcup_{v \notin \Gamma'} N(E_v)}.$$

The subgraphs of Γ resulting by removing the $\mathcal{L}$-nodes and adjacent edges from Γ are called the **Tyurina components** of Γ (following [Spi90, Definition III.3.1]).

Let $\Gamma_1, \ldots, \Gamma_s$ denote the Tyurina components of Γ which are not bamboos, and by $\Gamma'_1, \ldots, \Gamma'_r$ the maximal connected subgraphs in $\Gamma \setminus \bigcup_{j=1}^s \Gamma_j$. Therefore each Γ'_i consists of an $\mathcal{L}$-node and any attached bamboos and strings.

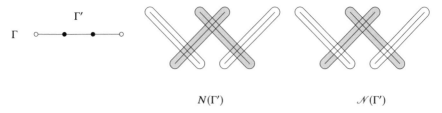

Fig. 7.14 The sets $N(\Gamma')$ and $\mathcal{N}(\Gamma')$

Assume that ϵ_0 is sufficiently small such that $\pi^{-1}(X \cap B_{\epsilon_0})$ is included in $N(\Gamma)$. For each $i = 1, \ldots, r$, define

$$Y_i := \pi(N(\Gamma'_i)) \cap B_{\epsilon_0},$$

and for each $j = 1, \ldots, s$, define

$$Z_j := \pi(\mathcal{N}(\Gamma_j)) \cap B_{\epsilon_0}.$$

Notice that the Y_i are in one-to-one correspondence with the $\mathcal{L}$-nodes.

Theorem 7.4.17 ([BNP14, Section 2, Proposition 5.1, Proposition 6.1])

1. *For each $i = 1, \ldots, r$, $(Y_i, 0)$ is thick;*
2. *For each $j = 1, \ldots, s$, $(Z_j, 0)$ is thin;*
3. *The decomposition $(X, 0) = \bigcup(Z_j, 0) \cup \bigcup(Y_i, 0)$ is a minimal thick-thin decomposition of $(X, 0)$.*

The proof of (2) is easy:

Proof Choose an embedding $(X, 0) \subset (\mathbb{C}^n, 0)$ and let $e_0 : X_0 \to X$ be the blow-up of the origin. If $x \in \mathbb{C}^n \setminus \{0\}$, denote by L_x the class of x in $\mathbb{P}^{n-1}$, so L_x represents the line through 0 and x in $\mathbb{C}^n$. By definition X_0 is the closure in $\mathbb{C}^n \times \mathbb{P}^{n-1}$ of the set $\{(x, L_x) : x \in X \setminus \{0\}\}$.

The semi-algebraic set Z_j is of real dimension 4. On the other hand, the strict transform of Z_j by e_0 meets the exceptional divisor $e_0^{-1}(0)$ at a single point (x, L_x), so the tangent cone at 0 to Z_j is the complex line L_x. Therefore $(Z_j, 0)$ is thin. $\square$

In the next section, we present the first part of the proof of the thickness of $(Y_i, 0)$, which consists of proving the following intermediate Lemma:

Lemma 7.4.18 *For each $\mathcal{L}$-node v, the subset $\pi(\mathcal{N}(v))$ of $(X, 0)$ is metrically conical.*

The rest of the proof of Point (1) of Theorem 7.4.17 is more delicate. In particular, it uses the key Polar Wedge Lemma [BNP14, Proposition 3.4] which is stated later in the present notes (Proposition 7.5.15) and a geometric decomposition of $(X, 0)$ into standard pieces which is a refinement of the thick-thin decomposition and

Fig. 7.15 The thick-thin
decomposition of the
singularity E_8

which leads to the complete classification of the inner Lipschitz geometry of $(X, 0)$
presented in Sect. 7.5.4. We refer to [BNP14] for the proofs.

The minimality (3) is proved in [BNP14, Section 8].

We now give several explicit examples of thick-thin decompositions. More
examples can be found in [BNP14, Section 15].

Example 7.4.19 Consider the normal surface singularity $(X, 0) \subset (\mathbb{C}^3, 0)$ with
equation $x^2 + y^3 + z^5 = 0$. This is the standard singularity E_8 (see [Dur79]). Its
minimal resolution has exceptional divisor a tree of eight $\mathbb{P}^1$ having self intersections
-2 and it factors through the blow-up of the point 0. The dual graph Γ is represented
on Fig. 7.15. It can be constructed with Laufer's method (see Appendix section 7.6).
The arrow represents the strict transform of a generic linear form $h = \alpha x + \beta y + \gamma z$
on $(X, 0)$. The vertex adjacent to it is the unique $\mathcal{L}$-node and Γ has two nodes which
are circled on the figure. The thick-thin decomposition of $(X, 0)$ has one thick piece
$(Y_1, 0)$ and one thin piece $(Z_1, 0)$. The subgraph Γ_1' of Γ such that $Y_1 = \pi(N(\Gamma_1'))$
is in black and the subgraph Γ_1 such that $Z_1 = \pi(\mathcal{N}(\Gamma_1))$ is in white.

Example 7.4.20 Consider the normal surface singularity $(X, 0) \subset (\mathbb{C}^3, 0)$ with
equation $x^2 + zy^2 + z^3 = 0$. This is the standard singularity D_4. Its minimal
resolution has exceptional divisor a tree of four $\mathbb{P}^1$'s having self intersections -2
and it factors through the blow-up of the point 0. The dual graph Γ is represented on
Fig. 7.16. It has one $\mathcal{L}$-node, which is the central vertex circled on the figure and no
other node. Therefore, the thick-thin decomposition of $(X, 0)$ has empty thin part
and $(X, 0)$ is metrically conical. The subgraph of Γ corresponding to the thick part
is the whole Γ.

Example 7.4.21 Consider the family of surface singularities in $(X_t, 0) \subset (\mathbb{C}^3, 0)$
with equations $x^5 + z^{15} + y^7 z + txy^6 = 0$ depending on the parameter $t \in \mathbb{C}$. This
is a μ-constant family introduced by Briançon and Speder in [BS75]. The thick-thin
decomposition changes radically when t becomes 0. Indeed, the minimal resolution
graph of every $(X_t, 0)$ is the first graph on Fig. 7.17 while the two other resolution
graphs describe the thick-thin decompositions for $t = 0$ and for $t \neq 0$. For $t \neq 0$ it
has three thick components and a single thin one. For $t = 0$, it has one component
of each type. We refer to [BNP14, Example 15.7] for further details.

Fig. 7.16 The thick-thin
decomposition of the
singularity D_4

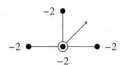

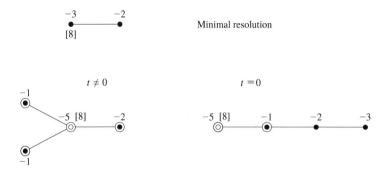

Fig. 7.17 The two thick-thin decompositions in the Briançon-Speder family $x^5 + z^{15} + y^7z + txy^6 = 0$

Exercise 7.4.22 Describe the thick-thin decomposition of every ADE surface singularity and show that among them, only A_1 and D_4 are metrically conical (Answer: [BNP14, Example 15.4]). The equations are:

- A_n: $x^2 + y^2 + z^{k+1} = 0$, $k \geq 1$
- D_n: $x^2 + zy^2 + z^{k-1} = 0$, $k \geq 4$
- E_6: $x^2 + y^3 + z^4 = 0$
- E_7: $x^2 + y^3 + yz^3 = 0$
- E_8: $x^2 + y^3 + z^5 = 0$

7.4.4 Generic Projection and Inner Metric: A Key Lemma

In this section, we state and prove Lemma 7.4.29 which is one of the key results which will lead to the complete classification of the inner metric of $(X, 0)$. We give two applications. The first one is the proof of Lemma 7.4.18. The second describes the inner contact between complex curves on a complex surface.

We first need to introduce the polar curves of generic projections and the Nash modification of $(X, 0)$. We refer to the lecture notes of Jawad Snoussi [Sno19] in the present volume for more details about Nash modification.

7.4.4.1 Polar Curves and Generic Projections

Let $(X, 0) \subset (\mathbb{C}^n, 0)$ be a normal surface singularity. We restrict ourselves to those $\mathcal{D}$ in $\mathbf{G}(n - 2, \mathbb{C}^n)$ such that the restriction $\ell_{\mathcal{D}}|_{(X,0)} \colon (X, 0) \to (\mathbb{C}^2, 0)$ is finite. The **polar curve** $\Pi_{\mathcal{D}}$ of $(X, 0)$ for the direction $\mathcal{D}$ is the closure in $(X, 0)$ of the critical locus of the restriction of $\ell_{\mathcal{D}}$ to $X \setminus \{0\}$. The **discriminant curve** $\Delta_{\mathcal{D}} \subset (\mathbb{C}^2, 0)$ is the image $\ell_{\mathcal{D}}(\Pi_{\mathcal{D}})$ of the polar curve $\Pi_{\mathcal{D}}$.

Proposition 7.4.23 ([Tei82, Lemme-clé V 1.2.2]) *An open dense subset* $\Omega \subset$ $\mathbf{G}(n-2, \mathbb{C}^n)$ *exists such that:*

1. *The family of curve germs* $(\Pi_{\mathcal{D}})_{\mathcal{D} \in \Omega}$ *is equisingular in terms of strong simultaneous resolution;*
2. *The curves* $\ell_{\mathcal{D}}(\Pi_{\mathcal{D}'})$, $(\mathcal{D}, \mathcal{D}') \in \Omega \times \Omega$ *form an equisingular family of reduced plane curves;*
3. *For each* $\mathcal{D}$, *the projection* $\ell_{\mathcal{D}}$ *is generic for its polar curve* $\Pi_{\mathcal{D}}$ *(Definition 7.3.3).*

Definition 7.4.24 The projection $\ell_{\mathcal{D}} \colon \mathbb{C}^n \to \mathbb{C}^2$ is **generic** for $(X, 0)$ if $\mathcal{D} \in \Omega$.

7.4.4.2 Nash Modification

Definition 7.4.25 Let $\lambda \colon X \setminus \{0\} \to \mathbf{G}(2, \mathbb{C}^n)$ be the map which sends $x \in X \setminus \{0\}$ to the tangent plane $T_x X$. The closure X_ν of the graph of λ in $X \times \mathbf{G}(2, \mathbb{C}^n)$ is a reduced analytic surface. By definition, the **Nash modification** of $(X, 0)$ is the morphism $\nu \colon X_\nu \to X$ induced by projection on the first factor.

Lemma 7.4.26 ([Spi90, Part III, Theorem 1.2]) *A morphism* $\pi \colon Y \to X$ *factors through Nash modification if and only if it has no base points for the family of polar curves of generic projections, i.e., there is no point* $p \in \pi^{-1}(0)$ *such that for every* $\mathcal{D} \in \Omega$, *the strict transform of* $\Pi_{\mathcal{D}}$ *by* π *passes through* p.

Definition 7.4.27 Let $(X, 0) \subset (\mathbb{C}^n, 0)$ be a complex surface germ and let $\nu \colon X_\nu \to X$ be the Nash modification of X. The **Gauss map** $\tilde{\lambda} \colon X_\nu \to \mathbf{G}(2, \mathbb{C}^n)$ is the restriction to X_ν of the projection of $X \times \mathbf{G}(2, \mathbb{C}^n)$ on the second factor.

Let $\ell \colon \mathbb{C}^n \to \mathbb{C}^2$ be a linear projection such that the restriction $\ell|_X \colon (X, 0) \to (\mathbb{C}^2, 0)$ is generic. Let Π and Δ be the polar and discriminant curves of $\ell|_X$.

Definition 7.4.28 The **local bilipschitz constant of** $\ell|_X$ is the map $K \colon X \setminus \{0\} \to \mathbb{R} \cup \{\infty\}$ defined as follows. It is infinite on the polar curve Π and at a point $p \in X \setminus \Pi$ it is the reciprocal of the shortest length among images of unit vectors in $T_p X$ under the projection $\ell|_{T_p X} \colon T_p X \to \mathbb{C}^2$.

Let Π^* denote the strict transform of the polar curve Π by the Nash modification ν. Set $B_\epsilon = \{x \in \mathbb{C}^n \colon \|x\|_{\mathbb{C}^n} \le \epsilon\}$.

Lemma 7.4.29 *Given any neighbourhood* U *of* $\Pi^* \cap \nu^{-1}(B_\epsilon \cap X)$ *in* $X_\nu \cap \nu^{-1}(B_\epsilon \cap X)$, *the local bilipschitz constant* K *of* $\ell|_X$ *is bounded on* $B_\epsilon \cap (X \setminus \nu(U))$.

Proof Let $\kappa \colon \mathbf{G}(2, \mathbb{C}^n) \to \mathbb{R} \cup \{\infty\}$ be the map sending $H \in \mathbf{G}(2, \mathbb{C}^n)$ to the bilipschitz constant of the restriction $\ell|_H \colon H \to \mathbb{C}^2$. The map $\kappa \circ \tilde{\lambda}$ coincides with $K \circ \nu$ on $X_\nu \setminus \nu^{-1}(0)$ and takes finite values outside Π^*. The map $\kappa \circ \tilde{\lambda}$ is continuous and therefore bounded on the compact set $\nu^{-1}(B_\epsilon \cap X) \setminus U$. $\square$

We will use "small" special versions of U called polar wedges, defined as follows.

Definition 7.4.30 Let Π_0 be a component of Π and let (u, v) be local coordinates in X_ν centered at $p = \Pi_0^* \cap v^{-1}(0)$ such that $v = 0$ is the local equation of $v^{-1}(0)$. For $\alpha > 0$, consider the polydisc $U_{\Pi_0}(\alpha) = \{(u, v) \in \mathbb{C}^2 : |u| \leq \alpha\}$. For small α, the set $W_{\Pi_0} = v(U_{\Pi_0}(\alpha))$ is called a **polar wedge** around Π_0 and the union $W = \bigcup_{\Pi_0 \subset \Pi} W_{\Pi_0}$ a **polar wedge** around Π.

7.4.4.3 Application 1

Proof of Lemma 7.4.18 We want to prove that for every $\mathcal{L}$-node ν, the germ $\pi(\mathcal{N}(\nu))$ is metrically conical.

Consider a polar wedge W around Π. A direct consequence of Lemma 7.4.29 is that the restriction of ℓ to $X \setminus W$ is a local bilipschitz homeomorphism for the inner metric. Therefore, for any metrically conical germ C in $(\mathbb{C}^2, 0)$, the intersection of the lifting $\ell^{-1}(C)$ with $\overline{X \setminus W}$ will be a metrically conical germ.

For each $j = 1, \ldots, s$, let $L_j \subset \mathbb{C}^n$ be the complex tangent line of the thin germ $(Z_j, 0)$ and let $L'_j \subset \mathbb{C}^2$ be image of L_j by the generic linear form $\ell \colon \mathbb{C}^n \to \mathbb{C}^2$. Assume L'_j has equation $y = a_j x$. For a real number $\alpha > 0$, we consider the conical subset $V_\alpha \subset \mathbb{C}^2$ defined as the union of the complex lines $y = \eta x$ such that $|\eta - a_j| \geq \alpha$, so V_α is the closure of a set obtained by removing conical neighbourhoods of the lines L'_j. Applying the above result, we obtain that for all $\alpha > 0$, the intersection of the lifting $\ell^{-1}(V_\alpha)$ with $\overline{X \setminus W}$ gives a metrically conical germ at 0. Since there exist two real numbers α_1, α_2 with $0 < \alpha_1 < \alpha_2$ such that inside a small ball B_ϵ, we have $\ell^{-1}(V_{\alpha_1}) \subset \pi(\mathcal{N}(\nu)) \subset \ell^{-1}(V_{\alpha_2})$, then the germ $\pi(\mathcal{N}(\nu)) \cap (\overline{X \setminus W})$ at 0 is also metrically conical.

If the strict transform of Π by π does not intersect the $\mathcal{L}$-curve E_ν, then the intersection $\pi(\mathcal{N}(\nu)) \cap (\overline{X \setminus W})$ is the whole $\pi(\mathcal{N}(\nu))$. Therefore $\pi(\mathcal{N}(\nu))$ is metrically conical.

If the strict transform of Π by π intersects the $\mathcal{L}$-curve E_ν, then we have to use a second generic projection $\ell' \colon (X, 0) \to (\mathbb{C}^2, 0)$ such that the strict transform of the polar curve Π' of ℓ' by π does not intersect U, and we prove that $\pi(\mathcal{N}(\nu)) \cap W$ is metrically conical using the above argument. Therefore $\pi(\mathcal{N}(\nu))$ is metrically conical as the union of two metrically conical sets. □

7.4.4.4 Application 2

Let $(X, 0)$ be a normal complex surface singularity.

Definition 7.4.31 Let $S_\epsilon = \{x \in \mathbb{C}^n : \|x\|_{\mathbb{C}^n} = \epsilon\}$. Let $(\gamma, 0)$ and $(\gamma', 0)$ be two distinct irreducible germs of complex curves inside $(X, 0)$. Let $q_{inn} = q_{inn}(\gamma, \gamma')$ be the rational number ≥ 1 defined by

$$d_i(\gamma \cap S_\epsilon, \gamma' \cap S_\epsilon) = \Theta(\epsilon^{q_{inn}}),$$

where d_i means inner distance in $(X, 0)$ as before.

We call $q_{inn}(\gamma, \gamma')$ the *the inner contact exponent* or *inner contact order* between γ and γ'.

The proof of the existence and rationality of the inner contact q_{inn} needs deep arguments of [KO97]. We refer to this paper for details.

Remark 7.4.32 One can also define the outer contact exponent q_{out} between two curves by using the outer metric instead of the inner one. In that case, the existence and rationality of q_{out} come easily from the fact that the outer distance d_o is a semialgebraic function. (While the inner distance d_i is not semi-algebraic.)

Definition 7.4.33 Let $\pi : Z \to X$ be a resolution of X and let E be an irreducible component of the exceptional divisor $\pi^{-1}(0)$. A *curvette* of E is a smooth curve $\delta \subset Z$ which is transversal to E at a smooth point of the exceptional divisor $\pi^{-1}(0)$.

Lemma 7.4.34 ([NPP20a, Lemma 15.1]) *Let $\pi : Z \to X$ be a resolution of $(X, 0)$ and let E be an irreducible component of the exceptional divisor $\pi^{-1}(0)$. Let $(\gamma, 0)$ and $(\gamma', 0)$ be the π-images of two curvettes of E meeting E at two distinct points. Then $q_{inn}(\gamma, \gamma')$ is independent of the choice of γ and γ'.*

Definition 7.4.35 We set $q_E = q_{inn}(\gamma, \gamma')$ and we call q_E the *inner rate* of E.

Remark 7.4.36 When $X = \mathbb{C}^2$, inner and outer metrics coincide and the result is well known and comes from classical plane curve theory: in that case, $q_{inn}(\gamma, \gamma')$ is the coincidence exponent between Puiseux expansions of the curves γ and γ' (see for example [GBT99, page 401]). The inner rate at each vertex of a sequence of blow-ups can be computed by using the classical dictionary between characteristic exponents of an irreducible curve and its resolution graph. We refer to [EN85, page 148] or [Wal04, Section 8.3] for details. As a consequence of this, the inner rates along any path from the root vertex to a leaf of T form a strictly increasing sequence.

Example 7.4.37 The dual tree T_0 of the minimal resolution $\sigma_0 : Y_0 \to \mathbb{C}^2$ of the curve γ with Puiseux expansion $y = z^{5/3}$ is obtained (Fig. 7.18) by computing the continued fraction development

$$\frac{5}{3} = 1 + \cfrac{1}{1 + \frac{1}{2}} =: [1, 1, 2]^+.$$

Since $1 + 1 + 2 = 4$, σ_0 consists of four successive blow-ups of points starting with the blow-up of the origin of $\mathbb{C}^2$ which correspond to the four vertices of T_0. The irreducible curves $E_1, \ldots, E_4$ are labelled in their order of appearance and the vertices of T_0 are also weighted by their self-intersections E_i^2.

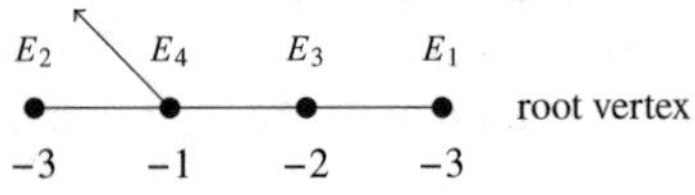

Fig. 7.18 The resolution tree T_0 of the curve $x^5 + z^{15} + y^7 z + txy^6 = 0$

Fig. 7.19 The inner rates in resolutions of the curve $y = x^{5/3}$

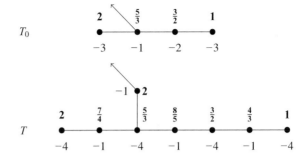

The inner rates are computed by using the approximation numbers associated with the sequence $[1, 1, 2]^+$: $q_{v_1} = [1]^+ = 1$, $q_{v_2} = [1, 1]^+ = 1 + \frac{1}{1} = 2$, $q_{v_3} = [1, 1, 1]^+ = \frac{3}{2}$ and $q_{v_3} = [1, 1, 2]^+ = \frac{5}{3}$

This gives the tree T_0 of Fig. 7.19 where each vertex is weighted by the self intersection of the corresponding exceptional curve E_i and with the inner rate q_{E_i} (in bold).

Let us blow up every intersection point between irreducible components of the total transform $\sigma_0^{-1}(\gamma)$. The resulting tree T is that used to compute the dual resolution graph of E_8: $x^2 + y^3 - z^5 = 0$ by Laufer's method (Appendix section 7.6). Again, the inner rates are in bold. Their computation is left to the reader as an exercise.

Proof of Lemma 7.4.34 Consider a generic projection $\ell: (X, 0) \to (\mathbb{C}^2, 0)$ which is also generic for the curve germ $(\gamma \cup \gamma', 0)$ (Definition 7.3.3). Then consider the minimal sequence of blow-ups $\sigma: Y \to \mathbb{C}^2$ such that the strict transforms $\ell(\gamma)^*$ and $\ell(\gamma')^*$ by σ do not intersect. Then $\ell(\gamma)^*$ and $\ell(\gamma')^*$ are two curvettes of the last exceptional curve C created by σ and we then have $q_{inn}(\ell(\gamma), \ell(\gamma')) = q_C$. Moreover, an easy argument using Hirzebruch–Jung resolution of surfaces (see [PP11] for an introduction to this resolution method) shows that σ does not depend on the choice of the curvettes γ^* and γ'^* of E. Now, since ℓ is generic for the curve $\gamma \cup \gamma'$, the strict transform of the polar curve Π of ℓ by π does not intersect the strict transform of $\gamma \cup \gamma'$, and then, $\gamma^* \cup \gamma'^*$ is outside any sufficiently small polar wedge of ℓ around Π. Therefore, by Lemma 7.4.29, we obtain $q_{inn}(\gamma, \gamma') = q_{inn}(\ell(\gamma), \ell(\gamma')) = q_C$ □

Example 7.4.38 The proof of Lemma 7.4.34 shows that the inner rates q_E can be computed by using inner rates in $\mathbb{C}^2$ through a generic projection $\ell: (X, 0) \to (\mathbb{C}^2, 0)$. Applying this, Fig. 7.20 shows the inner rate at each vertex of the minimal resolution graph of the surface singularity E_8: $x^2 + y^3 + z^5 = 0$. They are obtained by lifting the inner rates of the graph T of Example 7.4.37.

Fig. 7.20 The inner rates for
the singularity E_8

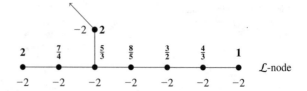

7.4.5 Fast Loops in the Thin Pieces

Consider a normal surface germ $(X, 0) \subset (\mathbb{C}^n, 0)$. We choose coordinates $(z_1 \ldots, z_n)$ in $\mathbb{C}^n$ so that z_1 and z_2 are generic linear forms and $\ell := (z_1, z_2): X \to \mathbb{C}^2$ is a generic linear projection. The family of Milnor balls we use in the sequel consists of standard "Milnor tubes" associated with the Milnor-Lê fibration for the map $\zeta := z_1|_X: X \to \mathbb{C}$ (see Section 3.5 of the lecture notes of Cisneros and Aguilar in chapter 1 of the present volume). Namely, for some sufficiently small ϵ_0 and some $R > 0$ we define for $\epsilon \le \epsilon_0$:

$$B_\epsilon := \{(z_1, \ldots, z_n) \in \mathbb{C}^n : |z_1| \le \epsilon, \|(z_1, \ldots, z_n)\| \le R\epsilon\} \quad \text{and} \quad S_\epsilon = \partial B_\epsilon.$$

By [BNP14, Proposition 4.1], on can choose ϵ_0 and R so that for $\epsilon \le \epsilon_0$:

1. $\zeta^{-1}(t)$ intersects the round sphere

$$S_{R\epsilon}^{2n-1} = \{(z_1, \ldots, z_n) \in \mathbb{C}^n : \|(z_1, \ldots, z_n)\| = R\epsilon\}$$

   transversely for $|t| \le \epsilon$;
2. the polar curve of the projection $\ell = (z_1, z_2)$ meets S_ϵ in the part $|z_1| = \epsilon$.

If $(A, 0)$ is a semialgebraic germ, we denote by $A^{(\epsilon)} = S_\epsilon \cap X$ its link with respect to the Milnor ball B_ϵ.

Theorem 7.4.39 ([BNP14, Theorem 1.7]) *Consider the minimal thick-thin decomposition*

$$(X, 0) = \bigcup_{i=1}^r (Y_i, 0) \cup \bigcup_{j=1}^s (Z_j, 0)$$

of $(X, 0)$. For $0 < \epsilon \le \epsilon_0$ and for each $j = 1, \ldots, s$, let $\zeta_j^{(\epsilon)}: Z_j^{(\epsilon)} \to S^1$ be the restriction to $Z_j^{(\epsilon)}$ of the generic linear form $h = z_1$. Then there exists $q_j > 1$ such that the fibers $\zeta_j^{-1}(t)$ have diameter $\Theta(\epsilon^{q_j})$.

Sketch of Proof of Theorem 7.4.39 The proof of Theorem 7.4.39 is based on two keypoints: Lemma 7.4.29, which implies that ℓ is an inner Lipschitz homeomorphism outside a polar wedge W, and the so called Polar Wedge Lemma [BNP14,

Proposition 3.4] which describes the geometry of a polar wedge. The idea is to use a generic linear projection $\ell = (z_1, z_2): (X, 0) \to (\mathbb{C}^2, 0)$ and to describe $(Z_j, 0)$ as a component of the lifting by ℓ of some semi-algebraic germ $(V_j, 0)$ in $\mathbb{C}^2$ which has the properties described in the Theorem, i.e., for small $\epsilon > 0$, $V_j^{(\epsilon)}$ fibers over S^1 with fibers having diameter $\Theta(\epsilon^{q_j})$ for some $q_j > 1$.

Consider a sequence $\sigma: Y \to \mathbb{C}^2$ of blow-ups of points which resolves the base points of the family of projected polar curves $\ell(\Pi_{\mathcal{D}})_{\mathcal{D} \in \Omega}$ and let T be its dual tree. Notice that the strict transforms of the curves $\ell(\Pi_{\mathcal{D}})$, $\mathcal{D} \in \Omega$ form an equisingular family of complex curves, but that these curves are not necessarily smooth, i.e., σ is not, in general, a resolution of $\ell(\Pi_{\mathcal{D}})$.

Denote by v_1 the root vertex of T, i.e., the vertex corresponding to the exceptional curve created by the first blow-up and by T_0 the subtree of T consisting of v_1 union any adjacent string or bamboo. Then Z_j is a component of $\ell^{-1}(V_j)$ where $V_j = \sigma(\mathcal{N}(T_j))$ and where T_j is a component of $T \backslash T_0$. Let v_j be the vertex of T_j adjacent to T_0. By classical curve theory, V_j is a set of the form $V_j = \{z_2 = \lambda z_1^{q_j}, a \leq |\lambda| \leq b\}$, where q_j is the inner rate of the exceptional curve represented by the vertex v_j. In particular, the 3-manifold $V_j^{(\epsilon)} = V_j \cap \{|z_1| = \epsilon\}$ is fibered over the circle S_ϵ^1 by the projection $z_1: V_j^{(\epsilon)} \to S_\epsilon^1$ and the fibers have diameter $\Theta(\epsilon^{q_j})$.

Let W be a polar wedge around Π. By Lemma 7.4.29, we know that ℓ is a locally inner bilipschitz homeomorphism outside W. Therefore, the fibers of the restriction $\zeta_j^{(\epsilon)}: Z_j^{(\epsilon)} \backslash W^{(\epsilon)} \to S^1$ have diameter $\Theta(\epsilon^{q_j})$. Moreover the Polar Wedge Lemma [BNP14, Proposition 3.4] guarantees that the fibers of the restriction of $\zeta_j^{(\epsilon)}$ to the link of a component of a polar wedge inside $(Z_j, 0)$ have diameter at most $\Theta(\epsilon^{q_j})$.

□

In [BNP14, Section 7], it is proved that each $Z_j^{(\epsilon)}$ contains loops which are essential in $X^{(\epsilon)}$. As a consequence of Theorem 7.4.39, we obtain the existence of families of fast loops γ_ϵ inside each $Z_j^{(\epsilon)}$.

7.5 Geometric Decompositions of a Surface Singularity

In this part, we explain how to break the thin pieces of the thick-thin decomposition into standard pieces which are still invariant by bilipschitz change of the inner metric. The resulting decomposition of $(X, 0)$ is what we call the inner geometric decomposition of $(X, 0)$. Then, we will define the outer geometric decomposition of $(X, 0)$, which is a refinement of the inner one, and which is invariant by bilipschitz change of the outer metric.

The inner and outer geometric decompositions will lead to several key results:

1. The complete classification of the inner Lipschitz geometry of a normal surface germ (Theorem 7.5.30);

2. A refined geometric decomposition which is an invariant of the outer Lipschitz geometry (Theorem 7.5.36);
3. A list of analytic invariants of the surface which are in fact invariants of the outer Lipschitz geometry (Theorem 7.5.38);

7.5.1 The Standard Pieces

In this section, we introduce the standard pieces of our geometric decompositions. We refer to [BNP14, Sections 11 and 13] for more details.

The pieces are topologically conical, but usually with metrics that make them shrink non-linearly towards the cone point. We will consider these pieces as germs at their cone-points, but for the moment, to simplify notation, we suppress this.

7.5.1.1 The B-Pieces

Let us start with a prototype which already appeared earlier in these notes (Example 7.4.7). Choose $q > 1$ in $\mathbb{Q}$ and $0 < a < b$ in $\mathbb{R}$. Let $Z \in \mathbb{C}^2$ be defined as the semi-algebraic set

$$Z := \{(x, y) \subset \mathbb{C}^2 : y = \lambda x^q, a \leq |\lambda| \leq b\}.$$

Then for all $\epsilon > 0$, the intersection $Z^{(\epsilon)} = Z \cap \{|x| = \epsilon\}$ is a 3-manifold (namely a thickened torus) and the restriction of the function x to $Z^{(\epsilon)}$ defines a locally trivial fibration $x \colon Z^{(\epsilon)} \to S^1_\epsilon$ whose fibers are annuli with diameter $\Theta(\epsilon^q)$.

Definition 7.5.1 ($B(q)$-Pieces) Let F be a compact oriented 2-manifold, $\phi \colon F \to F$ an orientation preserving diffeomorphism, and M_ϕ the mapping torus of ϕ, defined as:

$$M_\phi := ([0, 2\pi] \times F)/((2\pi, x) \sim (0, \phi(x)))\,.$$

Given a rational number $q > 1$, we will define a metric space $B(F, \phi, q)$ which is topologically the cone on the mapping torus M_ϕ.

For each $0 \leq \theta \leq 2\pi$ choose a Riemannian metric g_θ on F, varying smoothly with θ, such that for some small $\delta > 0$:

$$g_\theta = \begin{cases} g_0 & \text{for } \theta \in [0, \delta]\,, \\ \phi^* g_0 & \text{for } \theta \in [2\pi - \delta, 2\pi]\,. \end{cases}$$

Then for any $r \in (0, 1]$ the metric $r^2 d\theta^2 + r^{2q} g_\theta$ on $[0, 2\pi] \times F$ induces a smooth metric on M_ϕ. Thus

$$dr^2 + r^2 d\theta^2 + r^{2q} g_\theta$$

defines a smooth metric on $(0, 1] \times M_\phi$. The metric completion of $(0, 1] \times M_\phi$ adds a single point at $r = 0$. Denote this completion by $B(F, \phi, q)$. We call a metric space which is bilipschitz homeomorphic to $B(F, \phi, q)$ a $B(q)$-**piece** or simply a B-**piece**.

A $B(q)$-piece such that F is a disc is called a $D(q)$-**piece** or simply a D-**piece**.

A $B(q)$-piece such that F is an annulus $S^1 \times [0, 1]$ is called an $A(q, q)$-piece.

Example 7.5.2 The following is based on classical theory of plane curve singularities and is a generalization of the prototype given before Definition 7.5.1. Let $\sigma : Y \to \mathbb{C}^2$ be a sequence of blow-ups of points starting with the blow-up of the origin and let E_i be a component of $\sigma^{-1}(0)$ which is not the component created by the first blow-up. Then the inner rate q_{E_i} is strictly greater than 1, $B_i = \sigma(\mathcal{N}(E_i))$ is a $B(q_{E_i})$-piece fibered by the restriction of a generic linear form and the fiber consists of a disc minus a finite union of open discs inside it.

This is based on the fact that in suitable coordinates (x, y), one may construct such a piece B_i as a union of curves $\gamma_\lambda : y = \sum_{k=1}^{m} a_k x^{p_k} + \lambda x^{q_{E_i}}$, where $p_1 < \ldots < p_m < q_{E_i}$. Here $y = \sum_{k=1}^{m} a_k x^{p_k}$ is the common part of their Puiseux series and the coefficient $\lambda \in \mathbb{C}^*$ varies in a compact disc minus a finite union of open discs inside it.

Notice that if E_i intersects exactly one other exceptional curve E_j, then one gets a $D(q_{E_i})$-piece. If E_i intersects exactly two other curves E_j and E_k, one gets an $A(q_{E_i}, q_{E_i})$-piece.

7.5.1.2 The A-Pieces

Again, we start with a prototype. Choose $1 \leq q < q'$ in $\mathbb{Q}$ and $0 < a$ in $\mathbb{R}$ and let $Z \subset \mathbb{C}^2$ be defined as the semi-algebraic set

$$Z := \{(x, y) \subset \mathbb{C}^2 : y = \lambda x^s, |\lambda| = a, q \leq s \leq q'\}.$$

Then for all $\epsilon > 0$, the intersection $Z^{(\epsilon)} = Z \cap \{|x| = \epsilon\}$ is a thickened torus whose restriction of the function x to $Z^{(\epsilon)}$ defines a locally trivial fibration $x : Z^{(\epsilon)} \to S^1_\epsilon$ whose fibers are flat annuli with outer boundary of length $\Theta(\epsilon^q)$ and inner boundary of length $\Theta(\epsilon^{q'})$.

Definition 7.5.3 ($A(q, q')$-Pieces) Let q, q' be rational numbers such that $1 \leq q \leq q'$. Let A be the Euclidean annulus $\{(\rho, \psi) : 1 \leq \rho \leq 2, 0 \leq \psi \leq 2\pi\}$ in polar coordinates and for $0 < r \leq 1$ let $g_{q,q'}^{(r)}$ be the metric on A:

$$g_{q,q'}^{(r)} := (r^q - r^{q'})^2 d\rho^2 + ((\rho - 1)r^q + (2 - \rho)r^{q'})^2 d\psi^2 .$$

Endowed with this metric, A is isometric to the Euclidean annulus with inner and outer radii $r^{q'}$ and r^q. The metric completion of $(0, 1] \times \S^1 \times A$ with the metric

$$dr^2 + r^2 d\theta^2 + g_{q,q'}^{(r)}$$

compactifies it by adding a single point at $r = 0$. We call a metric space which is bilipschitz homeomorphic to this completion an $A(q, q')$-**piece** or simply an A-**piece**.

Notice that when $q = q'$, this definition of $A(q, q)$ coincides with that introduced in Definition 7.5.1.

Example 7.5.4 Let $\sigma : Y \to \mathbb{C}^2$ be as in Example 7.5.2 and let T be its dual tree. As already mentioned in Remark 7.4.36, the inner rates along any path from the root vertex to a leaf of T form a strictly increasing sequence. In particular, any edge e in T joins two vertices v and v', with inner rates respectively q and q' with $1 \leq q < q'$. Moreover, the semialgebraic set $Z = \sigma(N(v) \cap N(v'))$ is an $A(q, q')$-piece fibered by the restriction of a generic linear form and is bounded by the $B(q)$- and $B(q')$-pieces $\sigma(\mathcal{N}(v))$ and $\sigma(\mathcal{N}(v'))$.

More generally, let $S \subset T$ be a string in T which does not contain the root vertex of T. let $1 < q < q'$ be the two inner rates associated with the two vertices adjacent to S. Then $Z = \sigma(N(S))$ is an $A(q, q')$-piece fibered by the restriction of a generic linear form.

Definition 7.5.5 (Rate) The rational number q is called the **rate** of $B(q)$ or $D(q)$. The rational numbers q and q' are the two **rates** of $A(q, q')$.

7.5.1.3   Conical Pieces (or $B(1)$-Pieces)

Definition 7.5.6 (Conical Pieces) Given a compact smooth 3-manifold M, choose a Riemannian metric g on M and consider the metric $dr^2 + r^2 g$ on $(0, 1] \times M$. The completion of this adds a point at $r = 0$, giving a **metric cone on** M. We call a metric space which is bilipschitz homeomorphic to a metric cone a **conical piece** or a $B(1)$-**piece** (they were called CM-pieces in [BNP14]).

Example 7.5.7 Let $\sigma : Y \to X$ and T be as in Example 7.5.2 and let v_1 be the root vertex of T. Then $\sigma(\mathcal{N}(v_1))$ is a conical piece.

7.5.2 Geometric Decompositions of $\mathbb{C}^2$

A geometric decomposition of a semi-algebraic germ $(Y, 0)$ consists of a decomposition of $(Y, 0)$ as a union of A, B and conical pieces glued along their boundary components in such a way that the fibrations of B and A pieces coincide on the gluing.

Examples 7.5.2, 7.5.4, and 7.5.7 show that any sequence $\sigma : Y \to \mathbb{C}^2$ of blow-ups of points starting with the blow-up of the origin defines a geometric decomposition of $(\mathbb{C}^2, 0)$ whose B-pieces are in bijection with the exceptional curves E_i in $\sigma^{-1}(0)$ and the intermediate $A(q, q')$-pieces, $q < q'$ with the intersection points $E_i \cap E_j$.

Definition 7.5.8 We call this geometric decomposition of $(\mathbb{C}^2, 0)$ the geometric decomposition associated with σ.

Example 7.5.9 Consider the minimal resolution σ of the curve germ γ with Puiseux expansion $y = x^{3/2} + x^{7/4}$. Its resolution tree T, with exceptional curves E_i labelled in order of occurrence in the sequence of blow-ups, is pictured on Fig. 7.21. Each vertex is also weighted by the corresponding self-intersection E_i^2 and by the inner rate q_{E_i} in bold. The inner rates $q_{E_1} = 1, q_{E_2} = 2$ and $q_{E_3} = \frac{3}{2}$ are computed as in Example 7.4.37 using the first characteristic exponent $\frac{3}{2} = [1, 2]^+$. The two last inner rates are computed using the characteristic Puiseux exponents $\frac{3}{2}$ and $\frac{7}{4}$ as follows. Set $\frac{p_1}{q_1} = \frac{3}{2}$ and $\frac{p_2}{q_2} = \frac{7}{4}$ and write $\frac{p_2}{q_2} = \frac{p_1}{q_1} + \frac{1}{q_1} \frac{p'_2}{q'_2}$. Then the two last inner rates are computed by using the continued fraction development $\frac{p'_2}{q'_2} = [a_1, \ldots, a_r]^+$. In our case, we have $\frac{7}{4} = \frac{3}{2} + \frac{1}{2} \cdot \frac{1}{2}$, so $\frac{p'_2}{q'_2} = \frac{1}{2} = [0, 2]^+$. This gives $q_{E_4} = \frac{3}{2} + \frac{1}{1} = \frac{5}{2}$ and $q_{E_5} = \frac{3}{2} + \frac{1}{2} = \frac{7}{4}$. (Again, we refer to [EN85] or [Wal04] for details on these computations).

The underlying geometric decomposition of $(\mathbb{C}^2, 0)$ consists of:

- Five B-pieces $\sigma(\mathcal{N}(E_i))$, $i = 1, \ldots, 5$ in bijection with the vertices of T having rates respectively $1, 2, \frac{3}{2}, \frac{5}{2}, \frac{7}{4}$. Notice that the B-pieces corresponding to E_2 and E_4 are respectively a $D(2)$- and a $D(\frac{5}{2})$-piece since the corresponding vertices have valence one.

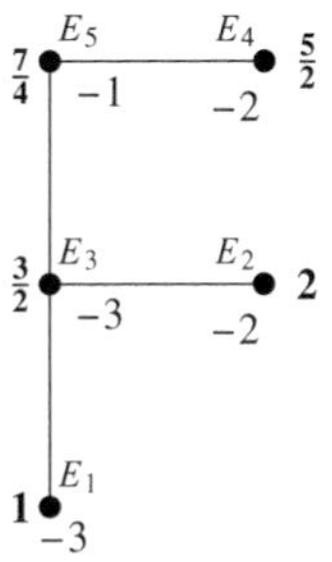

Fig. 7.21 Geometric decomposition of $(\mathbb{C}^2, 0)$ associated with the resolution of the curve $y = x^{3/2} + x^{7/4}$

- Four A-pieces in bijection with the edges of T: $\sigma(N(E_1) \cap N(E_3))$, $\sigma(N(E_3) \cap N(E_2))$, $\sigma(N(E_3) \cap N(E_5))$ and $\sigma(N(E_5) \cap N(E_4))$ which are respectively an $A(1, \frac{3}{2})$-piece, an $A(\frac{3}{2}, 2)$-piece, an $A(\frac{3}{2}, \frac{7}{4})$-piece and an $A(\frac{7}{4}, \frac{5}{2})$-piece.

Example 7.5.10 The trees T_0 and T in Example 7.4.37 describe two different geometric decompositions of $(\mathbb{C}^2, 0)$ associated with two resolutions of the curve $y = x^{5/3}$.

The following lemma shows that one can simplify a geometric decomposition by amalgamating pieces. In this lemma $\cong$ means bilipschitz equivalence and $\cup$ represents gluing along appropriate boundary components by an isometry. D^2 means the standard 2-disc.

Lemma 7.5.11 (Amalgamation Lemma)

1. $B(D^2, \phi, q) \cong B(D^2, id, q)$; $B(S^1 \times I, \phi, q) \cong B(S^1 \times I, id, q)$.
2. $A(q, q') \cup A(q', q'') \cong A(q, q'')$.
3. If F is the result of gluing a surface F' to a disk D^2 along boundary components then $B(F', \phi|_{F'}, q) \cup B(D^2, \phi|_{D^2}, q) \cong B(F, \phi, q)$.
4. $A(q, q') \cup B(D^2, id, q') \cong B(D^2, id, q)$.
5. Each $B(D^2, id, 1)$, $B(S^1 \times I, id, 1)$ or $B(F, \phi, 1)$ piece is a conical piece and a union of conical pieces glued along boundary components is a conical piece.□

Example 7.5.12 Consider again the geometric decomposition of $(\mathbb{C}^2, 0)$ introduced in Example 7.5.9. We can amalgamate the $D(2)$-piece union the $A(\frac{3}{2}, 2)$-piece to the neighbour $B(\frac{3}{2})$-piece. We can also amalgamate the $D(\frac{5}{2})$-piece union the adjacent $A(\frac{7}{4}, \frac{5}{2})$-piece to the neighbour $B(\frac{7}{4})$-piece. This produces a geometric decomposition of $(\mathbb{C}^2, 0)$ represented by the tree of Fig. 7.22, where we write inner rates only at the central vertices of B-pieces and not at the amalgamated pieces. This decomposition has five pieces: a conical $B(1)$ (black vertex), a $B(\frac{3}{2})$-piece (red vertices), a $B(\frac{7}{4})$-piece (blue vertices) and intermediate $A(1, \frac{3}{2})$- and $A(\frac{7}{4}, \frac{5}{2})$-pieces.

Remark 7.5.13 Notice that the new $B(\frac{7}{4})$-piece is now a D-piece. Then we could continue the amalgamation process by amalgamating iteratively all D-pieces. Of course, in the case of a geometric decomposition of $(\mathbb{C}^2, 0)$, an iterative amalgama-

Fig. 7.22 Amalgamated geometric decomposition

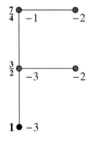

Fig. 7.23 Amalgamated
geometric decomposition for
the curve $y = x^{5/3}$

tion of the pieces always produces eventually a unique conical piece which is the whole $(\mathbb{C}^2, 0)$.

Example 7.5.14 In the tree T_0 of Example 7.5.10, the amalgamation of the $D(2)$-piece union the $A(\frac{5}{3}, 2)$-piece to the neighbour $B(\frac{5}{3})$-piece forms a bigger $B(\frac{5}{3})$-piece. The amalgamation of the $A(\frac{3}{2}, \frac{3}{2})$-piece with the two neighbour $A(1, \frac{3}{2})$- and $A(\frac{3}{2}, \frac{5}{3})$-pieces creates an intermediate $A(1, \frac{5}{3})$-piece between the $B(1)$- and the $B(\frac{5}{3})$-pieces. This creates a new geometric decomposition of $(\mathbb{C}^2, 0)$ with two B-pieces and one A-piece represented on Fig. 7.23. The red vertices correspond to the $B(\frac{5}{3})$-piece and the white one to the A-piece.

7.5.3   The Polar Wedge Lemma

Let $(X, 0) \subset (\mathbb{C}^2, 0)$ be a normal surface singularity. Consider a linear projection $\mathbb{C}^n \to \mathbb{C}^2$ which is generic for $(X, 0)$ (e.g. [NPP20a, Definition 2.4]) and denote again by $\ell: (X, 0) \to (\mathbb{C}^2, 0)$ its restriction to $(X, 0)$. Let Π be the polar curve of ℓ and let $\Delta = \ell(\Pi)$ be its discriminant curve.

Proposition 7.5.15 (Polar Wedge Lemma) *[BNP14, 3.4] Consider the resolution $\sigma: Y \to \mathbb{C}^2$ which resolves the base points of the family of projections of generic polar curves $(\ell(\Pi_{\mathcal{D}}))_{\mathcal{D} \in \Omega}$. Let Π_0 be an irreducible component of Π and let $\Delta_0 = \ell(\Pi_0)$. Let C be the irreducible component of $\sigma^{-1}(0)$ which intersects the strict transform of Δ_0^* by σ.*

Let W_{Π_0} be a polar wedge around Π_0 as introduced in Definition 7.4.30. Then W_{Π_0} is a $D(q_C)$-piece, and when $q_C > 1$, W_{Π_0} is fibered by its intersections with the real surfaces $\{h = t\} \cap X$, where $h: \mathbb{C}^n \to \mathbb{C}$ is a generic linear form.

7.5.4   The Geometric Decomposition and the Complete Lipschitz Classification for the Inner Metric

Let $(X, 0)$ be a surface germ, let $\ell: (X, 0) \to (\mathbb{C}^2, 0)$ be a generic linear projection with polar curve Π and let W be a polar wedge around Π. Let $\sigma: Y \to \mathbb{C}^2$ be the minimal sequence of blow-ups which resolves the base points of the family of projected polar curves $(\ell(\Pi_{\mathcal{D}}))_{\mathcal{D} \in \Omega}$ and consider the geometric decomposition of $(\mathbb{C}^2, 0)$ associated with σ (Definition 7.5.8).

Definition 7.5.16 Let T be the resolution tree of σ. We call Δ-*curve* any component of $\sigma^{-1}(0)$ which intersects the strict transform of the discriminant curve Δ of ℓ, and we call Δ-*node* of T any vertex representing a Δ-curve.

We call *node* of T any vertex which is either the root-vertex or a Δ-node or a vertex with valence ≥ 3.

Using Lemma 7.5.11, we amalgamate iteratively all the D-pieces of the geometric decomposition of $(\mathbb{C}^2, 0)$ associated with σ with the rule that we never amalgamate a piece containing a component of the discriminant curve Δ of ℓ. We then obtain a geometric decomposition of $(\mathbb{C}^2, 0)$ whose pieces are in bijection with the nodes of T.

Definition 7.5.17 We call this decomposition *the geometric decomposition of* $(\mathbb{C}^2, 0)$ *associated with the projection* $\ell \colon (X, 0) \to (\mathbb{C}^2, 0)$.

Example 7.5.18 Consider again the germ $(X, 0)$ of the surface E_8 with equation $x^2 + y^3 + z^5 = 0$ and the projection $\ell \colon (x, y, z) \to (y, z)$. In order to compute the geometric decomposition of $(\mathbb{C}^2, 0)$ associated with ℓ, we need to compute a resolution graph of $\sigma \colon Y \to \mathbb{C}^2$ as defined above with its inner rates. We will first compute the minimal resolution of $(X, 0)$ which factors through Nash modification.

We first consider the graph Γ of the minimal resolution π of E_8 as computed in the appendix of the present notes. We add to Γ decorations by arrows corresponding to the strict transforms of the coordinate functions x, y and $z \colon (X, 0) \to (\mathbb{C}, 0)$ and we denote the exceptional curves by $E_i, i = 1, \ldots, 8$ (the order is random). All the self-intersections of the exceptional curves equal -2 so we do not write them on the graph. We obtain the graph of Fig. 7.24.

Let $h \colon (X, 0) \to (\mathbb{C}, 0)$ be an analytic function, and let $(h \circ \pi) = \sum_{j=1}^{8} m_j E_j + h^*$ be its total transform by π, so m_j denotes the multiplicity of h along E_j and h^* its strict transform by π. Then, for all $j = 1, \ldots, 8$, we have $(h \circ \pi).E_j = 0$ ([Lau71, Theorem 2.6]). Using this, we compute the total transforms by π of the coordinate functions x, y and z:

$$(x \circ \pi) = 15E_1 + 12E_2 + 9E_3 + 6E_4 + 3E_5 + 10E_6 + 5E_7 + 8E_8 + x^*$$

$$(y \circ \pi) = 10E_1 + 8E_2 + 6E_3 + 4E_4 + 2E_5 + 7E_6 + 4E_7 + 5E_8 + y^*$$

$$(z \circ \pi) = 6E_1 + 5E_2 + 4E_3 + 3E_4 + 2E_5 + 4E_6 + 2E_7 + 3E_8 + z^*$$

Set $f(x, y, z) = x^2 + y^3 + z^5$. The polar curve Π of a generic linear projection $\ell \colon (X, 0) \to (\mathbb{C}^2, 0)$ has equation $g = 0$ where g is a generic linear combination of

Fig. 7.24 Resolution of the coordinates functions on the E_8 singularity

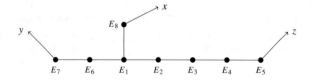

the partial derivatives $f_x = 2x$, $f_y = 3y^2$ and $f_z = 5z^4$. The multiplicities of g are given by the minimum of the compact part of the three divisors

$$(f_x \circ \pi) = 15E_1 + 12E_2 + 9E_3 + 6E_4 + 3E_5 + 10E_6 + 5E_7 + 8E_8 + f_x^*$$

$$(f_y \circ \pi) = 20E_1 + 16E_2 + 12E_3 + 8E_4 + 4E_5 + 14E_6 + 8E_7 + 10E_8 + f_y^*$$

$$(f_z \circ \pi) = 24E_1 + 20E_2 + 16E_3 + 12E_4 + 8E_5 + 16E_6 + 8E_7 + 12E_8 + f_z^*$$

We then obtain that the total transform of g is equal to:

$$(g \circ \pi) = 15E_1 + 12E_2 + 9E_3 + 6E_4 + 3E_5 + 10E_6 + 5E_7 + 8E_8 + \Pi^*.$$

In particular, Π is resolved by π and its strict transform Π^* has just one component, which intersects E_8.

Exercise 7.5.19

1. Prove that since the multiplicities $m_8(f_x) = 8$, $m_8(f_y) = 10$ and $m_8(z) = 12$ along E_8 are distinct, the family of polar curves, i.e., the linear system generated by f_x, f_y and f_z, has a base point on E_8.
2. Prove that one must blow up twice to get an exceptional curve E_{10} along which $m_{10}(f_x) = m_{10}(f_y)$, which resolves the linear system and, that this gives the resolution graph Γ' of Fig. 7.25.

Now, consider the computation of the resolution of E_8 by Laufer's method (see Appendix section 7.6) which consists of computing the double over $\ell: (X, 0) \to (\mathbb{C}^2, 0)$ branched over the discriminant curve $\Delta: y^3 + z^5 = 0$. We start with the minimal resolution $\sigma': Y' \to \mathbb{C}^2$ of Δ, and we see from the computation of self-intersections given in Appendix section 7.6 that we need to blow up five times the strict transform Δ^* in order to get the resolution graph Γ'. The resulting map is the morphism $\sigma: Y \to \mathbb{C}^2$ which resolves the base points of the family of projected polar curves $(\ell(\Pi_{\mathcal{D}}))_{\mathcal{D} \subset \Omega}$. The morphism σ is a composition of blow-ups of points and the last exceptional curve created in the process is the Δ-curve. Its inner rate is $\frac{5}{3} + 5.\frac{1}{3} = \frac{10}{3}$.

Fig. 7.25 The graph Γ'

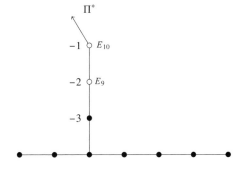

Fig. 7.26 Geometric
decomposition of $(\mathbb{C}^2, 0)$
associated with ℓ

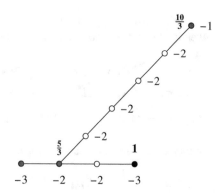

The geometric decomposition of $\mathbb{C}^2$ associated with ℓ is described by the resolution tree of σ with nodes weighted by the inner rates (Fig. 7.26).

Notice that the inner rate of the Δ-curve, which is also the inner rate of the curve $E_{v_{10}}$ of $\pi^{-1}(0)$ can also be computing using the equations as follows. For a generic $(a, b) \in \mathbb{C}^2$, $x + ay^2 + bz^4 = 0$ is the equation of the polar curve $\Pi_{a,b}$ of a generic projection. The image $\ell(\Pi_{a,b}) \subset \mathbb{C}^2$ under the projection $\ell = (y, z)$ has equation

$$y^3 + a^2 y^4 + 2aby^2 z^4 + z^5 + b^2 z^8 = 0$$

The discriminant curve $\Delta = \ell(\Pi_{0,0})$ has Puiseux expansion $y = (-z)^{\frac{5}{3}}$, while for $(a, b) \neq (0, 0)$, we get for $\ell(\Pi_{a,b})$ a Puiseux expansion $y = (-z)^{\frac{5}{3}} - \frac{a^2}{3} z^{\frac{10}{3}} + \cdots$. So the discriminant curve Δ has highest characteristic exponent $\frac{5}{3}$ and its contact exponent with a generic $\ell(\Pi_{a,b})$ is $\frac{10}{3}$.

By construction, the projection $\ell(W)$ of a polar wedge W is a union of D-pieces which refines the geometric decomposition of $(\mathbb{C}^2, 0)$ associated with ℓ. By Lemma 7.4.29, which guarantees that ℓ is a local bilipschitz homeomorphism for the inner metric outside W, any piece of this geometric decomposition outside the polar wedge W lifts to a piece of the same type. We obtain a geometric decomposition of $\overline{X \setminus W}$. Finally, the Polar Wedge Lemma 7.5.15 says that W is a union of D-pieces whose fibrations match with those of its neighbour B-pieces in $\overline{X \setminus W}$. We obtain the following result:

Proposition 7.5.20 *Each $B(q)$-piece (resp. $A(q, q')$-piece) of the geometric decomposition of $(\mathbb{C}^2, 0)$ associated with ℓ lifts by ℓ to a union of $B(q)$-pieces (resp. $A(q, q')$-pieces) in $(X, 0)$ (with the same rates).*

Therefore, we obtain a geometric decomposition of $(X, 0)$ into a union of B-pieces and A-pieces obtained by lifting by ℓ the A- and B-pieces of the geometric decomposition of $(\mathbb{C}^2, 0)$ associated with ℓ.

Definition 7.5.21 We call this decomposition the *initial geometric decomposition* of $(X, 0)$.

Fig. 7.27 Initial geometric decompostion for the singularity E_8

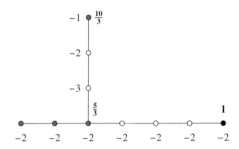

Example 7.5.22 The initial geometric decomposition of the surface germ E_8 is represented by the graph of Fig. 7.27. The vertices corresponding to the $B(\frac{5}{3})$-piece and the $B(\frac{10}{3})$-piece are in red, the $\mathcal{L}$-node is in black, the white vertices correspond to the A-pieces.

We will now amalgamate some pieces to define the *inner geometric decomposition* of $(X, 0)$. We first need to specify some special vertices in the resolution graph.

Definition 7.5.23 (Nodes) Let $\pi : Z \to X$ be a resolution of $(X, 0)$ which factors through the blow-up of the maximal ideal $e_0 : X_0 \to X$ and through the Nash modification. Let Γ be the dual resolution graph of π.

We call $\mathcal{L}$-*curve* any component of $\pi^{-1}(0)$ which corresponds to a component of $e_0^{-1}(0)$ and $\mathcal{L}$-*node* any vertex of Γ which represents an $\mathcal{L}$-curve.

We call *special $\mathcal{P}$-curve* any component E_i of $\pi^{-1}(0)$ which corresponds to a component of $\nu^{-1}(0)$ (i.e., it intersects the strict transform of the polar curve Π) and such that

1. The curve E_i intersects exactly two other components of E_j and E_k of $\pi^{-1}(0)$;
2. The inner rates satisfy: $max(q_{E_j}, q_{E_k}) < q_{E_i}$.

We call *special $\mathcal{P}$-node* any vertex of Γ which represents a special $\mathcal{P}$-curve.

We call *inner node* any vertex of Γ which has at least three incident edges or which represents a curve with genus > 0 or which is an $\mathcal{L}$- or a special $\mathcal{P}$-node.

Using Lemma 7.5.11, we now amalgamate iteratively D and A-pieces but with the rule that we never amalgamate the special A-pieces with a neighbouring piece.

Definition 7.5.24 We call this decomposition the *inner geometric decomposition* of $(X, 0)$.

The following is a straightforward consequence of this amalgamation rule. The pieces of the inner geometric decomposition of $(X, 0)$ can be described as follows (we refer to the lecture notes of Walter Neumann in the present volume for the notions of Seifert manifolds, graph decomposition and their relations with resolution graphs and plumbing).

Fig. 7.28 Inner geometric decompostion for the singularity E_8

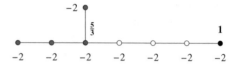

Proposition 7.5.25 *For each inner node (i) of Γ, let Γ_i be the subgraph of Γ consisting of (i) union any attached bamboo.*

1. *The B-pieces are the sets $B_i = \pi(\mathcal{N}(\Gamma_i))$, in bijection with the inner nodes of Γ. Moreover, for each node (i), B_i is a $B(q_i)$-piece, where q_i is the inner rate of the exceptional curve represented by (i) and the link $B_i^{(\epsilon)}$ is a Seifert manifold.*
2. *The A-pieces are the sets $A_{i,j} = \pi(N(S_{i,j}))$ where $S_{i,j}$ is a string or an edge joining two nodes (i) and (j) of Γ. Moreover, $A_{i,j}$ is an $A(q_i, q_j)$-piece and the link $A_{i,j}^{(\epsilon)}$ is a thickened torus having a common boundary component with both $B_i^{(\epsilon)}$ and $B_j^{(\epsilon)}$.*

In particular, the inner geometric decomposition of $(X, 0)$ induces a graph decomposition of the link $X^{(\epsilon)}$ whose Seifert components are the links $B_i^{(\epsilon)}$ and the separating tori are in bijection with the thickened tori $A_{i,j}^{(\epsilon)}$.

Remark 7.5.26 The inner geometric decomposition is a refinement of the thick-thin decomposition. Indeed, the thick part is the union of the $B(1)$-pieces and adjacent $A(1, q)$-pieces, and the thin part is the union of the remaining pieces.

Example 7.5.27 The inner geometric decomposition of the surface germ E_8 is represented by the graph of Fig. 7.28. The vertices corresponding to the $B(\frac{5}{3})$-piece are in red, the $\mathcal{L}$-node is in black, the white vertices correspond to the A-piece.

Exercise 7.5.28 Draw the resolution graph with inner rates at inner nodes representing the inner geometric decomposition of the surface germ $z^2 + f(x, y) = 0$ where $f(x, y) = 0$ is the plane curve with Puiseux expansion $y = x^{\frac{3}{2}} + x^{\frac{7}{4}}$.

Example 7.5.29 Here is an example with a special $\mathcal{P}$-node. This is a minimal surface singularity (see [Kol85]). Minimal singularities are special rational singularities which play a key role in resolution theory of surfaces, and they also share a remarkable metric property, as shown in [NPP20b]: they are Lipschitz normally embedded, i.e., their inner and outer metrics are Lipschitz equivalent. We refer to [NPP20b] for details on minimal singularities and for the computations on this particular example.

Consider the minimal surface singularity given by the minimal resolution graph of Fig. 7.29. The $\mathcal{L}$-nodes are the black vertices.

As shown in [NPP20b], one has to blow up once to obtain the minimal resolution which factors through Nash modification, creating the circled vertex on the graph of Fig. 7.30. The arrows on this graph correspond to the components of the polar curve. The inner rates (in bold) are computed in [NPP20b]. We obtain two special $\mathcal{P}$-nodes (the blue vertices).

Fig. 7.29 Minimal resolution of a minimal surface singularity

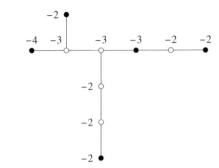

Fig. 7.30 $\mathcal{P}$-nodes and resolution of the generic polar curve

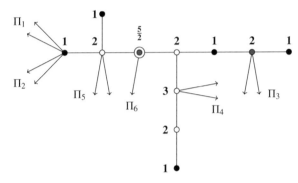

Fig. 7.31 Minimal resolution which factors through Nash transform

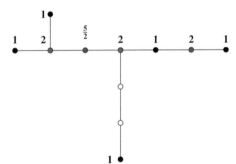

We then obtain the inner geometric decomposition described on Fig. 7.31. There are nine inner nodes, which correspond to five $B(1)$-pieces (in black), two special A-pieces with rates $\frac{5}{2}$ and 2 (in blue) and two $B(2)$-pieces (in red).

The terminology *inner geometric decomposition* comes from the following result:

Theorem 7.5.30 ([BNP14] Complete Classification Theorem for Inner Lipschitz Geometry) *The inner Lipschitz geometry of $(X, 0)$ determines and is uniquely determined by the following data:*

1. The graph decomposition of $X^{(\epsilon)}$ as the union of the links $B_i^{(\epsilon)}$ and $A_{i,j}^{(\epsilon)}$.

2. for each $B_i^{(\epsilon)}$, the inner rate $q_i \geq 1$.
3. for each $B_i^{(\epsilon)}$ such that $q_i > 1$, the homotopy class of the foliation by fibers of the fibration $z_1 \colon B_i^{(\epsilon)} \to S_\epsilon^1$.

Moreover, these data are completely encoded in the resolution graph Γ whose nodes are weighted by the rates q_i and by the multiplicities of a generic linear form h along the exceptional curves E_i up to a multiplicative constant. The latter is equivalent to the data of the maximal ideal Z_{max} (see [Nem99]) up to a multiple.

7.5.5 The Outer Lipschitz Decomposition

We now define on $(X, 0)$ a geometric decomposition of $(X, 0)$ which is a refinement of the inner geometric decomposition.

Definition 7.5.31 We use again the notations of Definition 7.5.23. We call $\mathcal{P}$-curve any component of $\pi^{-1}(0)$ which corresponds to a component of $\nu^{-1}(0)$ and $\mathcal{P}$-node any vertex of Γ which represents a $\mathcal{P}$-curve. We call *outer node* any vertex of Γ which has at least three incident edges or which represents a curve with genus > 0 or which is an $\mathcal{L}$- or a $\mathcal{P}$-node.

We start again with the initial geometric decomposition of $(X, 0)$ (Definition 7.5.21). Using Lemma 7.5.11, we amalgamate iteratively D and A-pieces but with the rule that we never amalgamate any B-pieces corresponding to a $\mathcal{P}$-node.

Definition 7.5.32 We call this decomposition the *outer geometric decomposition* of $(X, 0)$.

Let us now state an analog of Proposition 7.5.25:

Proposition 7.5.33 *The pieces of the outer geometric decomposition of $(X, 0)$ can be described as follows. For each outer node (i) of Γ, let Γ_i be the subgraph of Γ consisting of (i) and any attached bamboo.*

1. *The B-pieces are the sets $B_i = \pi(\mathcal{N}(\Gamma_i))$, in bijection with the outer nodes of Γ, and B_i is a $B(q_i)$-piece.*
2. *The A-pieces are the sets $A_{i,j} = \pi(N(S_{i,j}))$ where $S_{i,j}$ is a string or an edge joining two outer nodes i and j of Γ. Moreover, $A_{i,j}$ is an $A(q_i, q_j)$-piece.*

Example 7.5.34 The outer decomposition of the minimal singularity of Example 7.5.29 is described on Fig. 7.32. There is exactly one outer node which is not an inner node. So the outer decomposition is a refinement of the inner one: there is an extra $B(3)$-piece.

Example 7.5.35 The outer geometric decomposition of the E_8 singularity coincides with the initial geometric decomposition. So its graph is the one of Example 7.5.22.

Fig. 7.32 Outer Lipschitz
decomposition

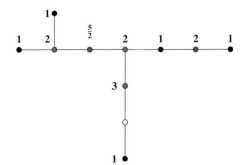

Notice that the E_8 example is very special. In general the outer geometric decomposition has much less pieces than the initial geometric decomposition.

Theorem 7.5.36 *The outer Lipschitz geometry of a normal surface singularity $(X, 0)$ determines the geometric decomposition of $(X, 0)$ up to self-bilipschitz homeomorphism.*

Moreover, these data are completely encoded in the resolution graph Γ where each outer node is weighted by the inner rate q_{E_i} and the self-intersection E_i^2 of the corresponding exceptional curve E_i and by the multiplicity m_i of a generic linear form h along E_i.

Notice that the latter is equivalent to the datum of the maximal ideal cycle $Z_{max} := \sum_i m_i E_i$ in the resolution π. (see [Nem99, 2.1] for details on Z_{max}).

The statement of Theorem 7.5.36 has some similarities with that of Theorem 7.5.30, but the proof is radically different. The proof of the Lipschitz invariance of the outer geometric decomposition is based on a bubble trick which enables one to recover first the B-pieces of the decomposition which have highest inner rate. Then the whole decomposition is determined by an inductive process based again on a second bubble trick by exploring the surface with bubbles having radius ϵ^q, with decreasing rates q. The proof is delicate. We refer to [NP12] for details.

Remark 7.5.37 As a byproduct of the bilipschitz invariance of the maximal ideal cycle Z_{max} stated in Theorem 7.5.36 we obtain that the multiplicity $m(X, 0)$ is an invariant of the Lipschitz geometry of a complex normal surface germ. Indeed, $m(X, 0)$ is nothing but the sum of the multiplicities of Z_{max} at the $\mathcal{L}$-nodes of Γ.

In [FdBFS18], the authors prove a broad generalization of this fact: the outer Lipschitz geometry of a surface singularity (not necessarily normal) determines its multiplicity

The Lipschitz invariance of the multiplicity is no longer true in higher dimension as proved in [BFSV18]. Actually, the proofs of the bilipschitz invariance in [NP12] and [FdBFS18] deeply use the classification of three-dimensional manifolds up to diffeomorphisms.

Using again bubble tricks, we can prove that beyond the weighted graph Γ and the maximal cycle Z_{max}, the outer Lipschitz geometry determines a large amount of other classical analytic invariants. These invariants are of two types. The first is related to the generic hyperplane sections and the blow-up of the maximal ideal, and the second is related to the polar and discriminant curves of generic plane projections and the Nash modification:

Theorem 7.5.38 ([NP12]) *If $(X, 0)$ is a normal complex surface singularity, then the outer Lipschitz geometry on X determines:*

1. *Invariants from generic hyperplane sections:*

   (a) *the decoration of the resolution graph Γ by arrows corresponding to the strict transform of a generic hyperplane section (these data are equivalent to the maximal ideal cycle Z_{max});*
   (b) *for a generic hyperplane H, the outer Lipschitz geometry of the curve $(X \cap H, 0)$.*

2. *Invariants from generic plane projections:*

   (a) *the decoration of the resolution graph Γ by arrows corresponding to the strict transform of the polar curve of a generic plane projection;*
   (b) *the embedded topology of the discriminant curve of a generic plane projection;*
   (c) *the outer Lipschitz geometry of the polar curve of a generic plane projection.*

7.6   Appendix: The Resolution of the E8 Surface Singularity

In this appendix, we explain how to compute the minimal resolution graph of a singularity with equation of the form $z^2 + f(x, y) = 0$ by Laufer's method, described in [Lau71, Chapter 2] (page 23 to 27 for the E_8 singularity). Here we will just introduce the method and perform it in the particular case of E_8. We invite the reader to study it in [Lau71] or in the course of Jawad Snoussi [Sno19] in the present volume and to compute further examples.

Laufer's method is based on the Hirzebruch–Jung algorithm which resolves any surface singularity.

7.6.1   Hirzebruch–Jung Algorithm

We refer to the paper [PP11] of Patrick Popescu-Pampu for more details on this part. The Hirzebruch–Jung algorithm consists in considering a finite morphism $\ell \colon (X, 0) \to (\mathbb{C}^2, 0)$. Then one takes a resolution $\sigma \colon Y \to \mathbb{C}^2$ of the discriminant curve Δ of ℓ, one resolves the singularities of Δ and one considers the pull-back $\widetilde{\sigma} \colon Z \to X$ of σ by ℓ. We then also have a finite morphism $\widetilde{\ell} \colon Z \to Y$ such that $\sigma \circ \widetilde{\ell} = \widetilde{\sigma} \circ \ell$. Let $n \colon Z_0 \to Z$ be the normalization of Z.

The singularities of Z_0 are quasi-ordinary singularities relative to the projection $\tilde{\ell} \circ n \colon Z_0 \to Y$ and with discriminant the singularities of the curve $\sigma^{-1}(\Delta)$, which are ordinary double points. Resolving these remaining singularities, one gets a morphism $\alpha \colon Z \to Z_0$. The composition $\pi = \tilde{\sigma} \circ \alpha \colon Z \to X$ is a resolution of $(X, 0)$ (in general far from being minimal).

7.6.2 Laufer's Method

It resolves the surface germ $(X, 0) \colon x^2 + f(y, z) = 0$ by applying Hirzebruch–Jung algorithm with the projection $\ell \colon (x, y, z) \mapsto (y, z)$ and then by giving an easy way to compute Z from a specific resolution tree T of the discriminant curve $\Delta \colon f(y, z) = 0$ of ℓ.

Let us explain it on the singularity E_8. The discriminant Δ of the projection $\ell \colon (X, 0) \to (\mathbb{C}^2, 0)$ has equation $f(y, z) = 0$ where $f(y, z) = y^3 + z^5$. We start with the minimal resolution $\sigma \colon Y \to \mathbb{C}^2$ of Δ. Its exceptional divisor consists in four curves $E_1, \ldots, E_4$ labelled in their order of occurrence in the sequence of blow-ups. Let m_i be the multiplicity of the function f along E_i. The integer m_i is defined as the exponent u^{m_i} appearing in the total transform of f by σ in coordinates centered at a smooth point of E_i, where $u = 0$ is the local equation of E_i (see the lectures notes of José Luis Cisneros in the present volume for details). So it can be computed when performing the sequence of blow-ups resolving $f = 0$.

By [Lau71, Theorem 2.6] these multiplicities can also be computed from the self-intersections E_j^2 using the fact that for each $j = 1, \ldots, 4$, the intersection $(\sigma^* f).E_j$ in Y equals 0, where $(\sigma^* f) = m_1 E_1 + \ldots + m_4 E_4 + f^*$, with f^* the strict transform of $f = 0$ by σ.

One obtains the following resolution tree T on which each vertex (i) is weighted by the self intersection E_i^2 and by the multiplicity m_i (into parenthesis). The arrow represents the strict transform of Δ.

Now, we blow up any intersection point between two components of the total transform $(\sigma^* f)$ having both even multiplicities. In the case of E_8, all the multiplicities are even, so we blow up every double point of $(\sigma^* f)$. We obtain the resolution tree T' of Fig. 7.34.

In the particular case where there are no adjacent vertices having both odd multiplicities (this is the case in the above tree T'), a resolution graph Γ of $(X, 0) \colon x^2 = f(y, z)$ is obtained as follows: Γ is isomorphic to T', and for any vertex (i) of T', the corresponding vertex of Γ carries self-intersection $2E_i^2$ if

Fig. 7.33 The minimal resolution tree of $y^3 + z^5 = 0$

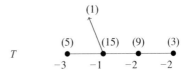

Fig. 7.34 The resolution tree T'

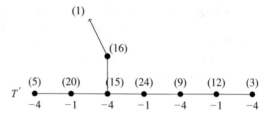

Fig. 7.35 The resolution graph Γ for E_8 with multiplicities of $y^3 + z^5$

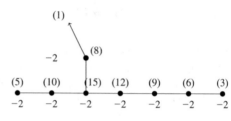

Fig. 7.36 The resolution minimal graph for E_8

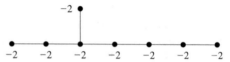

the multiplicity m_i is odd and $\frac{1}{2}E_i^2$ if it is even. Moreover, the multiplicity of the function $f \circ \ell \colon (X, 0) \to (\mathbb{C}, 0)$ is m_i if m_i is odd and $\frac{1}{2}m_i$ if m_i is even. In the case of E_8, we obtain the resolution graph Γ of Fig. 7.35, where the arrow represents the strict transform of $f \circ \ell \colon (X, 0) \to (\mathbb{C}, 0)$.

There are no -1-exceptional curves which could be blown down. So forgetting f and its multiplicities we get the well known graph of the minimal resolution of E_8 (Fig. 7.36).

In the case where some consecutive vertices have multiplicities which are odd, some vertices of T' may give two vertices in Γ. We refer to [Lau71] for details.

References

[BF08] L. Birbrair, A. Fernandes, Inner metric geometry of complex algebraic surfaces with isolated singularities. Commun. Pure Appl. Math. **61**(11), 1483–1494 (2008) 183

[BFN08] L. Birbrair, A. Fernandes, W.D. Neumann, Bi-Lipschitz geometry of weighted homogeneous surface singularities. Math. Ann. **342**(1), 139–144 (2008) 183

[BFN09] L. Birbrair, A. Fernandes, W.D. Neumann, Bi-Lipschitz geometry of complex surface singularities. Geom. Dedicata **139**, 259–267 (2009) 183

[BFN10] L. Birbrair, A. Fernandes, W.D. Neumann, Separating sets, metric tangent cone and applications for complex algebraic germs. Sel. Math. (N.S.) **16**(3), 377–391 (2010) 183

[BFSV18] L. Birbrair, A. Fernandes, J.E. Sampaio, M. Verbitsky, Multiplicity of singularities is not a bi-Lipschitz invariant (2018). arXiv preprint arXiv:1801.06849 172, 211

[BNP14] L. Birbrair, W.D. Neumann, An. ne Pichon, The thick-thin decomposition and the bilipschitz classification of normal surface singularities. Acta Math. **212**(2), 199–256 (2014) 167, 168, 183, 184, 186, 187, 188, 189, 190, 191, 196, 197, 198, 200, 203, 209

[BS75] J. Briançon, J.-P. Speder, La trivialité topologique n'implique pas les conditions de Whitney. C. R. Acad. Sci. Paris Sér. A-B **280**(6), Aiii, A365–A367 (1975) 190

[Dur79] A.H. Durfee, Fifteen characterizations of rational double points and simple critical points. Enseign. Math. (2) **25**(1–2), 131–163 (1979) 190

[Dur83] A.H. Durfee, Neighborhoods of algebraic sets. Trans. Am. Math. Soc. **276**(2), 517–530 (1983) 185

[EN85] D. Eisenbud, W. Neumann, in *Three-Dimensional Link Theory and Invariants of Plane Curve Singularities*. Annals of Mathematics Studies, vol. 110 (Princeton University Press, Princeton, 1985) 194, 201

[FdBFS18] J. Fernández de Bobadilla, A. Fernandes, J. Edson Sampaio, Multiplicity and degree as bi-Lipschitz invariants for complex sets. J. Topol. **11**(4), 958–966 (2018) 172, 211

[Fer03] A. Fernandes, Topological equivalence of complex curves and bi-Lipschitz homeomorphisms. Mich. Math. J. **51**(3), 593–606 (2003) 176

[GBGPPP19] E.R. García Barroso, P.D. González Pérez, P. Popescu-Pampu, The valuative tree is the projective limit of Eggers-Wall trees. Rev. R. Acad. Cienc. Exactas Fís. Nat. Ser. A Mat. RACSAM **113**(4), 4051–4105 (2019) 174

[GBT99] E. García Barroso, B. Teissier, Concentration multi-échelles de courbure dans des fibres de Milnor. Comment. Math. Helv. **74**(3), 398–418 (1999) 194

[KL77] T.C. Kuo, Y.C. Lu, On analytic function germs of two complex variables. Topology **16**(4), 299–310 (1977) 174

[KO97] K. Kurdyka, P. Orro, Distance géodésique sur un sous-analytique. **10**, 173–182 (1997). Real algebraic and analytic geometry (Segovia, 1995) 194

[Kol85] J. Kollár, Toward moduli of singular varieties. Compos. Math. **56**(3), 369–398 (1985) 208

[Lau71] H.B. Laufer, in *Normal Two-Dimensional Singularities*. Annals of Mathematics Studies, vol. 71 (Princeton University Press, Princeton; University of Tokyo Press, Tokyo, 1971) 204, 212, 213, 214

[Mos85] T. Mostowski, Lipschitz equisingularity. Dissertationes Math. (Rozprawy Mat.) **243**, 46 (1985) 172

[Nem99] A. Némethi, Five lectures on normal surface singularities, in *Low Dimensional Topology (Eger, 1996/Budapest, 1998)*. Bolyai Society Mathematical Studies, vol. 8 János Bolyai Math. Soc. (Budapest, 1999), pp. 269–351. With the assistance of Ágnes Szilárd and Sándor Kovács 210, 211

[NP12] W.D. Neumann, A. Pichon, Lipschitz geometry of complex surfaces: analytic invariants and equisingularity (2012). preprint arXiv:1211.4897 167, 168, 172, 176, 211, 212

[NP14] W.D. Neumann, A. Pichon, Lipschitz geometry of complex curves. J. Singul. **10**, 225–234 (2014) 167, 176, 179

[NPP20a] W.D. Neumann, H.M. Pedersen, A. Pichon, A characterization of lipschitz normally embedded surface singularities. J. Lond. Math. Soc. (2) **101**, 612–660 (2020) 194, 203

[NPP20b] W.D. Neumann, H.M. Pedersen, A. Pichon, Minimal surface singularities are lipschitz normally embedded. J. Lond. Math. Soc. (2) **101**, 641–658 (2020) 208

[Par88] A. Parusiński, Lipschitz properties of semi-analytic sets. Ann. Inst. Fourier (Grenoble) **38**(4), 189–213 (1988) 172

[Par94] A. Parusiński, Lipschitz stratification of subanalytic sets. Ann. Sci. École Norm. Sup. (4) **27**(6), 661–696 (1994) 172

[PP11] P. Popescu-Pampu, Introduction to Jung's method of resolution of singularities, in *Topology of Algebraic Varieties and Singularities*. Contemporary Mathematics, vol. 538 (American Mathematical Society, Providence, 2011), pp. 401–432 195, 212

[PT69]   F. Pham, B. Teissier, Fractions lipschitziennes d'une algèbre analytique complexe et saturation de zariski. Prépublications Ecole Polytechnique, (M17.0669) (1969) 176

[Sam17] J.E. Sampaio, Multiplicity, regularity and blow-spherical equivalence of complex analytic sets (2017). arXiv preprint arXiv:1702.06213 172

[Sno19] J. Snoussi, A quick trip into local singularities of complex curves and surfaces, in *The International School on Singularities and Lipschitz Geometry* (2019) 191, 212

[Spi90] M. Spivakovsky, Sandwiched singularities and desingularization of surfaces by normalized Nash transformations. Ann. Math. (2) **131**(3), 411–491 (1990) 188, 192

[Tei82] B. Teissier, Variétés polaires. II. Multiplicités polaires, sections planes, et conditions de Whitney, in *Algebraic geometry (La Rábida, 1981)*. Lecture Notes in Mathematics, vol. 961 (Springer, Berlin, 1982), pp. 314–491 174, 192

[TMW89] L.D. Trang, F. Michel, C. Weber, Sur le comportement des polaires associées aux germes de courbes planes. Compos. Math. **72**(1), 87–113 (1989) 176

[Wal04] C.T.C. Wall, in *Singular Points of Plane Curves*. London Mathematical Society Student Texts, vol. 63 (Cambridge University Press, Cambridge, 2004) 174, 194, 201

Chapter 8
The biLipschitz Geometry of Complex Curves: An Algebraic Approach

Arturo Giles Flores, Otoniel Nogueira da Silva, and Bernard Teissier

Abstract The purpose of these notes is to explain why a generic projection to a plane of a reduced germ of complex analytic space curve is a bi-Lipschitz homeomorphism for the outer metric. This is related to the fact that all topologically equivalent germs of plane curves are exactly the generic projections of a single germ of a space curve. The analytic algebra of this germ is the algebra of Lipschitz meromorphic functions on any of its generic projections. An application to the Lipschitz geometry of polar curves is given.

8.1 Introduction

These are the lecture notes of the course given by Bernard Teissier during the second week of the "International School on Singularities and Lipschitz Geometry" which took place in Cuernavaca, Mexico from June 11 to June 22, 2018. The aim of the course was to explore the concept of "generic plane linear projection" of a complex analytic germ of curve in $\mathbf{C}^N$. The objects of our study will therefore be germs of curves $(X, 0) \subset (\mathbf{C}^N, 0)$, linear map germs $\pi : (\mathbf{C}^N, 0) \to (\mathbf{C}^2, 0)$, and the images $(\pi(X), 0) \subset (\mathbf{C}^2, 0)$.

Intuitively, a projection π is generic for $(X, 0)$ if a small variation of π does not change the "equisingularity type" (or embedded topological type) of the image

A. G. Flores
Departamento de Matemáticas y Física, Universidad Autónoma de Aguascalientes, Aguascalientes, México
e-mail: arturo.giles@cimat.mx

O. N. Silva
Instituto de Matemáticas, Universidad Nacional Autónoma de México, Cuernavaca, Morelos, Mexico
e-mail: otoniel@im.unam.mx

B. Teissier (✉)
Institut Mathématique de Jussieu-Paris Rive Gauche, Paris, France
e-mail: bernard.teissier@imj-prg.fr

217
W. Neumann, A. Pichon (eds.), *Introduction to Lipschitz Geometry of Singularities*, Lecture Notes in Mathematics 2280, https://doi.org/10.1007/978-3-030-61807-0_8

$(\pi(X), 0)$ in $(\mathbf{C}^2, 0)$. The main objective was to provide algebraic criteria for a projection to be generic and to use them to prove two results related to Lipschitz geometry:

1. That all equisingular (topologically equivalent) germs of reduced plane curves are, up to analytic isomorphism, images of a single space curve $(X, 0) \subset (\mathbf{C}^N, 0)$ by generic linear projections $\pi : \mathbf{C}^N \to \mathbf{C}^2$, and that the restriction $\pi|(X, 0) : (X, 0) \to (\pi(X), 0)$ to $(X, 0)$ of such a generic projection is a biLipschitz map for the metrics induced by the Hermitian metrics of their respective ambient spaces. In particular, all topologically equivalent germs of plane curves are biLipschitz equivalent.

2. Given a reduced equidimensional germ of a complex space $(X, 0) \hookrightarrow (\mathbf{C}^N, 0)$, with dimension d, we consider a "general" projection $\pi : \mathbf{C}^N \to \mathbf{C}^2$ and the **polar curve on X** associated to the projection π. It is the closure in X of the critical locus of the restriction of π to the smooth part of X. If it is not empty, it is a curve usually denoted by $P_{d-1}(X, \pi)$ which plays an important role in the study of the Lipschitz geometry of X. We can consider π as defining a plane projection of the space curve $(P_{d-1}(X, \pi), 0)$ which varies with π. The result is that if the projection π is sufficiently general, then it is a generic plane projection for the curve $(P_{d-1}(X, \pi), 0) \subset (\mathbf{C}^N, 0)$.

The course assumed a certain familiarity with algebraic or complex analytic geometry, such as the definition of a complex analytic space X, the fact that its local algebras of functions are analytic algebras, that is, quotients of rings of convergent power series with complex coefficients, that the singular locus $\mathrm{Sing}X$ consisting of points where the local algebra is not isomorphic to a ring of convergent power series, is a closed analytic subspace, etc. The reader is also encouraged to consult the article [Sno20] of Jawad Snoussi in this volume, which has some overlap with the content of these notes.

8.1.1 What Is a Germ of Complex Analytic Curve?

A complex analytic curve[1] X may be locally regarded as a family of points in an open subset U of the complex affine space $\mathbf{C}^N$ which is the union of finitely many sets of points depending analytically on one complex parameter. It can also be defined as the zero set of a finite number of holomorphic functions $f_1, \ldots, f_s$ on U satisfying certain algebraic conditions:

$$X = \{z \in U \mid f_1(z) = \cdots = f_s(z) = 0\}.$$

---

[1] For more details on what follows in this section, we refer the reader to [Tei07].

A **germ of curve** $(X, 0) \subset (\mathbf{C}^N, 0)$ at a point which we take to be the origin is an equivalence class of curves in open neighborhoods of the origin. Two such objects defined respectively in U and U' are equivalent if their restrictions to a third neighborhood of the origin $U'' \subset U \cap U'$ coincide. Of course, when we speak of germs, we think of representatives in some "sufficiently small" neighborhood of the origin. Because of analyticity, to give a germ is equivalent to giving the convergent power series of $f_1, \ldots, f_s$ around the origin with respect to some coordinate system.

This allows us to associate to the germ $(X, 0) \subset (\mathbf{C}^N, 0)$ the analytic algebra of germs of holomorphic functions on $(X, 0)$:

$$O_{X,0} := \mathbf{C}\{z_1, \ldots, z_N\}/\langle f_1, \ldots, f_s \rangle,$$

where $\mathbf{C}\{z_1, \ldots, z_N\}$ denotes the ring of convergent power series. In these notes we will only be interested in reduced germs, meaning that the ideal $J := \langle f_1, \ldots, f_s \rangle$ is radical and $O_{X,0}$ is a reduced analytic algebra of pure dimension 1.

In the case of plane curves ($N = 2$) the ideal $I = \langle f \rangle \mathbf{C}\{x, y\}$ is principal and f is square free, which means that f has a factorization of the form $f = f_1 \cdots f_r$, where each f_i is irreducible in $\mathbf{C}\{x, y\}$ and they are all different. The point is that the f_i's correspond to germs $(X_i, 0) \subset (\mathbf{C}, 0)$ of analytically irreducible curves called the **branches** of the curve:

$$(X, 0) = \bigcup_{i=1}^{r} (X_i, 0).$$

For arbitrary N, the branches $(X_i, 0)$ correspond to the prime ideals appearing in the primary decomposition of the ideal (0) in $O_{X,0}$

$$(0) = P_1 \cap \ldots \cap P_r, \text{ where each } P_i \text{ is a minimal prime in } O_{X,0}$$

A germ of curve $(X, 0) \subset (\mathbf{C}^N, 0)$ may also be described parametrically by r sets of power series

$$\varphi_1^i(t_i), \ldots, \varphi_N^i(t_i) \in \mathbf{C}\{t_i\}, \ 1 \le i \le r,$$

where again r is the number of branches. For each i, $z_k = \varphi_k^i(t_i)$, $1 \le k \le N$ defines a germ of map $(\mathbf{D}_i, 0) \longrightarrow (\mathbf{C}^N, 0)$ where $\mathbf{D}_i$ is a disk in $\mathbf{C}$. Together these r n-uples of series correspond to a multi-germ of map

$$\varphi : \bigsqcup_{i=1}^{r} (\mathbf{D}_i, 0) \longrightarrow (\mathbf{C}^N, 0); \ z_k = \varphi_k^i(t_i). \tag{8.1}$$

The connexion between these two definitions goes back to Newton, who showed that an equation $f(x, y) = 0$, with $f(0, 0) = 0$ has solutions $y(x)$ which are power series in x with **rational exponents with bounded denominators** and coefficients in the algebraic closure of the smallest field containing the coefficients of $f(x, y)$. For Newton $f(x, y)$ is a polynomial with real coefficients, but the method works for series over any field k of characteristic zero. Note that if $\frac{\partial f(x,y)}{\partial y}$ does not vanish at $(0, 0)$ the implicit function Theorem gives us a power series $y(x)$ with integers as exponents. In the general case, such a series $y(x) = \sum_{i \in \mathbf{N}} a_i x^{\frac{i}{n}}$ gives rise to a parametrization

$$x = t^n, \quad y = \sum_{i \in \mathbf{N}} a_i t^i.$$

of one of the branches of the curve over an algebraic extension of k.

8.1.2 Structuring a Parametrization

Suppose that we have an irreducible and reduced germ of curve in $(\mathbf{C}^N, 0)$, given by $z_k = \varphi_k(t) \in \mathbf{C}\{t\}$, $k = 1, \ldots, N$. For simplicity we shall write $z_k = \varphi_k(t) = \sum_i a_k^{(i)} t^i$. We assume that the group generated by the exponents is $\mathbf{Z}$, which means that they are coprime. Let n be the smallest exponent appearing in all the series $\varphi_k(t)$; up to reindexing the variables z_i we may assume that it is the order of $\varphi_1(t)$, so that we may write $\varphi_1(t) = a_1^{(n)} t^n (1 + \psi(t))$ with $\psi(0) = 0$. By making a homothetic change on the variable z_1 we may assume that $a_1^{(n)} = 1$. Since we are in characteristic zero, we may extract an n-th root of the unit $1 + \psi(t)$ so that $1 + \psi(t) = u(t)^n$ where $u(t)$ is again invertible in $\mathbf{C}\{t\}$. Now we make the change of parameter $t' = tu(t)$ so that $\varphi_1(t') = t'^n$. Now by making a linear change of the form $z_i - a_i z_1$ on the coordinates $z_2, \ldots, z_N$ we may assume that z_1 is the only variable where the lowest exponent n appears. Geometrically this means that our curve is tangent to the z_1-axis at the origin: its set-theoretic tangent cone is the z_1-axis. Similarly, by making now a non linear change of coordinates of the form $z_i - \sum_k a_k^{(i)} z_1^k$ we may assume that the first exponent appearing in each $\varphi_k(t')$ is not divisible by n. This is geometrically more subtle and corresponds to Hironaka's **maximal contact**. Since t' is now our uniformizing parameter, we call it t henceforth.

Let us now compare $z_1 = t^n$ with one of the other coordinates, which we may write (up to a homothetic change of variables) $z_i = \varphi_i(t) = t^{b_i} + \cdots$. It may be that the exponents appearing in $\varphi_i(t)$ and n are not coprime. As we shall see below it means that the projection of our curve to the (z_1, z_i)-plane is not reduced. If that is the case, we may begin by dividing all the exponents by their greatest common divisor. The interesting case is therefore that of two series expansions t^n, $\varphi(t)$ with coprime exponents: we are in the case $N = 2$ of a plane branch to which we now turn.

The Case of a Plane Branch As we saw, after a change of coordinates and of uniformizing parameter, we can describe our plane branch by: $z_1 = t^n$, $z_2 = \varphi(t) \in$

$\mathbf{C}\{t\}$ where the smallest exponent of t in $\varphi(t)$ is not divisible by n. This smallest exponent is traditionally denoted by β_1. We take the g.c.d. of n and β_1; set $e_1 = (n, \beta_1) < n$. If $e_1 = 1$, the series $\varphi(t)$ is of the form $t^{\beta_1} + \sum_{k \geq 1} a_k t^{\beta_1 + k}$. If $e_1 > 1$, since the exponents are coprime, there has to be a smallest exponent β_2 in the series $\varphi(t)$ which is not divisible by e_1. Then we set $e_2 = (e_1, \beta_2) < e_1$, and we continue in this manner. Since $n > e_1 > e_2 > \cdots$ there exists an integer g such that $e_g = (e_{g-1}, \beta_g) = 1$. Finally we have the following structure for $\varphi(t)$: its expansion is decomposed into segments corresponding to the divisibility properties of the exponents.

$$z_2 = t^{\beta_1} + \sum_{k=1}^{s_1} a_{\beta_1 + k e_1} t^{\beta_1 + k e_1} + a_{\beta_2} t^{\beta_2} + \sum_{k=1}^{s_2} a_{\beta_2 + k e_2} t^{\beta_2 + k e_2} + \cdots + a_{\beta_j} t^{\beta_j} + \sum_{k=1}^{s_j} a_{\beta_j + k e_j} t^{\beta_j + k e_j} +$$

$$\cdots + a_{\beta_g} t^{\beta_g} + \sum_{k=1}^{\infty} a_{\beta_g + k} t^{\beta_g + k},$$

where all a_{β_i} are $\neq 0$ and each sum has to stop before the g.c.d. of the exponents drops and only the last segment is possibly infinite. The set of integers $n, \beta_1, \beta_2, \ldots, \beta_g$, which is often also denoted by $\beta_0, \beta_1, \beta_2, \ldots, \beta_g$, is called the **Puiseux characteristic** of the branch and the β_i, or sometimes the $\frac{\beta_i}{n}$, are called the **characteristic exponents**. It determines and is determined by the **embedded topological type** of the branch (see [Zar71, §7], [Zar68, Theorem 2.1, pg. 983], [Lej73]). This means that if two germs of plane branches $(X_1, 0)$ and $(X_2, 0)$ have the same Puiseux characteristic there exists a homeomorphism $(U_1, 0) \rightarrow (U_2, 0)$ of neighborhoods of the origin mapping the representative $X_1 \subset U_1$ to $X_2 \subset U_2$, and conversely. The two germs are also said to be **equisingular**. We shall meet this Puiseux characteristic again after Example 8.5.25 below, where we shall see that it determines not only the topology but also the biLipschitz geometry of the branch.

After what we have seen, the expansion above can be reinterpreted as a Newton expansion in terms of $t = z_1^{\frac{1}{n}}$, but here we have to choose a n-th root of z_1. The algebraic interpretation is that $\varphi(z_1^{\frac{1}{n}}) \in \mathbf{C}\{\{z_1\}\}\{z_1^{\frac{1}{n}}\}$ determines a cyclic extension of the field $\mathbf{C}\{\{z_1\}\}$ of meromorphic functions in z_1 with Galois group equal to the group μ_n of n-th roots of 1. The n series $\varphi(\omega z_1^{\frac{1}{n}})$, $\omega \in \mu_n$, are the roots of a unitary polynomial $\prod_{\omega \in \mu_n} (z_2 - \varphi(\omega z_1^{\frac{1}{n}})) \in \mathbf{C}\{z_1\}[z_2]$ whose vanishing is an equation for our germ of curve in the sense we shall see in the next section.

The structure of the series gives rise to a filtration of the Galois group:

$$\mu_n \supset \mu_{e_1} \supset \mu_{e_2} \supset \cdots \supset \mu_{e_g} = \{1\},$$

with the characteristic property that if we set $n = e_0$ and denote by v_t the t-adic order of a series, then for $1 \leq k \leq g$, we have that

$$\omega \in \mu_{e_{k-1}} \setminus \mu_{e_k} \iff v_t(\varphi(\omega t) - \varphi(t)) = \beta_k.$$

Let us now refine the structure according to [Zar06, Chapters III, IV,V]. The parametrization of a branch by t^n, $y(t)$ as above presents its analytic algebra $O_{X,0}$ as a subalgebra of $\mathbf{C}\{t\}$. The t-adic orders of the series in t which are in $O_{X,0}$ form a numerical semigroup $\Gamma \subset \mathbf{N}$ since one can multiply them and stay in $O_{X,0}$. Since the exponents are coprime the complement of Γ in $\mathbf{N}$ is finite (Dickson's Lemma) and the semigroup Γ is finitely generated. The smallest element c of $\mathbf{N}$ such that all integers $\geq c$ are in Γ is called the **conductor** of the semigroup. It is not difficult to verify (see [Zar06, Chapter III, Lemma 1.1]) that if the order of a series $\xi(t) \in O_{X,0}$ is $> \beta_1$, then $\xi(t) \in \langle x, y \rangle^2$, and therefore if the order s of $\xi(t)$ is in Γ we can make a change of coordinates $x' = x$, $y' = y - \xi(t)$ to eliminate a term in t^s from the expansion of $y(t)$. Using this, and the fact that by definition any element of Γ is the order of a series in $O_{X,0}$, Zariski proved in [Zar06, Chapter III, Proposition 1.2]:

Proposition 8.1.1 (Zariski)

(1) Assume that $n > 2$. Then one has $c \geq \beta_1 + 1$. Let $s_1, \ldots, s_q$ be the integers of the set $\{\beta_1 + 1, \ldots, c\}$ which do not belong to Γ. The branch $(X, 0)$ is analytically isomorphic to a branch given parametrically by:

$$x'(t) = t^n, \quad y'(t) = t^{\beta_1} + \sum_{i=1}^{q} a'_{s_i} t^{s_i}.$$

(2) If $n = 2$ then β_1 is odd since our germ is irreducible and the conductor is β_1; our curve is analytically isomorphic to $x(t) = t^2$, $y(t) = t^{\beta_1}$.

Zariski calls this a **short representation**. There are more simplifications of the expansion of $y(t)$ one can make without changing the analytic type. See [Zar06, Chapters III, IV,V].

The next thing we need to know is that the semigroup Γ determines and is determined by the Puiseux characteristic of the branch: it is a complete invariant of the equisingularity class. See [Zar06, Chap. II, §3]. In particular, in the short expansion, the coefficients of the t^{β_i} are $\neq 0$.

With this description of branches, we are able to describe the **contact** of two branches, which plays a key role in the characterization of the topological (and biLipschitz) type of a reduced germ of plane curve.

We shall see below how, conversely, the image of a parametrization can be defined by equations.

The modern presentation of the parametrization of a curve goes through the normalization, which is the topic of the next section.

8.2 Normalization

The property of being normal has an algebraic aspect which has to do with integral extension of rings.

Definition 8.2.1 Let $R \subset S$ be rings.

- The inclusion $R \subset S$ is called a finite extension if S is a finitely generated R-module.
- An element $s \in S$ is called integral over R if and only if it satisfies an equation

$$s^h + a_1 s^{h-1} + \cdots + a_{h-1} s + a_h = 0$$

with all $a_i \in R$. The extension is called integral if every element $s \in S$ is integral over R. (Just as in field theory, if the extension $R \subset S$ is finite it is integral. See [deJP00, Lemma 1.5.2])
- The ring R is said to be integrally closed in S if every element in S which is integral over R already belongs to R.
- The ring R is called normal if it is reduced and integrally closed in its total quotient ring $Q(R)$.

Suppose that R is a reduced ring. Recall that the set of non-zero divisors of a ring R is a multiplicatively closed set and the corresponding ring of fractions $Q(R)$ is called the total ring of fractions. It has the property that the canonical morphism $R \to Q(R)$ is injective.

The **normalization of** R is defined as the set $\overline{R}$ of all elements of $Q(R)$ which are integral over R. It is a reduced ring, integrally closed in $Q(R)$ and whose total ring of fractions coincides with $Q(R)$. In particular, the normalization $\overline{R}$ is a normal ring. Moreover, for the rings appearing in analytic or algebraic geometry, the extension $R \subset \overline{R}$ is finite in the sense that $\overline{R}$ is a finitely generated R-module.[2]

So what about if we start with the analytic algebra $O_{X,0}$ of a germ of analytic space $(X, 0) \subset (\mathbf{C}^N, 0)$? We will say that the germ $(X, 0)$ is **normal** if $O_{X,0}$ is a normal ring.

- Unique factorization domains are normal ([deJP00, Thm 1.5.5]), so the ring of power series $\mathbf{C}\{z_1, \ldots, z_n\}$ and the corresponding smooth germ $(\mathbf{C}^N, 0)$ are normal.
- Noetherian normal local rings are integral domains ([deJP00, Thm 1.5.7]), so a normal germ $(X, 0)$ is irreducible.
- Suppose $(X, 0)$ is irreducible. Since $O_{X,0}$ and its normalization have the same total ring of fractions, which in this case is a field, it follows from what we have just seen that $\overline{O_{X,0}}$ is a local noetherian domain. Moreover, by [deJP00, Cor. 3.325] it is an analytic algebra and so we can associate to it a normal germ $(\overline{X}, 0)$.

---

[2]It is interesting to note that the term "integral" comes from algebraic number theory in the tradition of Dedekind and the definition of the ring of integers of an algebraic number field, while the term "normal" was used by Zariski (see [Zar39]) in the course of his studies in birational geometry and resolution of singularities to designate an algebraic variety which could **not** be presented as the image of a different one by a finite birational map. This is why the terms "integral closure in the total ring of fractions" and "normalization" are used in algebraic or analytic geometry as names for the algebraic and geometric aspects of the same operation.

In particular we have:

$$O_{\overline{X},0} = \overline{O_{X,0}}.$$

- Splitting of normalization ([deJP00, Thm. 1.5.20]) tells us that if we have the irreducible decomposition

$$(X, 0) = (X_1, 0) \cup \ldots \cup (X_s, 0),$$

then the normalization $\overline{O_{X,0}}$ is equal to a direct sum of analytic algebras which are the normalizations of the analytic algebras $O_{X_i,0}$ corresponding to the irreducible components $(X_i, 0)$:

$$\overline{O_{X,0}} = \bigoplus_{i=1}^{s} \overline{O_{X_i,0}}.$$

Note that this implies that $(X, 0)$ and $(\overline{X}, 0)$ have the same dimension.

A **multi-germ** of analytic spaces (X, x) is a finite disjoint union:

$$(X, x) := (X_1, x_1) \sqcup (X_2, x_2) \sqcup \ldots \sqcup (X_r, x_r)$$

of germs of analytic spaces. The ring $O_{X,x}$ by definition is equal to $\bigoplus_{i=1}^{r} O_{X_i,x_i}$. The multi-germ (X, x) is called normal if $O_{X,x}$ is a normal ring.

Let $(Y, y) = (Y_1, y_1) \sqcup \ldots \sqcup (Y_s, y_s)$ be another multi-germ. A map $\varphi : (X, x) \to (Y, y)$ of multi-germs is given by a system of maps

$$\varphi_i : (X_i, x_i) \to (Y_{\alpha(i)}, y_{\alpha(i)}), \quad i \in \{1, \ldots, r\}, \ \alpha(i) \in \{1, \ldots, s\}.$$

Such a map φ induces, and is induced by, a **C**-algebra map $\varphi^* : O_{Y,y} \to O_{X,x}$.

Definition 8.2.2 Let (X, x) be a germ of analytic space. A normalization of (X, x) is a normal multi-germ $(\overline{X}, \overline{x})$ together with a finite, generically 1-1 map

$$n : (\overline{X}, \overline{x}) \to (X, x).$$

With this definition at hand, for any germ of analytic space $(X, 0)$ with irreducible decomposition

$$(X, 0) = (X_1, 0) \cup \ldots \cup (X_s, 0),$$

we can now obtain a normal multi-germ

$$(\overline{X}, \overline{x}) = (\overline{X_1}, x_1) \sqcup \ldots \sqcup (\overline{X_s}, x_s)$$

with associated normal ring

$$\overline{O_{X,0}} = \bigoplus_{i=1}^{s} \overline{O_{X_i,0}} = \bigoplus_{i=1}^{s} O_{\overline{X}_i, x_i},$$

and it is not hard to prove that the inclusion map $O_{X,0} \hookrightarrow \overline{O_{X,0}}$ induces a finite and generically 1-1 map, proving thus the existence of normalization ([deJP00, Thm 4.4.8]). Note that, geometrically, the normalization of a germ separates the irreducible components and normalizes each of them separately.

Example 8.2.3 Let $(X, 0) \subset (\mathbf{C}^2, 0)$ be the germ of plane curve defined by $f(x, y) = x^2 - y^2$. It has two irreducible components $(X_1, 0)$ and $(X_2, 0)$ with associated analytic algebras

$$O_{X_1,0} = \mathbf{C}\{x, y\}/\langle x - y \rangle \qquad O_{X_2,0} = \mathbf{C}\{x, y\}/\langle x + y \rangle.$$

These two germs are smooth, in particular they are normal and we have:

$$O_{X,0} = \frac{\mathbf{C}\{x, y\}}{\langle x^2 - y^2 \rangle} \longrightarrow \frac{\mathbf{C}\{x, y\}}{\langle x - y \rangle} \bigoplus \frac{\mathbf{C}\{x, y\}}{\langle x + y \rangle} = \overline{O_{X,0}}$$

$$f \longmapsto (f + \langle x - y \rangle, f + \langle x + y \rangle)$$

Since the germs are smooth and of dimension 1, their analytic algebras are isomorphic to the ring of convergent power series $\mathbf{C}\{t\}$:

$$\mathbf{C}\{x, y\}/\langle x - y \rangle \to \mathbf{C}\{t\} \qquad x \mapsto t, \; y \mapsto t$$

$$\mathbf{C}\{x, y\}/\langle x + y \rangle \to \mathbf{C}\{u\} \qquad x \mapsto u, \; y \mapsto -u$$

This means that the resulting normalization map

$$n : (\mathbf{C}, 0) \sqcup (\mathbf{C}, 0) \to (X, 0)$$

is the parametrization of each of the branches $t_1 \mapsto (t, t)$ and $t_2 \mapsto (u, -u)$.

It is useful to consider a function-theoretic interpretation of normal spaces. A general result tells us that in a smooth germ $(\mathbf{C}^d, 0)$ if you have a meromorphic function which is (locally) bounded then it is actually holomorphic (See for example [GF02, IV.4]). The algebraic version is that a locally bounded meromorphic function h satisfies an integral dependence relation of the form:

$$h^m + c_1 h^{m-1} + \cdots + c_m = 0; \qquad c_j \in O_n := \mathbf{C}\{z_1, \ldots, z_n\},$$

and since O_n is normal then $h \in O_n$.

Now there are many more analytic spaces for which $O_{X,x}$ is normal than just the non singular ones.

Definition 8.2.4 Given a reduced germ of analytic space (X, x), we call a function $f : X \setminus \text{Sing } X \to \mathbf{C}$ **weakly holomorphic** at $x \in X$ if:

- f is holomorphic on $X \setminus \text{Sing } X$ in a neighborhood of x.
- f is (locally) bounded near x.

A function is weakly holomorphic on X if it is so at every point.

The key point is proving that the germs at $x \in X$ of weakly holomorphic functions on X form a ring which is canonically isomorphic to the normalization of $O_{X,x}$. That is, f is weakly holomorphic on X if and only if it is meromorphic and satisfies an integral dependence relation. This gives us the following characterization:

Theorem 8.2.5 ([deJP00, Thm 4.4.15])

(1) Let (X, x) be a germ of reduced analytic space. Then a function f is weakly holomorphic on X if and only if f is in the integral closure of $O_{X,x}$ in its total ring of quotients.

(2) The integral closure of $O_{X,x}$ in its total ring of quotients is a direct sum of analytic algebras.

(3) The reduced germ (X, x) is normal if and only if every weakly holomorphic function germ can be extended to a holomorphic function.

Remark 8.2.6 Since this fact is fundamental for what follows, here is an idea of why boundedness and polynomial equation are related: The roots of a polynomial are bounded in terms of its coefficients, so a solution of a polynomial equation with holomorphic coefficients is bounded because holomorphic functions are. In the other direction, let $h = \frac{f}{g}$, with f, g in the maximal ideal of $O_{X,0}$, be our meromorphic function, let $(Y, 0) \subset (X, 0)$ be the subset defined by the ideal $\langle f, g \rangle O_{(X,0)}$, and consider the analytic subspace X' of $X \times \mathbf{P}^1$ which is the closure of the graph of the map $X \setminus Y \to \mathbf{P}^1$ defined by $x \mapsto [f(x) : g(x)] \in \mathbf{P}^1$. It is contained in the hypersurface of $X \times \mathbf{P}^1$ defined by $T_2 f(x) - T_1 g(x) = 0$ where $[T_1 : T_2]$ are projective coordinates on $\mathbf{P}^1$. The first projection induces a holomorphic map $e \colon X' \to X$ (we are blowup the ideal $\langle f, g \rangle$). The fiber over 0 is a complex analytic subspace of $\mathbf{P}^1$ and therefore is either $\mathbf{P}^1$ or a finite subset of it. If our meromorphic function is bounded, the point $[1 : 0] \in \mathbf{P}^1$ is not in the fiber, so that by the Weierstrass preparation Theorem (see Theorem 8.2.8 below), for a small enough representative X of the germ $(X, 0)$ the map $X' \to X$ is finite and X' has to be a hypersurface in $X \times \mathbf{C}$: its equation is our integral dependence relation.

Example 8.2.7 For the germ $(X, 0) \subset (\mathbf{C}^2, 0)$ defined by $xy = 0$ we have

$$\overline{O_{X,0}} = \mathbf{C}\{x\} \oplus \mathbf{C}\{y\}.$$

The function $f = (1, 0)$, meaning it is the constant function 1 on the x-axis and the constant function 0 on the y-axis, is holomorphic on $X \setminus \text{Sing } X = X \setminus \{0\}$ and is certainly bounded so it is weakly holomorphic. Note that it can not be continuously extended to $(X, 0)$. As a meromorphic function it can be written as

$$f(x, y) = \frac{x}{x + y}.$$

Let us wrap up this discussion on normal spaces and normalization by stating several important properties of which you can find detailed expositions in [Loj91, GLS07] and [KK83].

1. If X is reduced, the **non normal locus** is the set of points $x \in X$ where the local algebra $O_{X,x}$ is not normal; it is the complement of the **normal locus** and is a closed analytic subspace contained in the singular locus $\text{Sing} X$ of X. It is defined by the **conductor sheaf** which is the annihilator of the coherent O_X-module $\overline{O_X}/O_X$ and thus a coherent sheaf of ideals.
2. If T is a normal space and X is reduced then any map $T \to X$ which does not map any irreducible component of T to the non-normal locus of X factors uniquely through the normalization $n : \overline{X} \to X$.
3. If X is normal then $\dim \text{Sing}(X) \leq \dim X - 2$ (Singular locus of codimension at least 2).
4. If X is normal, the polar locus of a meromorphic function is either of codimension 1 or empty.

Going back to the curve case, a classical result of commutative algebra ([deJP00, Thm 4.4.9]) states that a Noetherian local ring of dimension one is normal if and only if it is regular. This implies that if $(X, 0) = \bigcup_{i=1}^r (X_i, 0) \subset (\mathbf{C}^N, 0)$ is a germ of analytic curve with r branches then the normal ring $\overline{O_{X,0}}$ is isomorphic to a direct sum of r copies of $\mathbf{C}\{t\}$ and the corresponding normalization map is equal to the parametrization of each branch, thus recovering the description in (8.1).

For plane curves, this result can also be seen using algebraic field extensions, but first we need a couple of definitions and the Weierstrass preparation Theorem. A convergent power series $f \in \mathbf{C}\{z_1, \ldots, z_N\}$ is called regular of order b in z_N if the power series $f(0, \ldots, 0, z_N)$ in the variable z_N has a zero of order b. A simple calculation shows that if f is of order b in the sense that $f \in \langle z_1, \ldots, z_N \rangle^b \setminus \langle z_1, \ldots, z_N \rangle^{b+1}$, then after a general linear change of coordinates, f is regular of order b in z_N (see [deJP00, Lemma 3.2.2]). Geometrically this means that if we consider the germ of hypersurface $(X, 0) \subset (\mathbf{C}^{N-1} \times \mathbf{C}, 0)$ defined by f and the first projection $p : X \to \mathbf{C}^{N-1}$, then for a small enough representative the fiber $p^{-1}(0)$ is the single point 0.

Theorem 8.2.8 (Weierstrass Preparation Theorem (See [deJP00, Thm 3.2.4]))
Let $f \in \mathbf{C}\{z_1, \ldots, z_N\}$ be regular of order b in z_N. Then there exists a unique monic polynomial $P \in \mathbf{C}\{z_1, \ldots, z_{N-1}\}[z_N]$

$$P(z_1, \ldots, z_N) = z_N^b + a_1(z_1, \ldots, z_{N-1})z_N^{b-1} + \cdots + a_N(z_1, \ldots, z_{N-1})$$

with $a_i(0) = 0$, and a unit $u \in \mathbf{C}\{z_1, \ldots, z_N\}$ such that we have the equality of convergent power series

$$f = uP.$$

As a consequence of this result we deduce two important facts: if we choose adequate coordinates such that $f = uP$ then it is equivalent to seek solutions of $f(z_1, \ldots, z_N) = 0$ and of $P(z_1, \ldots, z_N) = 0$. As a geometric consequence of this we get that if we consider the first projection as before and $p^{-1}(0) = \{0\}$, then for any point $q = (q_1, \ldots, q_{N-1}) \in \mathbf{C}^{N-1}$ sufficiently close to the origin the points of the fiber $p^{-1}(q)$ correspond to the roots of the polynomial of degree b

$$P(q_1, \ldots, q_{N-1}, z_N) = z_N^b + a_1(q_1, \ldots, q_{N-1})z_N^{b-1} + \cdots + a_N(q_1, \ldots, q_{N-1}),$$

and so all nearby fibers are also finite. More generally one uses this result to prove that if a complex analytic map $p \colon X' \to X$ is such that for some point $0 \in X$ we have that $p^{-1}(0)$ is a finite set, then there exists a neighborhood U of 0 in X such that the restricted map $p^{-1}(U) \to U$ is finite. See [deJP00, Thm 3.4.24].

For curve singularities, there is a classical invariant which measures how far the singularity is from being normal, or non singular. It has several geometric interpretations, the classical one being "diminution of genus", and we shall see more about it below.

Definition 8.2.9 Let $(X, 0)$ be a reduced curve singularity. Its δ-**invariant** is

$$\delta = \dim_{\mathbf{C}} \overline{O_{X,0}}/O_{X,0}.$$

This quotient is a finite dimensional vector space because it is the stalk of a coherent sheaf supported at the origin. For plane, and more generally Gorenstein, branches we have the equality $c = 2\delta$, where c is the conductor defined before Proposition 8.1.1. See [Zar06, Chap. II, §1].

Going back to the plane curve case, that is curves $(X, 0) \subset (\mathbf{C}^2, 0)$ defined by a convergent power series $f \in \mathbf{C}\{x, y\}$, or according to the Weierstrass preparation Theorem and possibly after a linear change of coordinates, by a polynomial $P \in \mathbf{C}\{x\}[y]$. Now from an algebraic point of view, consider the field of fractions $\mathbf{C}\{\{x\}\}$ of the integral domain $\mathbf{C}\{x\}$; the irreducible polynomial $y^n - x \in \mathbf{C}\{\{x\}\}[y]$ defines an algebraic extension of degree n of $\mathbf{C}\{\{x\}\}$, denoted by $\mathbf{C}\{\{x^{\frac{1}{n}}\}\}$, which is a Galois extension with Galois group equal to the group $\boldsymbol{\mu}_n$ of n-th roots of unity in $\mathbf{C}$. The action of $\boldsymbol{\mu}_n$ is exactly the change in determination of $x^{\frac{1}{n}}$ determined by $x^{\frac{1}{n}} \mapsto \omega x^{\frac{1}{n}}$ for $\omega \in \boldsymbol{\mu}_n$. A series of the form $y = \sum a_i x^{\frac{i}{n}}$ such that the greatest common divisor of n and all the exponents i which effectively appear is 1 gives n different series as ω runs through $\boldsymbol{\mu}_n$.

Suppose now that our polynomial P is an irreducible element of $\mathbf{C}\{x\}[y]$ of degree n. Then the Newton polygon method (see for example [Tei07, Che78], or

[BK86, Section 8.3]) provides a series $y(x^{1/m}) \in \mathbf{C}\{x^{\frac{1}{n}}\}$ such that $P(x, y(x^{\frac{1}{n}})) = 0$ and we have the equality:

$$P(x, y) = \prod_{\omega \in \mu_n} \left(y - y(\omega x^{\frac{1}{n}}) \right).$$

In particular we have that

$$\mathbf{C}\{\{x\}\}^* := \bigcup_{n \in \mathbf{N}} \mathbf{C}\{\{x^{\frac{1}{n}}\}\}$$

is an algebraically closed field (See [Wal78, IV.3] or [Che78, Thm 8.2.1]), and so every polynomial $P \in \mathbf{C}\{x\}[y]$ has all its roots in $\mathbf{C}\{\{x\}\}^*$. Finally, the relation with the parametrizations given by the normalization is the following, if

$$y(x^{\frac{1}{n}}) \in \mathbf{C}\{x^{\frac{1}{n}}\} \subset \mathbf{C}\{\{x\}\}^*$$

is a root of $P(x, y)$, then by taking $x = t^n$ we get the parametrization

$$t \mapsto (t^n, y(t)).$$

Let us finish this section by looking at plane projections from an algebraic perspective. For simplicity suppose $(X, 0) \subset (\mathbf{C}^N, 0)$ is a reduced and irreducible germ of complex analytic curve with $N \geq 3$. Let us write the associated analytic algebra

$$\mathcal{O}_{X,0} = \mathbf{C}\{z_1, \ldots, z_N\}/I,$$

where I is a prime ideal, and so $\mathcal{O}_{X,0}$ is an integral domain. If we choose a sufficiently general coordinate system (or if you prefer after a general linear coordinate change) the Noether normalization Theorem ([deJP00, Corollary 3.3.19]) tells us that we have a finite ring extension $\mathbf{C}\{z_1\} \hookrightarrow \mathcal{O}_{X,0}$. This implies that the we have an algebraic field extension

$$\mathbf{C}\{\{z_1\}\} \subset \mathrm{Quot}\left(\mathcal{O}_{X,0}\right),$$

and by the primitive element Theorem there exists an element $f \in \mathcal{O}_{X,0}$ such that $\mathrm{Quot}\left(\mathcal{O}_{X,0}\right) = \mathbf{C}\{\{z_1\}\}[f]$.

So if we denote by $\mathbf{C}\{z_1, f\}$ the analytic algebra obtained as the quotient of $\mathbf{C}\{x, y\}$ by the kernel of the map $\mathbf{C}\{x, y\} \to \mathcal{O}_{X,0}$ defined by $x \mapsto z_1 + J, y \mapsto f$ then we have finite ring extensions with the same field of fractions

$$\mathbf{C}\{z_1, f\} \hookrightarrow \mathcal{O}_{X,0} \hookrightarrow \mathbf{C}\{t\}.$$

Now $\mathbf{C}\{z_1, f\}$ is the analytic algebra of a plane curve $(X_1, 0) \subset (\mathbf{C}^2, 0)$ and it has the same normalization as $O_{X,0}$. We have used the primitive element Theorem as a substitute for the proof of the existence of a projection $\mathbf{C}^N \rightarrow \mathbf{C}^2$ sufficiently general for it to induce a " bimeromorphic" map $(X, 0) \rightarrow (X_1, 0)$. However the primitive element Theorem does not tell us much about the geometric nature of the projection. That is the object of the following sections.

8.3 Fitting Ideals: A Good Structure for the Image of a Finite Map

In this section, following [Tei73, §3] and [Rim72, Definition 5.6], [Tei77, §5], we will give the definitions of Fitting ideals, which we will use later to give a definition of the image, as a *complex analytic space*, of a finite map between complex analytic spaces.

Let A be a ring, and let M be an A-module of finite presentation, that is, there is an exact sequence, called a presentation of M:

$$A^q \xrightarrow{\Psi} A^p \longrightarrow M \longrightarrow 0,$$

where $p, q \in \mathbf{N}$. For each integer j we associate to M the ideal $F_j(M)$ of A generated by the $(p - j) \times (p - j)$ minors of the matrix (with entries in A) representing Ψ. Here we need the convention that if there are no $(p - j) \times (p - j)$ minors because j is too large, i.e., $j \geq p$, then $F_j(M) = A$ (the empty determinant is equal to 1) and if, at the other extreme, $p - j > q$, set $F_j(M) = 0$ (the ideal generated by the empty set is 0).

A Theorem of Fitting (see [Tou72, Chap. I, §2], [Eis95, Chap. 20, §2]) asserts that the ideals $F_j(M)$ depend only on the A-module M and not on the choice of a presentation. We call $F_j(M)$ the j-th Fitting ideal of M.

More generally, if (X, O_X) is a ringed space, and $\mathcal{M}$ a coherent sheaf of O_X-modules, we can define a sheaf of ideals $\mathcal{F}_i(\mathcal{M})$ of O_X, by defining $\mathcal{F}_i(\mathcal{M})$ locally as above, and then by uniqueness the ideals found locally patch up into a sheaf of ideals. Remark also that since $\mathcal{F}_i(\mathcal{M})$ is locally finitely generated, $\mathcal{F}_i(\mathcal{M})$ will be a coherent sheaf of ideals as soon as O_X is coherent, which is the case for a complex analytic space by Oka's Theorem (see [Loj91, Chap. VI, 1.3]).

Let now $f : (X, O_X) \rightarrow (Y, O_Y)$ be a map of complex analytic spaces. We would like to define the image of f as a complex analytic *subspace* of (Y, O_Y). This is not always possible, and in particular if one hopes to get a closed complex subspace of Y it is better to assume f is proper, and here we will consider only the case where f is *finite* (that is, proper with finite fibres).

The first sheaf of ideals that comes to mind as a candidate to define $f(X)$ is the sheaf of functions g on Y such that $g \circ f = 0$ on X, i.e., the annihilator sheaf of the sheaf of O_Y-modules f_*O_X, which is coherent by a theorem of Grauert-Remmert:

$$\text{Ann}_{O_Y}(f_*(O_X)) = \text{sheaf\{functions } g \text{ on } Y \text{ such that } g \cdot f_*O_X = 0 \}.$$

This is *not* a good choice because its formation does not commute with base extension, as we will show by an example below (Example 8.3.3).

The second option is the 0-th Fitting ideal of f_*O_X, which set theoretically also defines the image of f, since as a set the subspace of Y defined by it is $\{y \in Y \mid \dim_{\mathbb{C}}(f_*O_X) > 0\} = \{y \in Y \mid (f_*O_X)_y \neq 0\}$. Indeed, since the O_Y-module f_*O_X is coherent, it has locally on suitable open sets U of Y a presentation by an exact equence of $O_Y(U)$-modules:

$$O_Y(U)^q \xrightarrow{\Psi} O_Y(U)^p \longrightarrow f_*O_X(U) \longrightarrow 0.$$

The sheaf of ideals $\mathcal{F}_0(f_*O_X)$ is then generated on U by the $p \times p$ minors of a matrix representing Ψ.

Since both the formation of direct images and the formation of Fitting ideals commute with base change (see Proposition 8.3.2 below), this definition of the image will also have this property. So we set:

Definition 8.3.1 Let $f : X \rightarrow Y$ be a finite morphism of complex analytic spaces. The image $\text{im}(f)$ of f is the subspace of Y defined by the coherent sheaf of ideals $\mathcal{F}_0(f_*O_X)$. It is sometimes called the **Fitting image** of f to distinguish it from the one defined by the annihilator.

Proposition 8.3.2

1. *The formation of* $\text{im}(f)$ *commutes with base change: Given a complex analytic map* $\phi: T \rightarrow Y$, *consider the map* $f_T: X \times_Y T \rightarrow T$ *obtained by base extension, where* $X \times_Y T$ *is the fiber product. Then* $\text{im}(f_T) = \phi^{-1}(\text{im}(f))$ *as analytic spaces.*
2. *We have the inclusion* $\mathcal{F}_0(f_*O_X) \subset \text{Ann}(f_*O_X)$ *and the equality* $\sqrt{\mathcal{F}_0(f_*O_X)} = \sqrt{\text{Ann}(f_*O_X)}$.

Proof

(1) Since O_X is a finitely generated O_Y-module the O_T-module $O_{X \times_Y T}$ is equal to $O_X \otimes_{O_Y} O_T$ and if M is a finitely presented A-module as above and $A \rightarrow B$ is a map of algebras, then

$$B^q \xrightarrow{\Psi \otimes_A 1} B^p \longrightarrow M \otimes_A B \longrightarrow 0$$

is a presentation of $M \otimes_A B$ as a B-module and the matrix of $\Psi \otimes_A 1$ is the matrix of Ψ so that $F_j(M \otimes_A B) = F_j(M).B$.

(2) The inclusion follows directly from Cramer's rule and the equality from the definition of the Fitting ideal as defining the set of points where the cokernel of the second arrow is not zero.   □

Example 8.3.3 Let $f : (\mathbf{C}, 0) \to (\mathbf{C}^2, 0)$ be given by $x = t^{2k}$, $y = t^{3k}$ for some integer k. The set-theoretic image of f is the curve $y^2 - x^3 = 0$. However, we wish to obtain an ideal defining a space supported on that curve, but possibly with nilpotent functions. Let us compute $\mathcal{F}_0(f_*(\mathcal{O}_\mathbf{C}))_0$ as the 0-th Fitting ideal of $\mathbf{C}\{t\}$ considered as $\mathbf{C}\{x, y\}$-module via the map of rings $\mathbf{C}\{x, y\} \to \mathbf{C}\{t\}$ sending x to t^{2k} and y to t^{3k}. We must write a presentation of $\mathbf{C}\{t\}$ as $\mathbf{C}\{x, y\}$-module. Let $e_0 = 1$, $e_1 = t, \ldots, e_{2k-1} = t^{2k-1}$. It is easily seen that they form a system of generators of $\mathbf{C}\{t\}$ as $\mathbf{C}\{x, y\}$-module, and that between them we have the following $2k$ relations:

$$
\begin{aligned}
xe_k - ye_0 &= 0, & x^2e_0 - ye_k &= 0 \\
xe_{k+1} - ye_1 &= 0, & x^2e_1 - ye_{k+1} &= 0 \\
&\ \vdots & &\ \vdots \\
xe_{2k-1} - ye_{k-1} &= 0, & x^2e_{k-1} - ye_{2k-1} &= 0
\end{aligned}
$$

which are independent. Hence we have a sequence of $\mathbf{C}\{x, y\}$-modules:

$$
0 \longrightarrow \overset{2k-1}{\underset{i=0}{\bigoplus}} \mathbf{C}\{x, y\}e_i \overset{\psi}{\to} \overset{2k-1}{\underset{i=0}{\bigoplus}} \mathbf{C}\{x, y\}e_i \overset{\varphi}{\to} \mathbf{C}\{t\} \longrightarrow 0
$$

with $\varphi(e_i) = t^i$, and ψ is given by the $2k \times 2k$ matrix

$$
\psi =
\begin{bmatrix}
-y & 0 & \cdots & 0 & x & 0 & \cdots & 0 \\
0 & -y & \cdots & 0 & 0 & x & \cdots & 0 \\
\vdots & & \ddots & \vdots & & & \ddots & 0 \\
0 & 0 & \cdots & -y & 0 & 0 & \cdots & x \\
x^2 & 0 & \cdots & 0 & -y & 0 & \cdots & 0 \\
0 & x^2 & \cdots & 0 & 0 & -y & \cdots & 0 \\
\vdots & & \ddots & \vdots & & & \ddots & 0 \\
0 & 0 & \cdots & x^2 & 0 & 0 & \cdots & -y
\end{bmatrix}
$$

It is not hard to see that the sequence is exact, which means that the independent relations we have found must generate all relations between the e_i. Indeed, there is a general reason why $\mathbf{C}\{t\}$ must have a resolution of length 1 as $\mathbf{C}\{x, y\}$-module: the $\mathbf{C}\{x, y\}$-module $\mathbf{C}\{t\}$ is of homological dimension one (see [MP89]) and therefore the module of relations between the e_i is a free submodule of $\bigoplus_{i=0}^{2k-1} \mathbf{C}\{x, y\}$ and thus of rank $\leq 2k - 1$.

By permuting rows and columns of ψ one checks that $\det(\psi) = (y^2 - x^3)^k$ i.e., we have shown that

$$\mathcal{F}_0(f_* \mathcal{O}_C)_0 = (y^2 - x^3)^k \mathbf{C}\{x, y\}$$

Let us now calculate $\mathrm{Ann}_{\mathbf{C}\{x,y\}}\mathbf{C}\{t\}$; the annihilator is just the kernel of the map $\mathbf{C}\{x, y\} \to \mathbf{C}\{t\}$, which is the ideal generated by $(y^2 - x^3)$, certainly different from our Fitting ideal if $k > 1$.

Let us now make a base change by restricting our map over the x-axis, i.e., by the inclusion $\{y = 0\} \subset (\mathbf{C}^2, 0)$ or algebraically by $\mathbf{C}\{x, y\} \to \mathbf{C}\{x\}$ sending y to 0. Then the annihilator of $\mathbf{C}\{t\} \otimes_{\mathbf{C}\{x,y\}} \mathbf{C}\{x\} = \mathbf{C}\{t\}/(t^{3k})$ viewed as $\mathbf{C}\{x\}$-module is $(x^2)\mathbf{C}\{x\}$ while the image in $\mathbf{C}\{x\}$ of $(y^2 - x^3)\mathbf{C}\{x, y\}$ is $(x^3)\mathbf{C}\{x\}$. This shows that the formation of the annihilator does not commute with base change.

8.3.1 Equations Versus Parametrizations

As we said in Sect. 8.1.1, a germ of curve $(X_0, 0)$, abstractly, is a germ of a purely 1-dimensional analytic space, hence it is described by an analytic algebra $\mathcal{O}_{X_0,0}$ of pure dimension 1. Geometrically, $(X_0, 0)$ can be effectively given in two ways:

By equations: By giving an ideal $I = \langle f_1, \ldots, f_m \rangle$ in $\mathbf{C}\{x_1, \ldots, x_N\}$ such that $\mathcal{O}_{X_0,0} \simeq \mathbf{C}\{x_1, \ldots, x_N\}/I$. Saying that $\mathcal{O}_{X_0,0}$ is purely one-dimensional means that the ideal $\langle 0 \rangle$ has a primary decomposition $\langle 0 \rangle = Q_1 \cap \ldots \cap Q_r$ where $\sqrt{Q_i} = P_i$ is a minimal prime ideal in $\mathcal{O}_{X_0,0}$, and $\dim(\mathcal{O}_{X_0,0}/I) = 1$.

By a parametrization: By giving ourselves a germ of finite map $p : \bigsqcup_{i=1}^{r}(\mathbf{C}, 0) \to (\mathbf{C}^N, 0)$.

Here one has to be very careful: except when $n = 2$, it is not true, even if $r = 1$ and p is generically 1-to-1 so that the image (given by the Fitting structure) of this mapping is a reduced curve: it will have "embedded components" concentrated at the singular points, as will be shown in Example 8.3.4. The analysis of this phenomenon is beyond the scope of these notes. The case where $n = 2$ is explained in Proposition 8.3.6 in the next section.

Example 8.3.4 Consider the curve $(X_0, 0)$ parametrized by $n(t) = (t^4, t^6, t^7)$ which is a complete intersection (with the reduced structure) with ideal

$$\langle y^2 - x^3, z^2 - x^2 y \rangle \mathbf{C}\{x, y, z\}.$$

We have that $\mathbf{C}\{t\}$ is generated as a $\mathbf{C}\{x, y, z\}$-module by $e_0 = 1, e_1 = t, e_2 = t^2$ and $e_3 = t^3$ and it is not difficult to see that the relations are described by the following matrix

$$\Psi = \begin{bmatrix} y & 0 & -x & 0 \\ 0 & y & 0 & -x \\ -x^2 & 0 & y & 0 \\ 0 & -x^2 & 0 & y \\ z & 0 & 0 & -x \\ -x^2 & z & 0 & 0 \\ 0 & -x^2 & z & 0 \\ 0 & 0 & -x^2 & z \end{bmatrix}$$

that is, Ψ is the matrix of a presentation

$$\mathbf{C}\{x, y, z\}^8 \xrightarrow{\Psi} \mathbf{C}\{x, y, z\}^4 \longrightarrow \mathbf{C}\{t\} \longrightarrow 0.$$

of the $\mathbf{C}\{x, y, z\}$-module $\mathbf{C}\{t\}$. Computing the 4×4 minors of Ψ we find that:

$$F_0(\mathbf{C}\{t\}) =$$

$$\langle y^2 - x^3, z^2 - x^2 y \rangle \cap \langle z^2, xy^3, y^4, xy^2 z - x^4 z, y^3 z, x^4 y, x^3 y^2, x^3 yz, x^6, x^5 z \rangle \mathbf{C}\{x, y, z\},$$

where $\sqrt{\langle z^2, xy^3, y^4, xy^2 z - x^4 z, y^3 z, x^4 y, x^3 y^2, x^3 yz, x^6, x^5 z \rangle} = \langle x, y, z \rangle \mathbf{C}$ $\{x, y, z\}$.

The ring $\mathbf{C}\{x, y, z\}/F_0(\mathbf{C}\{t\})$ is not purely one-dimensional: it has an **embedded component**, i.e., an ideal of the primary decomposition of the ideal (0) which defines a subspace of strictly lower dimension, in this case dimension zero.

8.3.2 Deformations of Equations vs. Deformations of Parametrizations

In this subsection we consider deformations of a curve. We will follow the presentation given in [BG80]. The results in this subsection are due to B. Teissier (see [Tei77]).

Let $(X_0, 0) \subset (\mathbf{C}^N, 0)$ be a germ of a reduced curve and $X_0 \subset \mathbf{B}_0$ a representative, where $\mathbf{B}_0 \subset \mathbf{C}^N$ is a small open ball with center 0. Let

$$\varphi_0 : \overline{X}_0 = \bigsqcup_{j=1}^{r} \mathbf{D}_j \to X_0 \subset \mathbf{B}_0,$$

$$\varphi_0(\underline{t}) = (\varphi_0(t_1), \dots, \varphi_0(t_r)), \quad \underline{t} = (t_1, \dots, t_r),$$

be a representative of the normalization of X_0, where $\bigsqcup_{j=1}^{r} \mathbf{D}_j$ is the disjoint union of r open discs centered at the origin in $\mathbf{C}$, φ_0 is a r-uple of distinct maps $\varphi_0|\mathbf{D}_j : (\mathbf{D}_j, 0) \to \mathbf{B}_0$, and for each j the restriction $\varphi_0|\mathbf{D}_j$ is a homeomorphism $(\mathbf{D}_j, 0) \to (X_0^j, 0)$, where $(X_0^j, 0)$ is the j-th branch of $(X_0, 0)$. It induces an analytic isomorphism $(\mathbf{D}_j, 0) \setminus \{0\} \to (X_0^j, 0) \setminus \{0\}$ because an analytic map $\mathbf{C} \to \mathbf{C}$ which is a homeomorphism is an isomorphism.

Definition 8.3.5 Let $\mathbf{D} \subset \mathbf{C}^q$ be a small disc with center 0. A deformation of the normalization of X_0 is a holomorphic mapping

$$\varphi : \overline{X}_0 \times \mathbf{D} = \bigsqcup_{j=1}^{r} (\mathbf{D}_j \times \mathbf{D}) \to \mathbf{B}_0,$$

such that $\varphi(\underline{t}, v) = \varphi_0(\underline{t}) + v\psi(\underline{t}, v)$, $\underline{t} \in \overline{X}_0$, $v \in \mathbf{D}$ and $\psi(\underline{t}, v) = (\psi(t_1, v), \ldots, \psi(t_r, v))$ with $\psi(t_j, v) : (\mathbf{D}_j \times \mathbf{D}, 0) \to \mathbf{B}_0$. Note that we are dealing with representatives of germs of deformations of germs.

Then for sufficiently small $\mathbf{D}_j$ and $\mathbf{D}$ we have that $\phi = (\varphi, v) : \overline{X}_0 \times \mathbf{D} \to \mathbf{B}_0 \times \mathbf{D}$ is a finite mapping and therefore

$$X = \phi(\overline{X}_0 \times \mathbf{D}) \subset \mathbf{B}_0 \times \mathbf{D}$$

is a $q + 1$-dimensional analytic subset.

Proposition 8.3.6 *Given a germ of finite map* $n : \bigsqcup_{i=1}^{r} (\mathbf{D}_j \times \mathbf{D}, 0) \to (\mathbf{C}^2, 0) \times \mathbf{D}$ *with* $\mathbf{D} \subset \mathbf{C}^q$ *as above, corresponding to a map of analytic* $\mathbf{C}$*-algebras* $\mathcal{O}_{\mathbf{C}^q,0}\{x, y\} \to \bigoplus_{i=1}^{r} \mathcal{O}_{\mathbf{C}^q,0}\{t_i\}$ *which is the identity on* $\mathcal{O}_{\mathbf{C}^q,0}$ *and makes the second algebra a finite module over the first, the Fitting ideal* $F_0(\bigoplus_{i=1}^{r} \mathcal{O}_{\mathbf{C}^q,0}\{t_i\})$ *is a non zero principal ideal of* $\mathcal{O}_{\mathbf{C}^q,0}\{x, y\}$.

Proof The argument goes back to [Tei73, Chap.III, 3.4] (see also [Tei77, Section 5], [GLS07, Exercise 1.6.4], [MP89, Proposition 3.1]): the depth of the $\mathcal{O}_{\mathbf{C}^q,0}\{x, y\}$-module $\bigoplus_{i=1}^{r} \mathcal{O}_{\mathbf{C}^q,0}\{t_i\}$ is $q + 1$ because it is Cohen–Macaulay (see [GLS07, Theorem B.8.11]) of dimension $q + 1$, so by the Auslander-Buchsbaum formula its homological dimension is one (see [GLS07, Theorem B.9.3]), which implies that its minimal presentations are exact sequences of $\mathcal{O}_{\mathbf{C}^q,0}\{x, y\}$-modules

$$(0) \to \mathcal{O}_{\mathbf{C}^q,0}\{x, y\}^p \to \mathcal{O}_{\mathbf{C}^q,0}\{x, y\}^p \to \bigoplus_{i=1}^{r} \mathcal{O}_{\mathbf{C}^q,0}\{t_i\} \to (0).$$

Therefore the 0-th Fitting ideal is generated by the determinant of a $p \times p$-matrix. See also [MP89, Lemma 2.1]. $\qquad\square$

Applying Proposition 8.3.6 with $q = 0$, we see that none of the plane images we consider has embedded components. The Fitting image of a parametrization is reduced if and only if for $i = 1, \ldots, r$ the set of all exponents appearing in the series $x(t_i)$ or in $y(t_i)$ is coprime. For given i, the gcd of those exponents is the degree of the map $(\mathbf{D}, 0) \to (\mathbf{C}^2, 0)$ defined by $t_i \mapsto (x(t_i), y(t_i))$ and it is also the degree at which the equation $f_i(x, y) = 0$ of the *reduced* image curve is raised to give a generator of the Fitting ideal of the $\mathbf{C}\{x, y\}$-module $\mathbf{C}\{t_i\}$. Compare with Example 8.3.3 and see [MP89, Proposition 3.1] for a more general result.

Going back to the case $q > 0$, we see that if the parametrization for $v = 0$ has a reduced image X_0, then the hypersurface X in $\mathbf{D} \times \mathbf{C}^2$ which is the image of the parametrization is reduced. Otherwise, since it is the Fitting image, applying the compatibility with base change to $0 \in \mathbf{D}$, we find that the special fiber X_0 would not be reduced.

But then each fiber is reduced and its parametrization is an isomorphism outside of the singularities; it is a bimeromorphic map. This implies that the map $p : \bigsqcup_{i=1}^{r} (\mathbf{D}_j \times \mathbf{D}, 0 \times 0) \to (X, 0)$ is the normalization of the hypersurface X because the source is normal, and the map p is finite and bimeromorphic. It is a non singular normalization of the hypersurface X which induces the normalization of each fiber of the projection map $X \to \mathbf{D}$. So we have proved:

Corollary 8.3.7 *The parametrization of the total space of the family of plane curves* $n: (X, 0) \to (\mathbf{D}, 0)$ *obtained by deforming the parametrization of the germ of reduced plane curve* $(X_0, 0)$ *is a simultaneous normalization.*

Moreover, we can observe that the $O_{\mathbf{C}^q, 0}$-modules $O_{X,0}$ and $\overline{O}_{X,0} = \bigoplus_{i=1}^{r} O_{\mathbf{C}^q, 0}\{t_i\}$ are flat. The first one because a hypersurface is Cohen–Macaulay (see [GLS07, Theorem B.8.11]) and the second one because it is a sum of flat modules (see [GLS07, Corollary I 1.88]).

We have the exact sequence of $O_{\mathbf{C}^q, 0}$-modules:

$$(0) \to O_{X,0} \to \overline{O}_{X,0} \to \overline{O}_{X,0}/O_{X,0} \to (0),$$

and the flatness of the first two modules implies that of the third.

Now by the Weierstrass Preparation Theorem, the singular locus of X is finite over $\mathbf{D}, 0)$ and is the support of the $O_{\mathbf{C}^q, 0}$-module $\overline{O}_{X,0}/O_{X,0}$, which is flat and thus locally free. So the dimensions of its fibers over points $v \in \mathbf{D}$ is constant. Provided $\mathbf{D}$ and $\mathbf{B}_0$ are sufficiently small, this dimension is the sum of the δ-invariants (see Definition 8.2.9) of the finitely many singularities of the curve $X_v = n^{-1}(v)$, which all tend to) as $v \to 0$. So we have:

Corollary 8.3.8 *In a small enough representative of family of reduced plane curves obtained by deformation of the parametrization of the special fiber, the sum of the* δ*-invariants of the singularities of the fibers is constant.*

We shall use this in the proof of Proposition 8.4.6 below.

We have just seen that the space X defined by deforming a parametrization of a plane curve can also be described as the space defined by deforming an equation for the reduced plane curve X_0, since the equation for X has to reduce to the equation for X_0 when setting $v = 0$. The situation is more delicate for parametrized curves in $\mathbf{C}^N$ with $N > 2$, not only because of the behavior of the Fitting ideal, but also because in general deforming equations does not produce a flat family. The general definition of a deformation of a germ $(X_0, 0) \subset (\mathbf{C}^N, 0)$ is a germ $(X, 0) \subset (\mathbf{C}^N, 0) \times (S, 0)$, **flat over** S and defined by equations which, when restricted over $0 \in S$ give a set of equations of X_0, up to isomorphism. It is therefore a reasonable question to ask whether any such deformation can be obtained by a deformation of the parametrization of X_0 in the sense that X is the reduced image of such a deformation.

The answer is a converse to Corollary 8.3.8, as follows:

Proposition 8.3.9 (See [Tei80, Théorème 1, page 80], and [GLS07, §2.6] for plane curves) *Let $p \colon (X, 0) \to (S, 0)$ be a flat morphism where all the fibers are reduced curves and S is non singular. Then, for suitable representatives, the following conditions are equivalent:*

- *The normalization $\overline{X}$ is non singular, the composed map $\overline{X} \overset{n}{\to} X \overset{p}{\to} S$ is flat and for each $s \in S$ the map of fibers $(\overline{X})_s \to X_s$ is the normalization.*
- *The sum δ_s of the δ-invariants of the singular points of the fibers X_s is independent of $s \in S$.*

Since the map $p \circ n$ is flat with non singular fiber, at every point of $\overline{X}$ lying above $0 \in X$, the space $\overline{X}$ is locally isomorphic over S to a product of a disk by S. This shows that the map n is a deformation of the parametrization of X_0. The assumption that the fibers are reduced is necessary, as evidenced by the following example.

Example 8.3.10 Consider the of curve X_0 in $\mathbf{C}^3$ given by the equations $x = 0$, $z^2 - y^3 = 0$. The normalization of X_0 is given by

$$\varphi(t) = (0, t^2, t^3).$$

Consider the deformation $\Phi(v, t) = (vt, t^2, t^3, v)$. So, the reduced image X_{red} of Φ is given by the following equations:

$$x^2 - v^2 y = 0, \quad xy - vz = 0, \quad xz - vy^2 = 0, \quad z^2 - y^3 = 0.$$

Now, when we consider the projection $f : X_{red} \to \mathbf{D}$, the fiber $f^{-1}(0) = X_0$ is given by

$$x^2 = 0, \quad xy = 0, \quad xz = 0, \quad z^2 - y^3 = 0.$$

Note that it is not reduced at the origin, hence there is no deformation of the equations $x = z^2 - y^3 = 0$ which defines X_{red}. One can understand this as follows: while the special fiber of our family of curves has embedding dimension two, the general fiber has embedding dimension three. In an analytic family the embedding dimension of the fibers can only increase by specialization so that in our analytic family $f : X_{red} \to \mathbf{D}$ the ideal defining the special fiber has in its primary decomposition an infinitesimal embedded component with ideal $\langle x^2, y, z \rangle$ sticking out of the $x = 0$ plane, which makes the embedding dimension of $f^{-1}(0)$ equal to three as it must be. This fact was stressed also in [Tei77, §3, section 3.5].

More material on the plane curve case is found in [GLS07, Chap. II, §2]. There are generalizations of these results to the cases where the fibers may be non reduced and have embedded components. There are definitions of the invariant δ which apply to this more general situation. We refer the reader to [Le15] and the references therein.

Remark 8.3.11 We note that one can use Mond-Pellikaan's algorithm in [MP89] to find a presentation matrix of a finite analytic map germ $g : (X, 0) \to (\mathbf{C}^{d+1}, 0)$, where (X, x) is a germ of Cohen–Macaulay analytic space of dimension d. For the computations one can use also the software Singular [DGPS19] and the implementation of Mond-Pellikaan's algorithm given by Hernandes et al. in [HMP18]. At the web page of Miranda [Mir19] one can find a Singular library to compute presentation matrices based on the results of [HMP18].

8.4   General Projections

For a reduced and equidimensional germ of complex analytic variety $(X, 0) \subset (\mathbf{C}^N, 0)$ Whitney gave 6 possible definitions of tangent vectors [Whi65], the sets of which constitute tangent cones:

$$C_1(X, 0) \subset C_2(X, 0) \subset C_3(X, 0) \subset C_4(X, 0) \subset C_5(X, 0) \subset C_6(X, 0),$$

and when the germ $(X, 0)$ is smooth they all coincide with the tangent space $T_0 X$.

What is usually known as the **tangent cone** $C_{X,0}$ is what Whitney defined as the cone $C_3(X, 0)$ and is constructed by taking limits of secants through the origin. This means that if we take a representative $(X, 0) \subset (\mathbf{C}^N, 0)$, then a vector $\mathbf{v} \in \mathbf{C}^N$ is in $C_3(X, 0)$ if there exists a sequence of points $\{p_i\} \subset X \setminus \{0\}$ tending to 0 and a sequence of complex numbers $\{\lambda_i\} \subset \mathbf{C}^*$ such that

$$\lambda_i p_i \to \mathbf{v}.$$

Algebraically it is constructed by blowing up the point

$$e_0 : \mathrm{Bl}_0 X \to X$$

and the fiber over the origin is the **projectivized tangent cone** $e_0^{-1}(0) = \mathbf{P}C_3(X, 0)$. In particular it is a pure d-dimensional algebraic cone where d is the dimension of $(X, 0)$.

If $(X, 0)$ is a curve then the cone $C_3(X, 0)$ is a finite number of lines, one for each branch of X. By abuse of language they are called the tangents to X at 0. Of course different branches may have the same tangent.

Definition 8.4.1 A linear projection $\pi : (\mathbf{C}^N, 0) \rightarrow (\mathbf{C}^M, 0)$ with kernel D is called C_3-**general** (with respect to X) if it is transversal to the tangent cone. That is

$$D \cap C_3(X, 0) = \{0\}.$$

Note that by the Weierstrass Preparation Theorem (see 8.2.8 and [deJP00, Thm 3.4.24]) the condition is *equivalent* to the fact that the map

$$\pi|C_3(X, 0) \colon C_3(X, 0) \rightarrow C_3(\mathbf{C}^M, 0) = T_{\mathbf{C}^M, 0}$$

is finite (proper with finite fibers), which implies $M \geq d$. The restriction of a C_3-general projection to X

$$\pi|X : (X, 0) \rightarrow (\mathbf{C}^M, 0)$$

satisfies $\pi^{-1}(0) = \{0\}$ since otherwise the tangent cone to $\pi^{-1}(0)$, which is contained in $C_3(X, 0)$, would be contained in D. Again by the Weierstrass Preparation Theorem, this is equivalent to $\pi|X$ being a finite map. However the finiteness of $\pi|X$ does not imply that the projection is C_3-general; consider the projection of $y^3 - x^2 = 0$ to the x-axis.

Since $C_3(X, 0)$ is of dimension d, almost all (an open dense set of) linear projections $\pi : (\mathbf{C}^N, 0) \rightarrow (\mathbf{C}^{d+1}, 0)$ are C_3-general for $(X, 0)$. Since we assume X to be equidimensional, this tells us that $\pi(X) \subset \mathbf{C}^{d+1}$ is a hypersurface. In the curve case $(d = 1)$ this guarantees the existence of linear projections with image a plane curve.

By [Chi89, Cor. 8.2] we have that $C_3(\pi(X), 0) = \pi(C_3(X, 0))$ in $\mathbf{C}^{d+1}$. We leave it as an exercise for the reader to verify that this last equality is an equality of Fitting images. Hint: use the specialization spaces $\mathcal{X}$ and $\mathcal{Y}$ to the tangent cones for X and $\mathbf{C}^{d+1}$ respectively (see [GT18, §2, 2.4]) and the fact that the natural map $\mathcal{X} \rightarrow \mathcal{Y}$ is finite by the Weierstrass Preparation Theorem because the genericity assumption is equivalent to the finiteness of the map $C_3(X, 0) \rightarrow T_{\mathbf{C}^{d+1}, 0}$, and apply [MP89, Prop. 1.6].

Finally, a projection $\pi : (\mathbf{C}^N, 0) \rightarrow (\mathbf{C}^d, 0)$ is C_3-general for $(X, 0)$ if and only if the map $C_3(X, 0) \rightarrow C_3(\mathbf{C}^d, 0) = T_{\mathbf{C}^d, 0}$ which it induces is finite and surjective, and thus a ramified covering. These C_3-general maps all induce on $(X, 0)$ ramified analytic coverings $(X, 0) \rightarrow (\mathbf{C}^d, 0)$ of degree equal to the multiplicity of $(X, 0)$.

The cone $C_4(X, 0)$ is constructed by taking limits of tangent vectors at smooth points. One can prove that it is equivalent to taking limits at 0 in the appropriate

Grasmannian of tangent spaces at non singular points of X and so it is determined by the fiber over 0 of the Semple-Nash modification of a representative X of $(X, 0)$. Of course there is an analogous definition of a C_4-general linear projection and they do have interesting equisingularity properties. However, since in the curve case the cones C_3 and C_4 coincide we will skip this part and ask the interested reader to look at [Chi89, Stu72a] and [Stu72b].

The cone $C_5(X, 0)$ is constructed by taking limits of secants. This means that if we take a representative $X \subset \mathbf{C}^N$ then a vector $\mathbf{v} \in \mathbf{C}^N$ is in $C_5(X, 0)$ if there exist sequences of pairs of distinct points $\{p_i\}, \{q_i\} \subset X \setminus \{0\}$ tending to 0 as $i \to \infty$ and a sequence of complex numbers $\{\lambda_i\} \subset \mathbf{C}^*$ such that

$$\lambda_i (p_i - q_i) \to \mathbf{v}.$$

To prove that $C_5(X, 0)$ is an algebraic cone and has a bound for its dimension, take a small representative $X \subset \mathbf{C}^n$, consider the (closed) diagonal embedding $\delta : X \hookrightarrow X \times X$ and blow up its image Δ:

$$e_\Delta : \mathrm{Bl}_\Delta(X \times X) \to X \times X.$$

If we choose coordinates $(z_1, \ldots, z_N, w_1, \ldots, w_N)$ of the ambient space $\mathbf{C}^{2N}$, then we can obtain the space $\mathrm{Bl}_\Delta(X \times X)$ as the closure of the graph of the secant map defined away from the diagonal Δ by:

$$X \times X \setminus \Delta \longrightarrow \mathbf{P}^{N-1}$$

$$(z, w) \longmapsto [z_1 - w_1 : \cdots : z_N - w_N].$$

So we have $\mathrm{Bl}_\Delta(X \times X)$ as a closed subspace of the product $X \times X \times \mathbf{P}^{N-1}$, the map e_Δ is induced by the projection to $X \times X$, and the exceptional fiber is the divisor $D := e_\Delta^{-1}(\Delta) \subset \Delta \times \mathbf{P}^{N-1}$ which comes with a map $D \to \Delta$ such that for every point $(q, q) \in \Delta$ the fiber is the projective subvariety corresponding to the projectivization of the C_5-cone of X at q, that is $\mathbf{P}C_5(X, q)$. This is roughly the way Whitney proved that the C_5-cone is an algebraic variety in [Whi65, Th. 5.1]. Now $C_5(X)$ is the analytic space obtained by deprojectivization of the (fibers of) the divisor D and ψ corresponds to the pullback of e_Δ by δ:

$$X \times \mathbf{C}^N \supset C_5(X) \dashrightarrow \mathrm{Bl}_\Delta(X \times X)$$
$$\psi \downarrow \qquad\qquad\qquad \downarrow e_\Delta$$
$$X \xrightarrow{\quad\delta\quad} X \times X$$

where the upper arrow is defined only outside of $X \times \{0\}$. Note that the dimension of $C_5(X)$ is $2d$, and the dimension of $\psi^{-1}(p) = C_5(X, p)$ for a smooth point $p \in X$

is equal to d since in this case we have $C_5(X, p) = T_pX$. By the semicontinuity of the dimensions of the fibers of an analytic morphism, this implies that:

$$d \leq \dim C_5(X, 0) \leq 2d$$

Definition 8.4.2 A linear projection $\pi : (\mathbf{C}^N, 0) \rightarrow (\mathbf{C}^M, 0)$ with kernel D is called **generic** (or C_5**-general**) with respect to X if it is transversal to the cone $C_5(X, 0)$. That is

$$D \cap C_5(X, 0) = \{0\}.$$

In other words, no limit at 0 of secants to X is contained in D.

Note that a generic projection is in particular C_3-general and C_4-general.

Proposition 8.4.3 *Let $(X, 0) \subset (\mathbf{C}^N, 0)$ be a reduced equidimensional germ of complex analytic variety of dimension d and $\pi : (\mathbf{C}^N, 0) \rightarrow (\mathbf{C}^M, 0)$ a linear projection.*

(a) *If π is generic then the restriction to X induces a homeomorphism with its image.*
(b) *$(X, 0)$ is smooth if and only if $\dim C_5(X, 0) = d$*

Proof First of all note that the transversality to the cone $C_5(X, 0)$ implies that the restriction $\pi|X$ is injective for a small enough representative of X. But then the induced map $\pi|X : X \rightarrow \mathbf{C}^M$ is injective, continues and the map $X \rightarrow \pi(X)$ is open since π is and so it should be a homeomorphism of X with its image $\pi(X)$.

Now for (b): sufficiency is clear since $(X, 0)$ smooth implies $C_5(X, 0) = T_0X$ and so it is of dimension d. Conversely, if the dimension of $C_5(X, 0)$ is d there exist generic linear projections of X to $\mathbf{C}^d$. By (a) this gives us a homeomorphism between $(X, 0)$ and $(\mathbf{C}^d, 0)$. Note that π is also C_3-general so it induces a ramified covering of degree equal to the multiplicity of $(X, 0)$, but the injectivity gives us multiplicity 1 and so $(X, 0)$ is smooth.   □

For more on this and more general results see [Stu72a, Stu77] and [Chi89, Section 9.4]

An important thing to notice is that in the reducible case the cone $C_5(X, 0)$ contains but is *not* equal to the union of the C_5-cones of its irreducible components. For instance if $(X, 0)$ is a curve consisting of two smooth branches X_1 and X_2 then both cones $C_5(X_i, 0)$ are one-dimensional but since $(X, 0)$ is singular then by the previous result $C_5(X, 0)$ can not have dimension 1.

So now we have that if $(X, 0)$ is singular then $d + 1 \leq \dim C_5(X, 0) \leq 2d$, and for curves this gives $\dim C_5(X, 0) = 2$. This guarantees the existence of generic projections of curves to $\mathbf{C}^2$.

Corollary 8.4.4 *Let $(X, 0) \subset (\mathbf{C}^N, 0)$ be a germ of reduced analytic curve. Then almost all (an open dense set of) linear projections $\pi : (\mathbf{C}^N, 0) \rightarrow (\mathbf{C}^2, 0)$ are generic and their Fitting images $\pi(X) \subset \mathbf{C}^2$ are reduced plane curves* homeomorphic to X. Moreover, π induces an analytic isomorphism $X \setminus \{0\} \to \pi(X) \setminus \{0\}$.

Proof This follows from Proposition 8.3.6 and the fact that an analytic map $\mathbf{C} \to \mathbf{C}$ which is a homeomorphism is an isomorphism.   □

8.4.1   The Case of Dimension 1

In the case of curves we have the following important results:

Proposition 8.4.5 (See [BGG80, Prop IV.1]) *Let* $(X, 0) \subset (\mathbf{C}^N, 0)$ *be a germ of reduced analytic curve. If* $(X, 0)$ *is singular then the cone* $C_5(X, 0)$ *is a finite union of 2-planes each one of them containing at least one tangent line to* $(X, 0)$.

Proof We will only give an idea of the proof.

By Proposition 8.4.3 the cone $C_5(X, 0)$ is two-dimensional and by the blowup construction it has a finite number of irreducible components. So what one has to prove is that all the irreducible components are 2-planes. Again, by this blowup construction, any (direction of) line contained in $C_5(X, 0)$ can be picked off by lifting an arc

$$(\psi_1, \psi_2) : (\mathbf{C}, 0) \to (X \times X, (0, 0))$$

to $\mathrm{Bl}_\Delta(X \times X)$ like $(\psi_1(t), \psi_2(t), [\psi_1(t) - \psi_2(t)])$. Now each $\psi_i(t)$ is an arc $(\mathbf{C}, 0) \to (X, 0)$ and can be obtained using the parametrization of one of the branches of $(X, 0)$. Once you see this, what you have to do is consider the different cases and work out the calculations.

The first case is when $(X, 0) \subset (\mathbf{C}^N, 0)$ is irreducible of multiplicity n so according to Sect. 8.1.2, in suitable coordinates we have a parametrization of the form:

$$\varphi(t) = \left(t^n, \sum_{i>n} a_{2,i} t^i, \ldots, \sum_{i>n} a_{N,i} t^i \right)$$

with the tangent line being the z_1-axis $[1 : 0 : \cdots : 0]$. For every n-th root of unity $\omega \neq 1$ the lifted arc

$$t \mapsto (\varphi(t), \varphi(\omega t), [\varphi(t) - \varphi(\omega t)]) \in X \times X \times \mathbf{P}^{N-1}$$

will define a limit line $\ell_\omega \in \mathbf{P}^{N-1}$ as $t \to 0$ which is distinct from the z_1-axis and if you define H_ω as the 2-plane generated by the z_1-axis and the line in $\mathbf{C}^N$ corresponding to ℓ_ω, then you can prove that

$$C_5(X, 0) = H_{\omega_1} \cup \ldots \cup H_{\omega_{n-1}},$$

by verifying that any line obtained by lifting an arc is contained in one of these 2-planes. We note that they are not necessarily all different.

For the reducible case it is enough to consider two branches $(X, 0) = (X_1, 0) \cup (X_2, 0)$. In this case you have that the C_5-cone of each irreducible component $(X_i, 0)$ will be contained in $C_5(X, 0)$ but you will have additional components that come from the configuration of these two branches. For instance if they have different tangent lines ℓ_1 and ℓ_2 then all you have to add is the plane H_{12} generated by these two lines. i.e.,

$$C_5(X, 0) = C_5(X_1, 0) \cup C_5(X_2, 0) \cup H_{12}.$$

When the two branches are tangent (have the same tangent line) then you have to play a game very similar to the irreducible case by reparametrizing your branches in such a way as to travel through them at the same "speed" and using roots of unity to find lines ℓ_ω in the $C_5(X, 0)$ that are different from the tangent line and these will give you the additional 2-planes. i.e.,

$$C_5(X, 0) = C_5(X_1, 0) \cup C_5(X_2, 0) \cup H_{\omega_1} \cup \ldots \cup H_{\omega_k}. \qquad \square$$

Proposition 8.4.6 (see [BGG80, Prop IV.2]) *Let $(X, 0) \subset (\mathbf{C}^N, 0)$ be a germ of reduced analytic curve, and let $\Omega \subset G(N - 2, N)$ be the non-empty Zariski open set of the Grassmannian of $(N - 2)$-planes of $\mathbf{C}^N$ which are transversal to $C_5(X, 0)$. Then:*

(a) For $H \in \Omega$ the plane curve $(\pi_H(X), 0)$ is reduced and of constant topological (equisingularity) type with Milnor number μ_0.

(b) If $H \notin \Omega$ then one of the following statements is verified:

- *0 is not an isolated point of $H \cap X$.*
- *0 is an isolated point of $H \cap X$ but the curve $(\pi_H(X), 0)$ is not reduced.*
- *0 is an isolated point of $H \cap X$, the curve $(\pi_H(X), 0)$ is reduced but its Milnor number is greater than μ_0.*

Proof Let $W' \subset G(N - 2, N)$ be the open subset of the Grassmannian of $(N - 2)$-planes of $\mathbf{C}^N$ defined by the condition that $H \in W'$ if and only if $0 \in \mathbf{C}^N$ is an isolated point of $H \cap X$. Let $W \subset \mathbf{C}^{2N}$ with coordinate system $(a_1, \ldots, a_N, b_1, \ldots, b_N)$ be the associated open subset, where $d = (\underline{a}, \underline{b}) \in W$ if and only if the linear forms

$$a_1 z_1 + \cdots + a_N z_N \quad \text{and} \quad b_1 z_1 + \cdots + b_N z_N$$

are linearly independent and the $N - 2$ plane $H_d \subset \mathbf{C}^N$ they define is in W'. Let π_d be the linear projection

$$\pi_d : \mathbf{C}^N \longrightarrow \mathbf{C}^2$$

$$(z_1, \ldots, z_N) \mapsto (a_1 z_1 + \cdots + a_N z_N, b_1 z_1 + \cdots + b_N z_N)$$

Note that for $d \in W$ the germ $\pi_d : (X, 0) \to (\mathbf{C}^2, 0)$ is finite, and if we denote by $(\pi_d(X), 0) \subset (\mathbf{C}^2, 0)$ the image germ with the Fitting structure then by [MP89, Lemma 2.1] it is a (not necessarily reduced, but without embedded component, by Proposition 8.3.6) plane curve.

We put all these projections in an analytic family by considering the map

$$\Pi : \mathbf{C}^N \times W \longrightarrow \mathbf{C}^2 \times W$$

$$(z_1, \ldots, z_N, d) \mapsto (\pi_d(z_1, \ldots, z_N), d)$$

Note that for every $d \in W$ the map germ

$$\Pi : (X \times W, (0, d)) \to \left(\mathbf{C}^2 \times W, (0, d) \right)$$

is finite. And since the analytic algebra $O_{X \times W, (0,d)}$ is Cohen–Macaulay again by [MP89, Lemma 2.1] we have a germ of hypersurface $(\Pi(X \times, W), (0, d)) \subset \left(\mathbf{C}^2 \times W, (0, d) \right)$. By projecting to $W \subset \mathbf{C}^{2N}$ we obtain (by [GLS07, Thm B.8.11]) a flat map:

$$G : (\Pi(X \times W), (0, d)) \to (W, d).$$

Since the Fitting structure commutes with base change we have that the germ $\left(G^{-1}(d), (0, d) \right)$ is isomorphic to $(\pi_d(X), 0)$, and so we have a flat deformation of $(\pi_d(X), 0)$ where all the fibers are plane curves.

Note that if $\varphi : (\mathbf{C}, 0) \to (\mathbf{C}^N, 0), t \mapsto (\varphi_1(t), \ldots, \varphi_N(t))$ is the normalization of a branch of $(X, 0)$ then the plane curve $(\pi_d(X), 0)$ is parametrized by:

$$t \mapsto (a_1 \varphi_1(t) + \cdots + a_N \varphi_N(t), b_1 \varphi_1(t) + \cdots + b_N \varphi_N(t)),$$

and by varying d we get that the deformation space of G admits a parametrization in family.

Proof of (a) When H_d is transversal to $C_5(X, 0)$ then for every d' in a small neighborhood of d the $(n-2)$-plane $H_{d'}$ is also transversal to $C_5(X, 0)$ and all the corresponding projections $\pi_{d'}$ are therefore generic. By Corollary 8.4.4 this tells us that $\pi_{d'} : X \setminus \{0\} \to G^{-1}(d') \setminus \{0\}$ is an analytic isomorphism for every d' sufficiently close to d. This implies:

- All the curves in the family $G^{-1}(d')$ have the same number of branches as X.
- The parametrization in family is actually a normalization in family and by Corollary 8.3.8, or [Tei77, §3], see also [GLS07, II, Thm 2.56] the family is δ constant.

By the Milnor formula $\mu = 2\delta - r + 1$ the family $G : (\Pi(X \times, W), (0, d)) \to (W, d)$ is μ-constant and thus equisingular by [LR76] or [BGG80, Thm II.4].

Proof of (b) For $H_d \in W \setminus \Omega$ we have that the map

$$\pi_d : (X, 0) \to (\mathbf{C}^2, 0)$$

is finite but if it is generically k to 1 then by [MP89, Prop. 3.1] the Fitting structure of $(\pi_d(X), 0)$ is not reduced.

When π_d is generically 1-1 then $(\pi_d(X), 0)$ is reduced but by assumption there is a line $\ell \subset H_d \cap C_5(X, 0)$. Take a sequence of secants ℓ_k going through the points $x_k, y_k \in X \setminus \{0\}$ such that ℓ_k converges (in direction) to ℓ, since Ω is Zariski open we can find a sequence d_k tending to d such that $H_{d_k} \in \Omega$ and it contains (the direction of) ℓ_k. Note that $\pi_{d_k}(\ell_k) = q_k \neq 0$ and so the plane curve $(G^{-1}(d_k), q_k)$ is singular which implies that $\mu((\pi_d(X), 0)) > \mu((\pi_{d_k}(X), 0))$. $\square$

Example 8.4.7 Let $(X, 0) \subset (\mathbf{C}^3, 0)$ the germ of irreducible curve parametrized by

$$t \mapsto (t^4, t^5, t^7)$$

then the tangent cone $C_3(X, 0)$ is the z_1-axis.

By taking other arcs $t \mapsto (t^4, \omega t^5, \omega^3 t^7)$ were $\omega \in \mu_4 \setminus \{1\}$ and taking the limit as $t \to 0$ of the difference $[0 : (1 - \omega)t^5 : (1 - \omega^3)t^7]$ we get the z_2-axis as a limit of secants and we can deduce that the cone $C_5(X, 0)$ is the $z_1 z_2$-plane.

For $d = (1, 0, 0, 0, 1, 0)$ the corresponding projection

$$\pi_d(z_1, z_2, z_3) = (z_1, z_2)$$

is C_5-general and its image $\pi_d(X, 0) \subset (\mathbf{C}^2, 0)$ is the reduced plane curve $y^4 - x^5 = 0$ with Milnor number $\mu = 12$.

On the other hand For $d_0 = (1, 0, 0, 0, 0, 1)$ the corresponding projection

$$\pi_{d_0}(z_1, z_2, z_3) = (z_1, z_3)$$

is not C_5-general and its image $\pi_{d_0}(X, 0) \subset (\mathbf{C}^2, 0)$ is the reduced plane curve $y^4 - x^7 = 0$ with Milnor number $\mu = 18$.

By taking $d_\alpha = (1, 0, 0, 0, -\alpha^2, 1)$ we get a sequence of C_5-general projections π_{d_α} converging to π_{d_0}

$$\pi_{d_\alpha}(z_1, z_2, z_3) = (z_1, z_3 - \alpha^2 z_2)$$

Note that the plane curve $X_\alpha := \pi_{d_\alpha}(X)$ has a singular point in $(\alpha^4, 0)$ coming from the image of the secant going through the points $(\alpha^4, \alpha^5, \alpha^7)$ and $(\alpha^4, -\alpha^5, -\alpha^7)$ in X. Moreover as α tends to 0 these secants $d_\alpha = [0 : 1 : \alpha^2]$ converge to the z_2-axis $[0 : 1 : 0]$ in $\mathbf{P}^2$ which is precisely the intersection $H_{d_0} \cap C_5(X, 0)$.

8.5   Main Result

We have just seen that all (C_5-)generic plane projections of a reduced analytic curve are equisingular. Now our objective is to prove that all equisingular germs of reduced plane curves are generic projections of a single space curve. As we shall see, given a reduced plane curve $(X, 0) \subset (\mathbf{C}^2, 0)$ this space curve corresponds to the one-dimensional analytic algebra which is the Lipschitz saturation $O_{X,0}^s$ of $O_{X,0}$ in the sense of [PT69]. In doing so we will also give another reason why (a) of Proposition 8.4.6 is true, since we shall see that a projection π is generic for a space curve $(X, 0) \subset (\mathbf{C}^N, 0)$ if and only if it induces an isomorphism of the saturated algebras $O_{X,0}^s$ and $O_{\pi(X),0}^s$. In particular, two germs of reduced plane curves are equisingular (topologically equivalent) if and only if their saturations are *analytically* isomorphic.

In order to define these saturations we need the theory of integral closure of ideals.

8.5.1   Integral Closure of Ideals

Our main references for this subsection are, [LT08, Lip82, Tei73] and [HS06].

Definition 8.5.1 Let I be an ideal in a ring R. An element $r \in R$ is said to be integral over I if there exists an integer h and elements $a_j \in I^j$, $j = 1, \ldots, h$, such that

$$r^h + a_1 r^{h-1} + a_2 r^{h-2} + \cdots + a_{h-1} r + a_h = 0.$$

The set of all elements of R that are integral over I is an ideal called the **integral closure** of I and denoted by $\bar{I}$. We say that I is integrally closed if $I = \bar{I}$. If $I \subset J$ are ideals we say that J is integral over I if $J \subset \bar{I}$.

Remark 8.5.2 The following properties are easily verified:

1. $I \subset \bar{I}$. For each $r \in I$ choose $n = 1$ and $a_1 = -r$.
2. If $I \subset J$ are ideals then $\bar{I} \subset \bar{J}$ since an integral dependence equation for r over I is also an integral dependence equation for r over J.
3. $\bar{I} \subset \sqrt{I}$ since the integral dependence equation implies $r^n \in \langle a_1, \ldots, a_n \rangle \subset I$.
4. Radical ideals are integrally closed.
5. If $\varphi : R \to S$ is a ring morphism and $I \subset S$ is an integrally closed ideal of S then $\varphi^{-1}(I)$ is an integrally closed ideal of R.

A related concept is that of **reduction**: For ideals $J \subset I \subset R$ we say that J is a **reduction of** I if there exists a non-negative integer n such that $I^{n+1} = JI^n$. This implies that $\bar{I} = \bar{J}$. We can express integral dependence using equalities of ideals and modules.

Proposition 8.5.3 (See [LT08, Chapter 1], [HS06, Prop 1.1.7, Cor. 1.1.8 & Cor. 1.2.2]) *For any element $r \in R$ and ideal $I \subset R$. The following are equivalent:*

(a) $r \in \overline{I}$.
(b) There exists an integer k such that $(I + r)^k = I(I + r)^{k-1}$.
(c) I is a reduction of $I+ <r>$.
(d) There exists a finitely generated R-module M such that $rM \subset IM$ and if there exists $a \in R$ such that $aM = 0$, then there exists an integer ℓ such that $ar^\ell = 0$.

A very important corollary of this Proposition is that $\overline{I} \subset R$ is an integrally closed ideal of R and you can find a complete proof of this fact in [HS06, Cor. 1.3.1].

We have that $I \subset \overline{I} \subset \sqrt{I}$, but in fact the integral closure is much "closer" to I than to the radical and a very good family of examples in which it is easy to calculate and compare is that of monomial ideals in $\mathbf{C}\{z_1, \ldots, z_d\}$, which are the ideals generated by monomials. We begin with an example:

Example 8.5.4 For the ideal $I = \langle x^4, xy^2, y^3 \rangle \mathbf{C}\{x, y\}$ we have that

$$\overline{I} = \langle x^4, x^3y, xy^2, y^3 \rangle$$

and

$$\sqrt{I} = \langle x, y \rangle.$$

The exponent set of I consists of all integer lattice points in the yellow region below:

Similarly, in $\mathbf{C}\{z_1, \ldots, z_d\}$ we have $\overline{\langle z_1^n, \ldots, z_d^n \rangle} = \langle z_1, \ldots, z_d \rangle^n$.

The **exponent vector** of a monomial $z_1^{m_1} \cdots z_d^{m_d}$ is $(m_1, \ldots, m_d) \in \mathbf{N}^d$. For any monomial ideal I, the set of all exponent vectors of all the monomials in I is called the **exponent set of** I. Since a monomial is in I if and only if it is a multiple in $\mathbf{C}\{z_1, \ldots, z_d\}$ of one of the monomial generators of I, the exponent set of I consists of all those points of $\mathbf{N}^d$ which are componentwise greater or equal than

Fig. 8.1 The point $(3, 1)$ representing the monomial x^3y is in the convex hull of the yellow region, whose integral points represent monomials in I. The integral dependence relation is $(x^3y)^2 - x^5.xy^2 = 0$

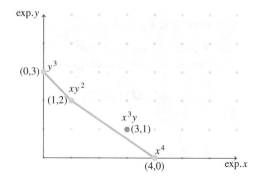

the exponent vector of one of the monomial generators of I. Moreover one can prove that $\overline{I}$ is monomial and its exponent set is equal to all the integer lattice points in the convex hull of the set of exponents of elements of I. (See [Tei82, Chap.1, §2], [Tei04, §3, §4], [HS06, Props 1.4.2 & 1.4.6]). To understand how this theory can be used in the setting of complex analytic geometry the following result is fundamental.

Theorem 8.5.5 ([LT08, Thm 2.1, p. 799]) *Let X be a reduced complex analytic space. Let $Y \subset X$ be a closed, nowhere dense, analytic subspace of X, and x a point in Y. Let $\mathcal{I} \subset \mathcal{O}_X$ be the coherent ideal defining Y, and let $\mathcal{J} \subset \mathcal{O}_X$ be another coherent ideal. Let I (resp. J) be the stalk of $\mathcal{I}$ (resp. $\mathcal{J}$) at x. Then the following statements are equivalent:*

1. *$J \subset \overline{I}$,*
2. *For every germ of morphism $\phi : (\mathbf{C}, 0) \to (X, x)$*

$$\phi^* J \cdot \mathcal{O}_{\mathbf{C},0} \subset \phi^* I \cdot \mathcal{O}_{\mathbf{C},0},$$

3. *For every morphism $\pi : X' \to X$ such that X' is a normal analytic space, π is proper and surjective, and $\mathcal{I} \cdot \mathcal{O}_{X'}$ is locally invertible, there exists an open subset $U \subset X$ containing x, such that:*

$$\mathcal{J} \cdot \mathcal{O}_{X'}|\pi^{-1}(U) \subset \mathcal{I} \cdot \mathcal{O}_{X'}|\pi^{-1}(U),$$

3*. *If $\Pi : \overline{Bl_I X} \to X$ denotes the normalized blowup of X along $\mathcal{I}$, then there exists an open subset $U \subset X$ containing x, such that:*

$$\mathcal{J} \cdot \mathcal{O}_{\overline{Bl_I X}}|\Pi^{-1}(U) \subset \mathcal{I} \cdot \mathcal{O}_{\overline{Bl_I X}}|\Pi^{-1}(U),$$

4. *Let $V \subset X$ be a neighborhood of x, where both $\mathcal{J}$ and $\mathcal{I}$ are generated by their global sections. Then for every system of generators $g_1, \ldots, g_m$ of $\Gamma(V, \mathcal{I})$ and every $f \in \Gamma(V, \mathcal{J})$, there exist a neighborhood V' of x in V and a constant C such that:*

$$|f(y)| \leq C \sup_{i=1,\ldots,m} |g_i(y)|$$

for every $y \in V'$.

Let us take a closer look at statement 2: For any arc $\varphi : (\mathbf{C}, 0) \to (X, 0) \subset (\mathbf{C}^N, 0)$ we have a corresponding morphism of analytic algebras

$$\varphi^* : \mathcal{O}_{X,0} = \mathbf{C}\{z_1, \ldots, z_N\}/\mathfrak{a} \longrightarrow \mathbf{C}\{t\}$$

$$z_i + \mathfrak{a} \mapsto \varphi_i(t) = t^{m_i} u_i(t)$$

where $m_i \geq 1$ and $u_i(t)$ is a unit in $\mathbf{C}\{t\}$. So if $I \subset \mathcal{O}_{X,0}$ is an ideal then $\varphi^*(I)\mathcal{O}_{\mathbf{C},0} = \langle t^k \rangle \mathbf{C}\{t\}$ for some integer k and an element $g \in \mathcal{O}_{X,0}$ is in $\overline{I}$ if and only if for any such arc $t \mapsto \varphi(t)$ the order of the series $g(\varphi_1(t), \ldots, \varphi_N(t))$ is greater or equal than this k.

The fact that the normalized blowup map is proper implies that the condition of statement 2 needs to be verified only for finitely many arcs. Since the general statement is somewhat cumbersome, let us illustrate how this works in the case where the ideal I a complete intersection defining the origin in $(X, 0)$. Let $I = \langle h_1, \ldots, h_d \rangle \subset O_{X,0}$. The blowup $Bl_I X$ of I in X is the subspace of $X \times \mathbf{P}^{d-1}$ defined by the $d - 1$ equations $\frac{h_1}{T_1} = \frac{h_2}{T_2} = \cdots = \frac{h_d}{T_d}$, again a complete intersection. The fiber of the natural projection $Bl_I X \to \mathbf{P}^{d-1}$ over a point $t \in \mathbf{P}^{d-1}$ with coordinates $(t_1 : t_2 : \cdots : t_d)$ is a curve in $Bl_I X$ which is isomorphic to its image in X defined by the equations $h_i t_j - h_j t_i = 0$. So we can view $Bl_I X$ as a family of curves C_t on X parametrized by $\mathbf{P}^{d-1}$, which is the exceptional divisor of the map $Bl_I X \to X$. When we pass to the normalization $n \colon \overline{Bl_I X} \to Bl_I X$, by general Theorems on normalization (see Proposition 8.3.9 and use the fact that by generic flatness there is a dense open $U \subset \mathbf{P}^{d-1}$ where δ is constant), there exists a Zariski dense open subset $U \subset \mathbf{P}^{d-1}$ such that $n^{-1}(U)$ is a non singular divisor in a non singular space $n^{-1}((X \times U) \cap Bl_I X)$, and for each point $t \in U$ the map n induces a normalisation of the curve C_t. This normalization is then a union of disks, one for each irreducible component of C_t, and each disk transversal to $n^{-1}(\mathbf{P}^{d-1})$ in $n^{-1}((X \times U) \cap Bl_I X)$. Because a meromorphic function on a normal space is holomorphic if it has no poles in codimension one, to verify that an element $g \in O_{X,0}$ is in $\bar{I}$, it suffices to verify that for some $t \in U$, the order of vanishing of g along each arc parametrizing a branch of C_t is larger than the order of vanishing of the ideal I. Because of what we have just seen, the order of vanishing along these arcs will, after lifting to $\overline{Bl_I X}$, translate as the order of vanishing along some irreducible component of the exceptional divisor in $\overline{Bl_I X}$. Since the ideal I is locally principal on $\overline{Bl_I X}$, to prove that $g \in \bar{I}$ it suffices to prove that after lifting to $\overline{Bl_I X}$ the function g becomes a multiple of the local equations of the exceptional divisor. But the polar set of the quotient of g by that equation is contained in that exceptional divisor and the inequalities of orders imply that there are no poles at a general point of each irreducible component. Because $\overline{Bl_I X}$ is normal, there are no poles anywhere and on $\overline{Bl_I X}$ the pull back of the function g is indeed in the pull back of the ideal I so that g is in $\bar{I}$.

We shall use this below to describe the saturation.

With this at hand we can now characterize C_3-general projections in terms of integral closure of ideals. Let $(X, 0) \subset (\mathbf{C}^N, 0)$ be a reduced germ of analytic space of pure dimension d. Let us choose coordinates $z_1, \ldots, z_N$ on $\mathbf{C}^N$, denote by L the linear subspace of $\mathbf{C}^N$ defined by $z_1 = \cdots = z_d = 0$ and let $\mathfrak{a}$ be the ideal of $O_{X,0}$ generated by the images of $z_1, \ldots, z_d$.

Proposition 8.5.6 *The restriction to $(X, 0)$*

$$\pi|(X, 0) : (X, 0) \to (\mathbf{C}^d, 0) \quad (z_1, \ldots, z_N) \mapsto (z_1, \ldots, z_d)$$

of the linear projection π with kernel L is C_3-general if and only if $\bar{\mathfrak{a}} = \mathfrak{m}$ where $\mathfrak{m} = \langle z_1, \dots, z_N \rangle O_{X,0}$ is the maximal ideal of the analytic algebra $O_{X,0}$.

Proof Recall that π is C_3-general if and only if $C_3(X,0) \cap L = \{0\}$. Let $\ell = [a_1 : \cdots : a_N] \in \mathbf{P}^{N-1}$ be a line in the (projectivized) tangent cone $C_3(X,0)$, then $\ell \not\subset L$ if and only if $a_i \neq 0$ for some $i \in \{1, \dots, d\}$. Note that any arc $\varphi : (\mathbf{C}, 0) \to (X, 0)$ determines a line in $C_3(X, 0)$, the limit as $t \to 0$ of

$$t \longrightarrow [\varphi_1(t) : \cdots : \varphi_N(t)] \in \mathbf{P}^{N-1},$$

and conversely any line in the tangent cone can be obtained through an arc since it corresponds to a point in the fiber over 0 of the blowup $Bl_0 X \to X$. On the other hand, for every arc $\varphi : (\mathbf{C}, 0) \to (X, 0)$ we have that

$$\varphi^*(\mathfrak{a})O_{\mathbf{C},0} = \langle \varphi_1(t), \dots, \varphi_d(t) \rangle \mathbf{C}\{t\} = \langle t^k \rangle \mathbf{C}\{t\},$$

where $k = \min\{\mathrm{ord}_0 \varphi_i(t) \mid i = 1, \dots, d\}$. Finally $a_i \neq 0$ for some $i \in \{1, \dots, d\}$ if and only if for all $j \in \{d+1, \dots, N\}$

$$\mathrm{ord}_0 \varphi_j(t) \geq k = \min\{\mathrm{ord}_0 \varphi_i(t) \mid i = 1, \dots, d\}$$

if and only if $\varphi^*(z_j) \in \varphi^*(\mathfrak{a})O_{\mathbf{C},0}$ if and only if $z_j \in \bar{\mathfrak{a}}$ for all $j \in \{d+1, \dots, N\}$, that is, $\bar{\mathfrak{a}} = \mathfrak{m}$. $\qquad\square$

By a linear change of coordinates in $\mathbf{C}^N$ we can always place ourselves in the setting of the previous result. But the theory of integral closure also gives us an algebraic way to prove that for a given germ $(X, 0)$ of pure dimension d almost all linear projections $\pi : (\mathbf{C}^N, 0) \to (\mathbf{C}^d, 0)$ are C_3-general as stated in the following result (For a proof see [Mat89, Thm 14.14])

Theorem 8.5.7 (Rees-Samuel) *Let $O_{X,0}$ be a d-dimensional analytic algebra with maximal ideal $\mathfrak{m} = \langle z_1, \dots, z_N \rangle$. For $i \in \{1, \dots, d\}$, let $y_i = \sum_{j=1}^{N} \lambda_{ij} z_j$ be a sufficiently general $\mathbf{C}$-linear combination of $z_1, \dots, z_N$. Then the ideal $\mathfrak{a} = \langle y_1, \dots, y_d \rangle$ satisfies $\bar{\mathfrak{a}} = \mathfrak{m}$.*

We can take this one step further by considering another important aspect of this theory, namely its relation with multiplicity. For a local Noetherian ring $(R, \mathfrak{m})$ and an $\mathfrak{m}$-primary ideal $\mathfrak{a} \subset R$ we can define a Hilbert Samuel function

$$k \in \mathbf{N} \mapsto \dim_{R/\mathfrak{m}} R/\mathfrak{a}^k.$$

The result is that for large enough k the Hilbert-Samuel function behaves like a polynomial of degree equal to the dimension of R and its leading coefficient is of the form $e(\mathfrak{a}) k^d / d!$, where $e(\mathfrak{a})$ is a positive integer called the multiplicity of the ideal $\mathfrak{a}$. In the case R is the analytic algebra $\mathcal{O}_{X,0}$ of a germ $(X, 0)$ and $\mathfrak{a} = \mathfrak{m}$ it IS the multiplicity of the germ. (See [deJP00, Section 4.2])

Theorem 8.5.8 (Rees (See [Ree61, Thm 3.2],[HS06, Thm 11.3.1])) *Let* $(\mathcal{O}_{X,0}, \mathfrak{m})$ *be a reduced and equidimensional analytic algebra and* $\mathfrak{a} \subset \mathfrak{b}$ *two* $\mathfrak{m}$-*primary ideals. Then* $\overline{\mathfrak{a}} = \overline{\mathfrak{b}}$ *if and only if* $e(\mathfrak{a}) = e(\mathfrak{b})$.

A geometric interpretation of this result is described by Lipman in [Lip82]. Let $(X, 0)$ be a germ of reduced and equidimensional singularity of dimension d with associated analytic algebra $(\mathcal{O}_{X,0}, \mathfrak{m})$. Every $\mathfrak{m}$-primary ideal is generated by at least d elements, and every d-tuple $(f_1, \ldots, f_d)$ of elements of $\mathfrak{m}$ defines a map-germ $F : (X, 0) \to (\mathbf{C}^d, 0)$.

Now, the ideal $\mathfrak{a} = \langle f_1, \ldots, f_d \rangle$ is $\mathfrak{m}$-primary if and only if F is finite. As we have mentioned before you can prove that such an $F : (X, 0) \to (\mathbf{C}^d, 0)$ is then a ramified analytic cover of degree equal to $e(\mathfrak{a})$ and by Rees' Theorem this degree will be the multiplicity of $(X, 0)$ $(= e(\mathfrak{m}))$ if and only if $\overline{\mathfrak{a}} = \mathfrak{m}$.

Moreover using Nakayama's Lemma one checks that $\mathfrak{a}$ is a reduction of $\mathfrak{m}$ (equivalently $\overline{\mathfrak{a}} = \mathfrak{m}$) if and only if in the graded $\mathbf{C}$-algebra

$$\operatorname{gr}_{\mathfrak{m}} \mathcal{O} = \bigoplus_{k \geq 0} \mathfrak{m}^k / \mathfrak{m}^{k+1}, \quad \text{with } \mathfrak{m}^0 = \mathcal{O}$$

(which is the homogeneous coordinate ring of the projectivized tangent cone $\mathbf{PC}_3(X, 0)$ see [GT18, Section 2.4]) the images $\overline{f_i}$ of the f_i in $\mathfrak{m}/\mathfrak{m}^2$ generate an irrelevant ideal (that is, an ideal containing all elements of $\operatorname{gr}_{\mathfrak{m}} \mathcal{O}$ of sufficiently large degree so that its zero locus in projective space is empty).

What this last condition means is that first of all the $\overline{f_i}$ are linearly independent over $\mathbf{C}$, so that there is an embedding of the germ $(X, 0)$ into $(\mathbf{C}^N, 0)$ for some N and a linear projection $\pi : (\mathbf{C}^N, 0) \to (\mathbf{C}^d, 0)$ such that its restriction to $(X, 0)$ is germwise the F associated above to $(f_1, \ldots, f_d)$ and secondly, since $\overline{\mathfrak{a}} = \mathfrak{m}$ by Proposition 8.5.6 the projection π is C_3-general.

We end this section by establishing a result analogous to Proposition 8.5.6 but with respect to generic projections of curves.

Definition 8.5.9 Let $\varphi_1 : R \to A_1$ and $\varphi_2 : R \to A_2$ be morphisms of $\mathbf{C}$-analytic algebras. There is a unique $\mathbf{C}$-analytic algebra, denoted $A_1 \widehat{\otimes}_R A_2$, together with morphisms $\theta_i : A_i \to A_1 \widehat{\otimes}_R A_2$, $i = 1, 2$, such that $\theta_1 \circ \varphi_1 = \theta_2 \circ \varphi_2$ and for every pair of morphisms of $\mathbf{C}$-analytic algebras $\psi_1 : A_1 \to B$, $\psi_2 : A_2 \to B$ satisfying $\psi_1 \circ \varphi_1 = \psi_2 \circ \varphi_2$ there is a unique morphism of $\mathbf{C}$-analytic algebras $\psi : A_1 \widehat{\otimes}_R A_2 \to B$ making the whole diagram commute. The algebra $A_1 \widehat{\otimes}_R A_2$ is called the analytic tensor product of A_1 and A_2 over R.

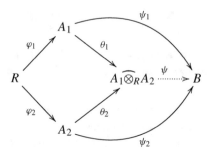

Geometrically this analytic tensor product is the operation on the analytic algebras that corresponds to the fibre product of analytic spaces. Given holomorphic maps $\phi_1 : (X_1, p_1) \to (Y, q)$ and $\phi_2 : (X_2, p_2) \to (Y, q)$ we have the fibre product:

$$
\begin{array}{ccc}
X_1 \times_Y X_2 & \xrightarrow{\ \Pi_1\ } & X_1 \\
\Big\downarrow{\scriptstyle \Pi_2} & & \Big\downarrow{\scriptstyle \phi_1} \\
X_2 & \xrightarrow[\ \phi_2\ ]{} & Y
\end{array}
$$

which induces the corresponding diagram of analytic algebras

$$
\begin{array}{ccc}
O_{Y,q} & \xrightarrow{\ \varphi_1\ } & O_{X_1, p_1} \\
\Big\downarrow{\scriptstyle \varphi_2} & & \Big\downarrow{} \\
O_{X_2, p_2} & \longrightarrow & O_{X_1 \times_Y X_2, (p_1, p_2)}
\end{array}
$$

that is, the analytic algebra $O_{X_1 \times_Y X_2, (p_1, p_2)}$ is isomorphic to $O_{X_1, p_1} \widehat{\otimes}_{O_{Y,q}} O_{X_2, p_2}$.

Remark 8.5.10 See [GP07, Def 1.28, Example 1.46.1 & Lemma 1.89] and [Ada12].

1. When $R = \mathbf{C}$ in the definition, the analytic tensor product $O_{X_1, p_1} \widehat{\otimes}_{\mathbf{C}} O_{X_2, p_2}$ is the analytic algebra corresponding to the product germ $(X_1 \times X_2, (p_1, p_2))$. Moreover if $O_{X_1, p_1} = \mathbf{C}\{\mathbf{z}\}/I$ and $O_{X_2, p_2} = \mathbf{C}\{\mathbf{w}\}/J$ with $\mathbf{z} = (z_1, \ldots, z_N)$ and $\mathbf{w} = (w_1, \ldots, w_M)$ then

$$
O_{X_1, p_1} \widehat{\otimes}_{\mathbf{C}} O_{X_2, p_2} = \frac{\mathbf{C}\{\mathbf{z}, \mathbf{w}\}}{\langle I\mathbf{C}\{\mathbf{z}, \mathbf{w}\} + J\mathbf{C}\{\mathbf{z}, \mathbf{w}\}\rangle}.
$$

2. In general if $(X_1, p_1) \subset (\mathbf{C}^N, 0)$, $(X_2, p_2) \subset (\mathbf{C}^M, 0)$ and $(Y, q) \subset (\mathbf{C}^k, 0)$ with $O_{Y,q} = \mathbf{C}\{y_1, \ldots, y_k\}/K$, denoting by $y_k(\mathbf{z})$ (resp. $y_k(\mathbf{w})$) a representative in $\mathbf{C}\{\mathbf{z}\}$ (resp. $\mathbf{C}\{\mathbf{w}\}$) of the image of y_k in O_{X_1, p_1} (resp. O_{X_2, p_2} by the structure

maps $O_{Y,q} \to O_{X_i,p_i}$ $i = 1, 2$, then

$$O_{X_1,p_1} \widehat{\otimes}_{O_{Y,q}} O_{X_2,p_2} = \frac{\mathbf{C}\{z\}}{I} \widehat{\otimes}_{O_{Y,q}} \frac{\mathbf{C}\{w\}}{J}$$

$$\cong \frac{\mathbf{C}\{z, w\}}{I\mathbf{C}\{z, w\} + J\mathbf{C}\{z, w\} + \langle y_1(z) - y_1(w), \ldots, y_k(z) - y_k(w)\rangle \mathbf{C}\{z, w\}}.$$

Let $(X, 0) \subset (\mathbf{C}^N, 0)$ be a germ of reduced singular curve. By Proposition 8.4.5 the C_5 cone is a finite union of 2-planes of $\mathbf{C}^n$

$$C_5(X, 0) = H_1 \cup \ldots \cup H_r.$$

If we let $\pi : (\mathbf{C}^N, 0) \to (\mathbf{C}^2, 0)$ denote the linear projection to the first two coordinates $(z_1, \ldots, z_N) \mapsto (z_1, z_2)$ then π is generic if and only if $\pi(H_i) = \mathbf{C}^2$ for $i = 1, \ldots, r$. Recall that the construction of $C_5(X, 0)$ goes through the blowup of the diagonal of $X \times X$, so let $I_\Delta \subset O_{X \times X, (0,0)}$ be the ideal defining this diagonal

$$I_\Delta = \langle z_1 - w_1, \ldots, z_N - w_N\rangle O_{X \times X, (0,0)}.$$

Proposition 8.5.11 *Let $I_{\Delta_2} \subset O_{X \times X, (0,0)}$ be the ideal coming from the projection π, that is, $I_{\Delta_2} = \langle z_1 - w_1, z_2 - w_2\rangle O_{X \times X, (0,0)}$. Then π is generic if and only if $\overline{I_{\Delta_2}} = \overline{I_\Delta}$.*

Proof The proof is now very similar to the C_3-general case, and since $I_{\Delta_2} \subset I_\Delta$ all we have to prove is that genericity is equivalent to the inclusion $I_\Delta \subset \overline{I_{\Delta_2}}$.

Let $L = V(z_1, z_2)$ be the kernel of π. Then π is generic if and only if $C_5(X, 0) \cap L = \{0\}$. Let $\ell = [a_1 : \cdots : a_N] \in \mathbf{P}^{N-1}$ be a line in the (projectivized) cone $C_5(X, 0)$, then $\ell \not\subset L$ if and only if $a_i \neq 0$ for some $i \in \{1, 2\}$. This time the lines in $C_5(X, 0)$ are determined by taking the limit as $t \to 0$ of the secants associated to pairs of arcs $(\varphi, \psi) : (\mathbf{C}, 0) \to (X \times X, (0, 0))$

$$t \longrightarrow [\varphi_1(t) - \psi_1(t) : \cdots : \varphi_N(t) - \psi_N(t)] \in \mathbf{P}^{N-1}$$

Again for every such pair of arcs $(\varphi, \psi) : (\mathbf{C}, 0) \to (X \times X, (0, 0))$ we have that

$$(\varphi, \psi)^*(I_{\Delta_2})O_{\mathbf{C},0} = \langle \varphi_1(t) - \psi_1(t), \varphi_2(t) - \psi_2(t)\rangle \mathbf{C}\{t\} = \langle t^k\rangle \mathbf{C}\{t\}$$

where $k = \min\{\mathrm{ord}_0(\varphi_1(t) - \psi_1(t)), \mathrm{ord}_0(\varphi_2(t) - \psi_2(t))\}$. Finally $a_i \neq 0$ for some $i \in \{1, 2\}$ if and only if for all $j \in \{3, \ldots, N\}$

$$\mathrm{ord}_0(\varphi_j(t) - \psi_j(t)) \geq k = \min\{\mathrm{ord}_0(\varphi_1(t) - \psi_1(t)), \mathrm{ord}_0(\varphi_2(t) - \psi_2(t))\}$$

if and only if $(\varphi, \psi)^*(z_j - w_j) \in (\varphi, \psi)^*(I_{\Delta_2})O_{\mathbf{C},0}$ if and only if $z_j - w_j \in \overline{I_{\Delta_2}}$ for all $j \in \{3, \ldots, N\}$, that is, $I_\Delta \subset \overline{I_{\Delta_2}}$. $\square$

8.5.2 Lipschitz Saturation

Let $n^*: O_{X,0} \hookrightarrow \overline{O_{X,0}}$ be the integral closure of a reduced complex analytic algebra which is a quotient of $\mathbf{C}\{z_1, \ldots, z_N\}$. Recall that $\overline{O_{X,0}}$ is a direct sum of normal analytic algebras (in particular integral domains), one for each irreducible component of the germ $(X, 0)$. By Definition 8.5.9 the following commutative diagram determines a unique morphism Ψ of direct sums of analytic algebras:

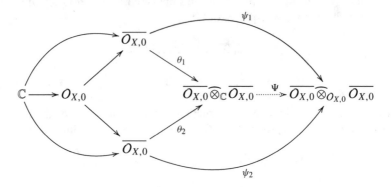

where $\theta_1(f) = f\widehat{\otimes}_\mathbf{C} 1$ and $\theta_2(f) = 1\widehat{\otimes}_\mathbf{C} f$. Note that the map Ψ : $\overline{O_{X,0}} \widehat{\otimes}_\mathbf{C} \overline{O_{X,0}} \to \overline{O_{X,0}} \widehat{\otimes}_{O_{X,0}} \overline{O_{X,0}}$ is the morphism of sums of analytic algebras corresponding to the inclusion $(\overline{X} \times_X \overline{X}, (0,0)) \hookrightarrow (\overline{X} \times \overline{X}, (0,0))$. By Remark 8.5.10 if we denote by $n : (\overline{X}, 0) \to (X, 0)$ the normalization map and $\overline{O_{X,0}} = \bigoplus_{i=1}^{r} \mathbf{C}\{\underline{t_i}\}/J_i(\underline{t_i})$ with $\underline{t_i} = (\underline{t}_{i,1}, \ldots, \underline{t}_{i,m_i})$, then

$$\overline{O_{X,0}} \widehat{\otimes}_\mathbf{C} \overline{O_{X,0}} = \bigoplus_{i,j} \frac{\mathbf{C}\{\underline{t_i}, \underline{u}_j\}}{\langle J_i(\underline{t_i}), J_j(\underline{u}_j)\rangle}$$

and Ψ is a surjection with kernel

$$I_\Delta = \langle z_1\widehat{\otimes}_\mathbf{C} 1 - 1\widehat{\otimes}_\mathbf{C} z_1, \ldots, z_N\widehat{\otimes}_\mathbf{C} 1 - 1\widehat{\otimes}_\mathbf{C} z_N\rangle \overline{O_{X,0}} \widehat{\otimes}_\mathbf{C} \overline{O_{X,0}}.$$

Definition 8.5.12 Let I_Δ be the kernel of the morphism Ψ above. We define the Lipschitz saturation $O_{X,0}^s$ of $O_{X,0}$ as the algebra

$$O_{X,0}^s := \left\{ f \in \overline{O_{X,0}} \,|\, \theta_1(f) - \theta_2(f) \in \overline{I_\Delta} \right\} = \left\{ f \in \overline{O_{X,0}} \,|\, f\widehat{\otimes}_\mathbf{C} 1 - 1\widehat{\otimes}_\mathbf{C} f \in \overline{I_\Delta} \right\}.$$

Example 8.5.13 Let $(X, 0) \subset (\mathbf{C}^3, 0)$ be the irreducible curve with normalization map:

$$\eta : (\mathbf{C}, 0) \longrightarrow (X, 0)$$

$$t \mapsto (t^4, t^6, t^7)$$

In this setting the map Ψ above is

$$\Psi : \overline{O_{X,0}} \,\widehat{\otimes}_{\mathbf{C}}\, \overline{O_{X,0}} \to \overline{O_{X,0}} \,\widehat{\otimes}_{O_{X,0}}\, \overline{O_{X,0}}$$

$$\Psi : \mathbf{C}\{t,u\} \longrightarrow \frac{\mathbf{C}\{t,u\}}{\langle t^4 - u^4, t^6 - u^6, t^7 - u^7 \rangle}.$$

The maps θ_i are just inclusions, $\mathbf{C}\{t\} \hookrightarrow \mathbf{C}\{t,u\}$, $\mathbf{C}\{u\} \hookrightarrow \mathbf{C}\{t,u\}$ and the ideal $I_\Delta = \langle t^4 - u^4, t^6 - u^6, t^7 - u^7 \rangle$. By definition $O^s_{X,0} :=$ $\{f \in \mathbf{C}\{t\} \mid f(t) - f(u) \in \overline{I_\Delta}\}$ and note that $O_{X,0} \subset O^s_{X,0}$. For example $t^5 \in \mathbf{C}\{t\}$ is in $O^s_{X,0}$ if and only if $t^5 - u^5$ is in $\overline{I_\Delta}$. By taking the arc $\phi(\tau) = (\tau, -\tau)$ we have that $\phi^* I_\Delta O_{\mathbf{C},0} = \langle \tau^7 \rangle$ and $\phi^*(t^5 - u^5) = 2\tau^5 \notin \phi^* I_\Delta O_{\mathbf{C},0}$, so by Theorem 8.5.5(2) the element $t^5 \in \mathbf{C}\{t\} = \overline{O_{X,0}}$ is not in the Lipschitz saturation $O^s_{X,0}$. For this particular arc we have $\phi^*(t^9 - u^9) = 2\tau^9 \in \phi^* I_\Delta O_{\mathbf{C},0}$, and one can actually prove that $t^9 \in O^s_{X,0}$. As we shall see later on, in fact $O^s_{X,0} = \mathbf{C}\{t^4, t^6, t^7, t^9\}$.

We are going to show that the Lipschitz saturation $O^s_{X,0}$ is always an analytic algebra, even if the germ $(X, 0)$ is not irreducible. To begin to understand why this is true, let's look at the irreducible case. Define the map

$$\alpha : \overline{O_{X,0}} \to \overline{O_{X,0}} \,\widehat{\otimes}_{\mathbf{C}}\, \overline{O_{X,0}}$$

$$f \mapsto \theta_1(f) - \theta_2(f) = f(z) - f(w)$$

It is not a ring map, however if $n^* : O_{X,0} \hookrightarrow \overline{O_{X,0}}$ denotes the inclusion coming from the normalization map $n : \overline{X} \to X$ then $\alpha(n^*(O_{X,0})) = \alpha(n^*(\mathfrak{m}_{X,0}))$ and $I = \mathrm{Ker}\,\Psi = \langle \alpha(n^*(\mathfrak{m}_{X,0})) \rangle$.

By Definition 8.5.9 $\overline{O_{X,0}} \,\widehat{\otimes}_{O_{X,0}}\, \overline{O_{X,0}}$ is an $O_{X,0}$-algebra, in particular an $O_{X,0}$-module. However, an interesting point is that since $n : \overline{X} \to X$ is a finite map, by [GP07, Lemma 1.89] this algebra is isomorphic to the algebraic tensor product $\overline{O_{X,0}} \otimes_{O_{X,0}} \overline{O_{X,0}}$, so for instance

$$\frac{\mathbf{C}\{t,u\}}{\langle t^2 - u^2, t^3 - u^3 \rangle} \cong \mathbf{C}\{t\} \otimes_{\mathbf{C}\{t^2,t^3\}} \mathbf{C}\{u\}$$

Lemma 8.5.14 *The map*

$$\Psi \circ \alpha : \overline{O_{X,0}} \longrightarrow \overline{O_{X,0}} \,\widehat{\otimes}_{O_{X,0}}\, \overline{O_{X,0}}$$

is a morphism of $O_{X,0}$-modules.

Proof Indeed for $r \in O_{X,0}$ and $f \in \overline{O_{X,0}}$:

$$rf \stackrel{\alpha}{\mapsto} r(z)f(z) - r(w)f(w) \stackrel{\Psi}{\mapsto} r(z)f(z) - r(w)f(w) + I$$

but $r(z) = r(w) \bmod(I)$ so $r(z)f(z) - r(w)f(w) = (r(z) + I)(f(z) - f(w) + I) = r(\Psi \circ \alpha)(f)$. $\square$

By definition $\overline{I_\Delta}$ is an ideal of $\overline{O_{X,0}} \widehat{\otimes}_\mathbb{C} \overline{O_{X,0}}$ and since Ψ is a surjective ring homomorphism we have that $\Psi(\overline{I_\Delta}) \subset \overline{O_{X,0}} \widehat{\otimes}_{O_{X,0}} \overline{O_{X,0}}$ is an ideal, in particular it is an $O_{X,0}$-module. But this implies that

$$(\Psi \circ \alpha)^{-1}(\Psi(\overline{I_\Delta})) = \alpha^{-1}(\overline{I_\Delta}) = O_{X,0}^s \subset \overline{O_{X,0}}$$

is an $O_{X,0}$-module. $\square$

Lemma 8.5.15 *The Lipschitz saturation $O_{X,0}^s$ is an $O_{X,0}$-algebra and a direct sum of analytic algebras.*

Proof Since $O_{X,0}^s$ is a submodule of the Noetherian module $\overline{O_{X,0}}$, it is a finitely generated $O_{X,0}$-module. Even more, you can easily check that $O_{X,0}^s$ is closed under multiplication, so it is an $O_{X,0}$-algebra and by [deJP00, Cor. 3.3.25 & 3.3.26] this implies that $O_{X,0}^s$ is a direct sum of analytic algebras.

Indeed, take $f_1, f_2 \in O_{X,0}^s$ then $(\Psi \circ \alpha)(f_1) = f_1(z) - f_1(w) + I_\Delta \in \Psi(\overline{I_\Delta})$, but it is an ideal so $(f_2(z) + I_\Delta)(f_1(z) - f_1(w) + I_\Delta) \in \Psi(\overline{I_\Delta})$. Analogously $(f_1(w) + I_\Delta)(f_2(z) - f_2(w) + I_\Delta) \in \Psi(\overline{I_\Delta})$ by taking their sum we get that $(\Psi \circ \alpha)(f_1 f_2) = f_1(z)f_2(z) - f_1(w)f_2(w) + I_\Delta \in \Psi(\overline{I_\Delta})$ which implies that $f_1 f_2 \in O_{X,0}^s$ as claimed. $\square$

Before proving that $O_{X,0}^s$ is actually an analytic algebra we would like to give an idea of how things work in the non-irreducible case so suppose there are two irreducible components $(X, 0) = (X_1, 0) \cup (X_2, 0)$. As we said before $\overline{X}$ is then a multi-germ $(\overline{X_1}, p) \sqcup (\overline{X_2}, q)$ and $\overline{O_{X,0}} = \overline{O_{X_1,0}} \oplus \overline{O_{X_2,0}} = O_{\overline{X_1},p} \oplus O_{\overline{X_2},q}$. Since the analytic tensor product should be the algebraic counterpart of the fibre product then we should consider/define

$$\overline{O_{X,0}} \widehat{\otimes}_{O_{X,0}} \overline{O_{X,0}} =$$

$$O_{\overline{X_1},p} \widehat{\otimes}_{O_{X,0}} O_{\overline{X_1},p} \oplus O_{\overline{X_1},p} \widehat{\otimes}_{O_{X,0}} O_{\overline{X_2},q} \oplus O_{\overline{X_2},q} \widehat{\otimes}_{O_{X,0}} O_{\overline{X_1},p} \oplus O_{\overline{X_2},q} \widehat{\otimes}_{O_{X,0}} O_{\overline{X_2},q}$$

and analogously for $\overline{O_{X,0}} \widehat{\otimes}_\mathbb{C} \overline{O_{X,0}}$. By componentwise taking the ring maps Ψ_{ij} coming from the universal property of the irreducible case, for example:

$$\Psi_{12} : O_{\overline{X_1},p} \widehat{\otimes}_\mathbb{C} O_{\overline{X_2},q} \to O_{\overline{X_1},p} \widehat{\otimes}_{O_{X,0}} O_{\overline{X_2},q}$$

we get the ring map Ψ as before with kernel $I_\Delta = I_{11} \oplus I_{12} \oplus I_{21} \oplus I_{22}$. The map α should now be defined as

$$\alpha : \overline{O_{X,0}} \longrightarrow \overline{O_{X,0}} \widehat{\otimes}_\mathbb{C} \overline{O_{X,0}}$$

$$(f_1, f_2) \mapsto (f_1(z) - f_1(w), f_1(z) - f_2(w), f_2(z) - f_1(w), f_2(z) - f_2(w))$$

and we get the same definition for the Lipschitz saturation

$$O^s_{X,0} := \left\{ f = (f_1, f_2) \in \overline{O_{X,0}} \,|\, \alpha(f) \in \overline{I_\Delta} \right\}.$$

More importantly both Lemmas remain valid. Note that in this context of two irreducible components we have $\alpha(f) \in \overline{I_\Delta}$ if and only if $f_1(z) - f_1(w) \in \overline{I_{11}}$, $f_1(z) - f_2(w) \in \overline{I_{12}}$, $f_2(z) - f_1(w) \in \overline{I_{21}}$ and $f_2(z) - f_2(w) \in \overline{I_{22}}$.

Proposition 8.5.16 (See [PT69, Theorem 1.2], [Tei82, Prop. 6.1.1]) *The algebra $O^s_{X,0}$ is the ring of germs of meromorphic functions on $(X, 0)$ which are locally Lipschitz with respect to the ambient metric.*

Proof Recall that $\overline{O_{X,0}}$ is the ring of meromorphic functions on $(X, 0)$ that are locally bounded and a Lipschitz meromorphic function is locally bounded. Now if $h \in O^s_{X,0}$ we need to prove that there exists a real positive constant $C > 0$ such that for every couple $(x_1, x_2) \in X \setminus \operatorname{Sing} X \times X \setminus \operatorname{Sing} X$ (in a small enough representative) we have

$$|h(x_1) - h(x_2)| \le C||x_1 - x_2||.$$

Let $n : \overline{X} \to X \subset \mathbf{C}^n$ be the normalization map. In the irreducible case where $O_{X,0} = \mathbf{C}\{z_1, \ldots, z_N\}/\langle f_1, \ldots, f_s \rangle$ and $\overline{O_{X,0}} = \mathbf{C}\{t_1, \ldots, t_m\}/J(\underline{t})$, the map n induces a morphism of analytic algebras which may be described by

$$n^* : O_{X,0} \longrightarrow \overline{O_{X,0}}$$

$$z_i \mapsto z_i(t_1, \ldots, t_m) = z_i(\underline{t})$$

and referring to the maps α and Ψ as above we have that

$$I_\Delta = \operatorname{Ker} \Psi = \langle z_1(\underline{t}) - z_1(\underline{u}), \ldots, z_N(\underline{t}) - z_N(\underline{u}) \rangle.$$

By definition, $h \in O^s_{X,0}$ if $\alpha(h) = h(\underline{t}) - h(\underline{u}) \in \overline{I_\Delta}$ and by Theorem 8.5.5 there exists a constant C such that

$$|h(\underline{t}) - h(\underline{u})| \le C \sup |z_i(\underline{t}) - z_i(\underline{u})| = C||z(\underline{t}) - z(\underline{u})||,$$

with $(z(\underline{t}), z(\underline{u})) \in X \times X$ and so h is Lipschitz. Reading the proof in the opposite sense gives that a meromorphic, locally Lipschitz function h is necessarily in $O^s_{X,0}$.

If $(X, 0)$ has r irreducible components then $\overline{X}$ is a multi-germ and then we have r maps $n_k : (\overline{X_k}, x_k) \to (X_k, 0) \subset (X, 0)$ with coordinate functions $z_1(\underline{t_k}), \ldots, z_N(\underline{t_k})$. Then for $h = (h_1, \ldots, h_r) \in \overline{O_{X,0}}$ we have that $\alpha(h) =$

$$\left(h_i(\underline{t_i})\widehat{\otimes}1 - 1\widehat{\otimes}h_j(\underline{u_j})\right)_{i,j} \in \bigoplus_{i,j=1}^{i,j=r} \overline{O_{X_i,0}} \widehat{\otimes}_{\mathbf{C}} \overline{O_{X_j,0}}, \text{ and}$$

$$I_\Delta = \bigoplus_{i,j=1}^{i,j=r} I_{ij} \text{ with } I_{ij} = \langle z_1(\underline{t_i})\widehat{\otimes}1 - 1\widehat{\otimes}z_1(\underline{u_j}), \ldots, z_N(\underline{t_i})\widehat{\otimes}1 - 1\widehat{\otimes}z_N(\underline{u_j})\rangle \overline{O_{X_i,0}} \widehat{\otimes}_{\mathbf{C}} \overline{O_{X_j,0}}$$

and $\alpha(h) \in \overline{I_\Delta}$ if and only if $h_i(\underline{t_i})\widehat{\otimes}1 - 1\widehat{\otimes}h_j(\underline{u_j}) \in \overline{I_{ij}}$ for all (i, j).

So in the spirit of Example 8.2.7 the "coordinate" h_i of h indicates you how to evaluate h in points of the corresponding irreducible component $(X_i, 0)$ of $(X, 0)$ and for $i \neq j$ the condition $h_i(\underline{t_i})\widehat{\otimes}1 - 1\widehat{\otimes}h_j(\underline{u_j}) \in \overline{I_{ij}}$ tells you that the Lipschitz condition must also be satisfied when you take points in different irreducible components.   □

Corollary 8.5.17 (See [PT69, Corollary 1.3]) *Let $(X, 0) \subset (\mathbf{C}^N, 0)$ be a reduced germ of complex analytic singularity. The ring $O_{X,0}^s$ is an analytic algebra.*

Proof We already proved in Lemma 8.5.15 that $O_{X,0}^s$ is a direct sum of analytic algebras, but if there were more than one, the function $(1, 0, \ldots, 0) \in O_{X,0}^s$ would not be Lipschitz, contradicting Proposition 8.5.16.   □

From Lemma 8.5.15 we have injective ring morphisms

$$O_{X,0} \hookrightarrow O_{X,0}^s \hookrightarrow \overline{O_{X,0}}.$$

Since $\overline{O_{X,0}}$ is contained in the total ring of fractions $Q(O_{X,0})$, the total ring of fractions of the Lipschitz saturation $O_{X,0}^s$ coincides with $Q(O_{X,0})$ and by transitivity of integral dependence the normalizations also coincides i.e., $\overline{O_{X,0}^s} = \overline{O_{X,0}}$. In terms of holomorphic maps we have:

$$\overline{X} \xrightarrow{n_s} X^s \xrightarrow{\zeta} X,$$

where X^s is the germ of complex analytic singularity corresponding to the analytic algebra $O_{X,0}^s$, the map $n_s : \overline{X} \to X^s$ is the normalization map of X^s, $\zeta : (X^s, 0) \to (X, 0)$ is finite and induces an isomorphism outside the non-normal locus of X, and $n = \zeta \circ n_s : \overline{X} \to X$ is the normalization map of X.

Definition 8.5.18 The germ $(X^s, 0)$ together with the finite map $\zeta : (X^s, 0) \to (X, 0)$ is called the Lipschitz saturation of $(X, 0)$.

Lemma 8.5.19 *Let $(X, 0) \subset (\mathbf{C}^n, 0)$ be a reduced germ of complex analytic singularity, then $\left(O_{X,0}^s\right)^s = O_{X,0}^s$.*

Proof Following the notation of Lemma 8.5.14 we have the maps:

$$O_{X,0} \hookrightarrow O_{X,o}^s \hookrightarrow \overline{O_{X,0}} \xrightarrow{\alpha} \overline{O_{X,0}} \widehat{\otimes}_{\mathbf{C}} \overline{O_{X,0}},$$

and this induces a map

$$\overline{O_{X,0}} \widehat{\otimes}_{O_{X,0}} \overline{O_{X,0}} \cong \frac{\overline{O_{X,0}} \widehat{\otimes}_{\mathbf{C}} \overline{O_{X,0}}}{\langle \alpha(O_{X,0}) \rangle} \longrightarrow \frac{\overline{O_{X,0}} \widehat{\otimes}_{\mathbf{C}} \overline{O_{X,0}}}{\langle \alpha(O_{X,0}^s) \rangle} \cong \overline{O_{X,0}} \widehat{\otimes}_{O_{X,0}^s} \overline{O_{X,0}}$$

that makes the following diagram commute

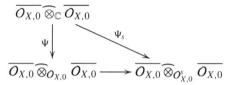

If we denote $I_\Delta = \mathrm{Ker}\,\Psi$ and $I_{\Delta^s} = \mathrm{Ker}\,\Psi_s$ we have $I_\Delta \subset I_{\Delta^s}$. Now by definition we have $O_{X,0}^s = \{h \in \overline{O_{X,0}} \,|\, \alpha(h) \in \overline{I_\Delta}\}$ so $\alpha(O_{X,0}^s) \subset \overline{I_\Delta}$ which implies $\overline{I_\Delta} = \overline{I_{\Delta^s}}$ and so $\left(O_{X,0}^s\right)^s = \{h \in \overline{O_{X,0}} \,|\, \alpha(h) \in \overline{I_{\Delta^s}} = \overline{I_\Delta}\} = O_{X,0}^s$. □

8.5.3 The Case of Dimension 1

Let $(X, 0) \subset (\mathbf{C}^2, 0)$ be a germ of reduced plane curve, and $\zeta : (X^s, 0) \to (X, 0) \subset (\mathbf{C}^2, 0)$ the finite map given by the Lipschitz saturation of $(X, 0)$. What we want to emphasize is that this map can always be realized as a linear projection on suitable representatives. Indeed, any representative of $(X^s, 0) \subset (\mathbf{C}^m, 0)$ can be re-embedded as the graph of ζ in $\mathbf{C}^{m+2}$, namely by the map $X^s \to \mathbf{C}^2 \times \mathbf{C}^m : p \mapsto (\zeta_1(p), \zeta_2(p), p)$. The map ζ is now the projection of $(X^s, 0)$ to $(X, 0)$ by the first two coordinates: $(z_1, \ldots, z_{m+2}) \mapsto (z_1, z_2)$.

Proposition 8.5.20 (See [Tei82, Proposition 6.2.1]) *For a germ of reduced plane curve $(X, 0) \subset (\mathbf{C}^2, 0)$ the Lipschitz saturation map $\zeta : (X^s, 0) \to (X, 0)$ is a generic projection.*

Proof Suppose first that $(X, 0)$ is irreducible, in this case we have the holomorphic maps

$$(\mathbf{C}, 0) \xrightarrow{\eta_s} (X^s, 0) \subset (\mathbf{C}^{m+2}, 0) \xrightarrow{\zeta} (X, 0) \subset (\mathbf{C}^2, 0)$$

$$t \mapsto (z_1(t), z_2(t), z_3(t), \ldots, z_{m+2}(t)) \mapsto (z_1(t), z_2(t))$$

By Proposition 8.5.11 we have to prove that the ideals $I_{\Delta^s} = \langle z_1 - w_1, \ldots, z_{m+2} - w_{m+2} \rangle$ and $I_{\Delta_2^s} = \langle z_1 - w_1, z_2 - w_2 \rangle$ have the same integral closure in $O_{X^s \times X^s, (0,0)}$. In this coordinate system we have the normalization map $\eta_s^* : O_{X^s, 0} \hookrightarrow \overline{O_{X,0}}$

given by

$$\frac{\mathbf{C}\{z_1,\ldots,z_{m+2}\}}{I} \hookrightarrow \mathbf{C}\{t\}$$

$$z_i \mapsto z_i(t) \quad i = 1, 2; \qquad z_{j+2} \mapsto z_{j+2}(t) \quad j = 1,\ldots m,$$

which induces the morphism

$$\theta : O_{X^s \times X^s, (0,0)} = \frac{\mathbf{C}\{z_1,\ldots,z_{m+2}, w_1,\ldots,w_{m+2}\}}{I(z) + I(w)} \hookrightarrow \mathbf{C}\{t, u\} = \overline{O_{X,0}} \widehat{\otimes}_{\mathbf{C}} \overline{O_{X,0}}$$

$$z_i \mapsto z_i(t) \quad i = 1, 2; \qquad z_{j+2} \mapsto z_{j+2}(t) \quad j = 1,\ldots m$$

$$w_i \mapsto z_i(u) \quad i = 1, 2; \qquad w_{j+2} \mapsto z_{j+2}(u) \quad j = 1,\ldots m.$$

But from the proof of Lemma 8.5.19 we have that the ideals $I_{\Delta_2^s} = \langle z_1(t) - z_1(u), z_2(t) - z_2(u)\rangle$ and $I_{\Delta^s} = \langle z_1(t) - z_1(u), z_2(t) - z_2(u), z_3(t) - z_3(u),\ldots, z_{m+2}(t) - z_{m+2}(u)\rangle$ have the same integral closure in $\mathbf{C}\{t, u\}$ and so by Remark 8.5.2(5) the ideals $\theta^{-1}(I_{\Delta^s})$ and $\theta^{-1}(I_{\Delta_2^s})$ have the same integral closure in $O_{X^s \times X^s, (0,0)}$, which is what we wanted.

In the reducible case the proof works exactly the same way, it is just a lot messier to write down. The only thing you have to keep track off is the following. Suppose $(X, 0)$ has two irreducible components $(X_1, 0) \cup (X_2, 0)$ then $(X^s, 0)$ also has two irreducible components and $\overline{O_{X,0}} \cong \mathbf{C}\{t_1\} \oplus \mathbf{C}\{t_2\}$. This implies that the normalization map $\eta_s^* : O_{X^s,0} \hookrightarrow \overline{O_{X,0}}$ is given by

$$\frac{\mathbf{C}\{z_1,\ldots,z_{m+2}\}}{I} \hookrightarrow \mathbf{C}\{t_1\} \oplus \mathbf{C}\{t_2\}$$

$$z_i \mapsto (z_i(t_1), z_i(t_2)) \quad i = 1, 2; \qquad z_{j+2} \mapsto (z_{j+2}(t_1), z_{j+2}(t_2)) \quad j = 1,\ldots m$$

In this case $\overline{O_{X,0}} \widehat{\otimes}_{\mathbf{C}} \overline{O_{X,0}} \cong \mathbf{C}\{t_1, u_1\} \oplus \mathbf{C}\{t_1, u_2\} \oplus \mathbf{C}\{t_2, u_1\} \oplus \mathbf{C}\{t_2, u_2\}$ and the induced morphism θ looks like:

$$\theta : O_{X^s \times X^s, (0,0)} = \frac{\mathbf{C}\{z_1,\ldots,z_{m+2}, w_1,\ldots,w_{m+2}\}}{I(z) + I(w)} \hookrightarrow \overline{O_{X,0}} \widehat{\otimes}_{\mathbf{C}} \overline{O_{X,0}}$$

$$z_i \mapsto (z_i(t_1), z_i(t_1), z_i(t_2), z_i(t_2)) \quad i = 1, 2$$

$$z_{j+2} \mapsto (z_{j+2}(t_1), z_{j+2}(t_1), z_{j+2}(t_2), z_{j+2}(t_2)) \quad j = 1,\ldots m$$

$$w_i \mapsto (z_i(u_1), z_i(u_2), z_i(u_1), z_i(u_2)) \quad i = 1, 2$$

$$w_{j+2} \mapsto z_{j+2}(u_1), z_{j+2}(u_2), z_{j+2}(u_1), z_{j+2}(u_2)) \quad j = 1,\ldots m,$$

then you have the map α as in the proof of Proposition 8.5.16 and the rest follows through. □

Remark 8.5.21

1. Since the Lipschitz saturation map $\zeta : (X^s, 0) \to (X, 0)$ is a generic projection, the multiplicity of $(X^s, 0)$ is equal to the multiplicity of $(X, 0)$.
2. Except if the plane branch $(X, 0)$ is non singular, the map $(\overline{X}, 0) \to (X, 0)$ is never obtained as a generic projection since the multiplicity changes. However, among all germs $(X', 0)$ which dominate $(X, 0)$ and are dominated by $(\overline{X}, 0)$, and in addition are such that the map $(X', 0) \to (X, 0)$ can be represented by a *generic* linear projection, there is a unique one, up to isomorphism, which dominates all the others: it is the saturation.

Corollary 8.5.22 *Let $(X, 0) \subset (\mathbf{C}^2, 0)$ be a reduced plane curve. The Lipschitz saturation map $\zeta : (X^s, 0) \to (X, 0)$ is a biLipschitz homeomorphism.*

Proof We already know that a generic projection induces a homeomorphism with its image (Proposition 8.4.3), so by Proposition 8.5.20 the map ζ is a homeorphism and since it is the restriction to X^s of the linear projection $(z_1, \ldots, z_{m+2}) \mapsto (z_1, z_2)$ it is Lipschitz. The inverse of ζ can be described on each irreducible component X_k by

$$(z_1(t_k), z_2(t_k)) \mapsto (z_1(t_k), z_2(t_k), z_3(t_k), \ldots, z_{m+2}(t_k)),$$

and since for all $j \in \{1, \ldots, m\}$, $z_{j+2}(\underline{t}) \in \mathcal{O}^s_{X,0}$ Proposition 8.5.16 tells us that it is also Lipschitz. □

Our main result now follows from the following theorem:

Theorem 8.5.23 (See [PT69, §4], [BGG80, Prop. VI.3.2]) *Let $\mathcal{O}_{X,0}$ be the analytic algebra of a germ of reduced plane curve $(X, 0) \subset (\mathbf{C}^2, 0)$. The Lipschitz saturation $\mathcal{O}^s_{X,0}$ determines and is determined by the characteristic exponents of its branches (irreducible components) and the intersection multiplicities $m_{ij} = (X_i, X_j)$ of each pair of branches. In particular the saturated curve $(X^s, 0)$ is an invariant (up to isomorphism) of the equisingularity class of $(X, 0)$.*

This implies that every member of the equisingularity class of a germ of reduced plane curve $(X, 0) \subset (\mathbf{C}^2, 0)$ can be obtained by a generic projection of the Lipschitz saturation $(X^s, 0)$ of any one of them. The proof of the Proposition involves a lot of calculations and can be found in the references. For this reason we would rather describe how to calculate the saturated curve $(X^s, 0)$. Let us start with the irreducible case:

Definition 8.5.24 Let $h \in \mathbf{C}\{t\}$ be a power series with coprime exponents. If

$$h = \sum_{j=0}^{\infty} a_j t^j,$$

we define the **set of exponents** of f as $Ex(f) = \{j \in \mathbf{N} \mid a_j \neq 0\}$. And for a germ of analytically irreducible plane curve $(X, 0) \subset (\mathbf{C}^2, 0)$ we define the **set of exponents** of $O_{X,0}$ as

$$E(O_{X,0}) = \bigcup_{h \in \mathfrak{m}} Ex(h),$$

where $\mathfrak{m}$ is the maximal ideal of $O_{X,0}$. Note that the semigroup $\Gamma(X)$ of the plane branch $(X, 0)$ is contained in $E(O_{X,0})$. (See [Tei07, Section 8]).

If $(X, 0) \subset (\mathbf{C}^2, 0)$ is irreducible then:

1. For every $j \in E(O_{X,0})$ we have that $t^j \in O^s_{X,0}$.
2. The analytic algebra $O^s_{X,0}$ is monomial, in particular:

$$O^s_{X,0} = \mathbf{C}\{t^j \mid j \in E(O^s_{X,0})\}.$$

For a numerical semigroup (i.e., a subsemigroup of $(\mathbf{N}, +)$ with finite complement) there is the concept of saturated semigroup (see [RG09, Chapter 3, Section 2]) which is defined as follows:

For $A \subset \mathbf{N}$ and $a \in A \setminus \{0\}$ define

$$d_A(a) = \gcd\{x \in A \mid x \leq a\}.$$

Then a non-empty subset $A \subset \mathbf{N}$ such that $0 \in A$ and $\gcd(A) = 1$ is a saturated numerical semigroup if and only if $a + k d_A(a) \in A$ for all $a \in A \setminus \{0\}$ and $k \in \mathbf{N}$.

The reader can verify that the condition indeed implies that A is a semigroup and that the intersection of two saturated semigroups is again saturated, so that any $A \subset \mathbf{N}$ such that $0 \in A$ and $\gcd(A) = 1$ is contained in a smallest saturated semigroup.

Example 8.5.25 Let $(X, 0) \subset (\mathbf{C}^2, 0)$ be the cusp singularity defined by $y^2 - x^3 = 0$. Its normalization map is $t \mapsto (t^2, t^3)$ and so its semigroup is generated by $\langle 2, 3 \rangle$. Since $\Gamma(X) = \mathbf{N} \setminus \{1\}$ then $E(O_{X,0}) = \Gamma(X)$ is a saturated numerical semigroup.

This definition also tells us how to obtain the smallest saturated semigroup containing any $A \subset \mathbf{N}$ with $0 \in A$ and $\gcd(A) = 1$, for example the set of exponents $E(O_{X,0})$.

Let $e_0 = \beta_0 = \min\{x \in A\}$ and define

$$\widetilde{A_0} := A \cup \beta_0 \cdot \mathbf{N}.$$

In the case of $E(O_{X,0})$ we have that $e_0 = \beta_0$ is the multiplicity of $(X, 0)$.

Let $\beta_1 := \min\{x \in A \mid e_0 \text{ does not divide } x\}$ and $e_1 = \gcd\{\beta_0, \beta_1\}$; note that $e_1 = d_A(\beta_1)$. Again define

$$\widetilde{A_1} := \widetilde{A_0} \cup \{\beta_1 + k e_1 \mid k \in \mathbf{N}\}.$$

Continuing this way we obtain two sequences of natural numbers $e_0 > e_1 > \cdots > e_g = 1 = \gcd(A)$ and $\beta_0 < \beta_1 < \cdots < \beta_g$ and an associated sequence of subsets of $\widetilde{A}_0 \subset \widetilde{A}_1 \subset \cdots \subset \widetilde{A}_g \subset \mathbf{N}$ where $\beta_{i+1} := \min\{x \in A \mid e_i \text{ does not divide } x\}$, $e_i := \gcd\{\beta_0, \ldots, \beta_i\} = d_A(\beta_i)$ and

$$\widetilde{A_{i+1}} := \widetilde{A}_i \cup \{\beta_{i+1} + k e_{i+1} \mid k \in \mathbf{N}\}.$$

Note that $\widetilde{A} := \widetilde{A}_g$ is a saturated semigroup which is completely determined by its **characteristic sequence** $\{\beta_0, \ldots, \beta_g\}$. Moreover if $t \mapsto \left(t^n, \sum_{i \geq n} a_i t^i\right)$ is a Puiseux parametrization of the plane branch $(X, 0) \subset (\mathbf{C}^2, 0)$, the characteristic sequence of $E(\mathcal{O}_{X,0})$ is the set of characteristic exponents of $(X, 0)$ and so it determines its equisingularity class.

Proposition 8.5.26 ((Pham-Teissier), See [PT69, §4], [BGG80, Thm VI.1.6])
For a germ of irreducible plane curve singularity $(X, 0) \subset (\mathbf{C}^2, 0)$ the Lipschitz saturation $\mathcal{O}_{X,0}^s$ is given by

$$\mathcal{O}_{X,0}^s = \mathbf{C}\{t^p \mid p \in \widetilde{E(\mathcal{O}_{X,0})}\}.$$

In particular $E\left(\mathcal{O}_{X,0}^s\right) = \widetilde{E(\mathcal{O}_{X,0})}$.

Let us give a sketch of the proof: we start from a structured parametrization $(t^n, y(t))$ of our branch X as in Sect. 8.1.2 and we have to study integral dependence over the ideal $I_\Delta = (t - u)\mathcal{N} := \langle t^n - u^n, y(t) - y(u) \rangle \subset \mathbf{C}\{t, u\}$. Here $\mathcal{N}$ is the primary ideal $\langle \frac{t^n - u^n}{t - u}, \frac{y(t) - y(u)}{t - u} \rangle \mathbf{C}\{t, u\}$. According to what we saw after Theorem 8.5.5, to verify that $g(t) - g(u)$ is integral over I, which is the same as $\frac{g(t) - g(u)}{t - u}$ being integral over $\mathcal{N}$, it suffices to verify that its order along any of the branches of a plane curve $C_T \subset \mathbf{C}^2$ defined by $T_1 \frac{t^n - u^n}{t - u} - T_2 \frac{y(t) - y(u)}{t - u} = 0$ is larger than that of the ideal I for $T = [T_1 : T_2]$ in the open set $U \subset \mathbf{P}^1$. Now we claim that the open set U is $T_1 \neq 0$. Indeed, since the order of $y(t)$ is $> n$ all the plane curves C_T with $T_1 \neq 0$ have a tangent cone consisting of $n - 1$ lines in general position. It is not difficult then to show (see [Tei73, Chap. II, Lemma 2.6, Proposition 2.7]) that they are equisingular with their tangent cone, and therefore are all equisingular, with simultaneous normalization. So the curve $\frac{t^n - u^n}{t - u} = 0$ is in U, and its branches are the lines $u = \omega t$, $\omega \in \mu_n \setminus \{1\}$, which means that a function $g(t) \in \mathbf{C}\{t\}$ is in the saturation if and only if we have

$$\operatorname{ord}_t(g(t) - g(\omega t)) \geq \operatorname{ord}_t(y(t) - y(\omega t)) \text{ for all } \omega \in \mu_n \setminus \{1\}.$$

The result now follows easily from what we saw at the end of Sect. 8.1.2 about the orders of the $y(t) - y(\omega t)$.

It may be interesting to remark here that this construction gives an intrinsic (coordinate free) definition of the Puiseux characteristic as the set of valuations (orders of vanishing) of the ideal $\mathcal{N}$ along the irreducible components of the exceptional divisor of the normalized blowup of N in $\overline{X} \times \overline{X}$. For more details, see [PT69, §4] and [BGG80, Thm VI.1.6].

Remark 8.5.27 It is shown in [Tei80, 5.2] that the multiplicity, in the sense we saw after Theorem 8.5.7, of the primary ideal N is equal to twice the invariant δ which appears in Propositions 8.3.9 and 8.4.6. It is also shown there that δ is the maximum number of different singular points (then necessarily ordinary double points) which can appear when deforming the parametrization of the plane branch. Both results extends to reducible curves. One can define an analogous ideal N for a non-plane branch but then, in view of Theorem 8.5.8 and Proposition 8.5.11, its multiplicity is twice the δ-invariant of a generic plane projection and no longer the classical $\dim_{\mathbf{C}} \frac{\overline{O_{X,0}}}{O_{X,0}}$ in this case.

Example 8.5.28 Let $(X, 0) \subset (\mathbf{C}^2, 0)$ be the plane branch with normalization map:

$$\eta : (\mathbf{C}, 0) \longrightarrow (X, 0)$$
$$t \mapsto (t^4, t^6 + t^7)$$

Then the exponent set $E(O_{X,0})$ contains the semigroup $\Gamma(X) = \langle 4, 6, 13 \rangle \mathbf{N}$ but by definition it also contains 7. Now $\beta_1 = 6$ and $e_1 = 2$ so

$$\widetilde{E_1} = E(O_{X,0}) \cup \{6 + 2k \mid k \in \mathbf{N}\}.$$

In the next step $\beta_2 = 7$ and $e_2 = 1$ so $g = 2$ and we get the saturated semigroup

$$\widetilde{E_2} = \widetilde{E_1} \cup \{7 + k \mid k \in \mathbf{N}\}.$$

Note that $\widetilde{E(O_{X,0})} = \langle 4, 6, 7, 9 \rangle \mathbf{N}$ and so we have the normalization map for the Lipschitz saturation $(X^s, 0) \subset (\mathbf{C}^4, 0)$ given by:

$$\eta^s : (\mathbf{C}, 0) \longrightarrow (X^s, 0)$$
$$t \mapsto (t^4, t^6, t^7, t^9)$$

By making the change of coordinates in $(\mathbf{C}^4, 0)$, $(x, y, z, w) \mapsto (x, y + z, z, w)$ we can view the Lipschitz saturation map

$$\zeta : (X^s, 0) \to (X, 0)$$

as the projection on the first two coordinates as before.

Remark 8.5.29 (See [Tei82, Chap. I, Theorem 6.3.1], [BGG80, Appendice]) A more concrete way of seeing that all plane branches with the same Puiseux characteristic are generic plane projections of a single space curve is to go back to the notations of Sect. 8.1.2 to write down explicitly the saturation of a plane branch

$(X, 0)$ with given characteristic $(n, \beta_1, \ldots, \beta_g)$: it is isomorphic to the monomial curve with analytic algebra

$$\mathbf{C}\{t^n, t^{2n}, \ldots, t^{\beta_1}, t^{\beta_1+e_1}, \ldots, t^{\beta_2}, t^{\beta_2+e_2}, \ldots, t^{\beta_3}, \ldots, t^{\beta_g}, t^{\beta_g+1}, \ldots\},$$

where $n = e_0 = \beta_0$ as above. The semigroup generated by these exponents, which are those of $\widetilde{E(O_{X,0})}$, is finitely generated by Dickson's Lemma and because the Puiseux exponents are coprime its complement in $\mathbf{N}$ is finite. For more details on the saturation of semigroups we refer to [RG09, Chapter 3, Section 2].

As we saw in Sect. 8.1.2, up to isomorphism, the image of this monomial curve by a generic linear projection can be parametrized by $x = t^n$, $y = \sum_{p \in \widetilde{E(O_{X,0})}\setminus\{n\}} a_p t^p$. Now we see that the generic projections are precisely those which are such that the coefficient of t^n is $\neq 0$ and for $p = \beta_1, \ldots, \beta_g$ we have $a_p \neq 0$, which means that the projection has characteristic $(n, \beta_1, \ldots, \beta_g)$.

Remark that, except if $n = 2$, the semigroup of integers generated by the exponents of the monomials belonging to the saturation $O^s_{X,0}$ is different from the semigroup Γ we saw in Sect. 8.1.2. This has the interesting consequence that the specialization of a plane branch to the monomial curve with the same semigroup, which is Whitney equisingular (see [GT18, Remark 4.1]; the argument there can be generalized to any plane branch), is not in general biLipschitz trivial.

When $(X, 0)$ is not irreducible it is a bit more complicated, nevertheless the Lipschitz saturation $O^s_{X,0}$ can be described in the following way:

Theorem 8.5.30 (See [PT69, §4] and [BGG80, Thm VI.2.2]) *Let $O_{X,0}$ be the analytic algebra of a reduced plane curve $(X, 0) = (X_1, 0) \cup \ldots \cup (X_r, 0)$ with normalization $\overline{O_{X,0}} = \mathbf{C}\{t_1\} \oplus \cdots \oplus \mathbf{C}\{t_r\}$. We may assume that the image of x in $\overline{O_{X,0}}$ is $(t_1^{n_1}, \ldots, t_r^{n_r})$ where n_i is the multiplicity of the branch $(X_i, 0)$. Let μ be the least common multiple of $\{n_1, \ldots, n_r\}$. Then the element $\underline{h} = (h_1, \ldots, h_r) \in \overline{O_{X,0}}$ is in the Lipschitz saturation $O^s_{X,0}$ if and only if the following two conditions are satisfied:*

1. *For every $j \in \{1, \ldots, r\}$ we have that $h_j \in O^s_{X_j,0}$.*
2. *For every μ-th root of unity ϵ and every couple $i \neq j$ we have the inequality*

$$m_{i,j,\epsilon}(h) \geq m_{i,j,\epsilon} := \inf_{g \in O_{X,0}} \left\{ v_\tau \left(g_i(\tau^{\mu/n_i}) - g_j([\epsilon\tau]^{\mu/n_j}) \right) \right\},$$

where $m_{i,j,\epsilon}(\underline{h}) = v_\tau \left(h_i(\tau^{\mu/n_i}) - h_j([\epsilon\tau]^{\mu/n_j}) \right)$ and v_τ is the valuation of $\mathbf{C}\{\tau\}$ given by the order of the series. The number $m_{i,j,\epsilon}$ depends only on the characteristic exponents and the intersection multiplicity of the branches X_i and X_j.

Example 8.5.31 Let $(X, 0) = (X_1, 0) \cup (X_2, 0)$ be the plane curve with normalization map:

$$\eta : (\mathbf{C}, 0) \sqcup (\mathbf{C}, 0) \longrightarrow (X, 0)$$

$$t_1 \mapsto (t_1^4, t_1^6 + t_1^7)$$

$$t_2 \mapsto (t_2^3, t_2^5)$$

In the previous example we already calculated the Lipschitz saturation $O_{X_1,0}^s = \mathbf{C}\{t_1^4, t_1^6, t_1^7, t_1^9\}$ and following the algorithm we get the Lipschitz saturation $O_{X_2,0}^s = \mathbf{C}\{t_2^3, t_2^5, t_2^7\}$. Since the branches are tangent, their intersection multiplicity is greater than the product of their multiplicities and is equal to the order of the series in t_1 obtained by substituting the parametrization of $(X_1, 0)$ in the equation $y^3 - x^5 = 0$ defining $(X_2, 0)$. In this case it is equal to 18.

By definition $\mu = \mathrm{lcm}\{3, 4\} = 12$ and it is not hard to prove that for any 12-th root of unity ϵ

$$m_{1,2,\epsilon} = v_\tau \left(y_1(\tau^3) - y_2([\epsilon\tau]^4) \right)$$

$$= v_\tau \left(\tau^{18} + \tau^{21} - \epsilon^8\tau^{20} \right) = 18.$$

So from the Theorem 8.5.30 we have that $h = (h_1(t_1), h_2(t_2))$ is in $O_{X,0}^s$ if and only if $h_1(t_1) \in O_{X_1,0}^s$, $h_2(t_2) \in O_{X_2,0}^s$ and $m_{1,2,\epsilon}(h) \geq 18$. For example if $h = (t_1^4, t_2^5)$ then

$$m_{1,2,\epsilon}(h) = v_\tau \left((\tau^3)^4 - ([\epsilon\tau]^4)^5 \right)$$

$$= v_\tau(\tau^{12} - \epsilon^8\tau^{20}) = 12 \Rightarrow h \notin O_{X,0}^s.$$

On the other hand if $h = (t_1^6 + t_1^7, t_2^5)$ then

$$m_{1,2,\epsilon}(h) = v_\tau \left((\tau^3)^6 + (\tau^3)^7 - ([\epsilon\tau]^4)^5 \right)$$

$$= v_\tau(\tau^{18} + \tau^{21} - \epsilon^8\tau^{20}) = 18 \Rightarrow h \in O_{X,0}^s.$$

We will end this section with the following consequence of the Theorem:

Corollary 8.5.32 (See [BGG80, VI.3.7]) *Let $(X, 0) = (X_1, 0) \cup \ldots \cup (X_r, 0)$ be a reduced plane curve with normalization $\overline{O_{X,0}} = \mathbf{C}\{t_1\} \oplus \cdots \oplus \mathbf{C}\{t_r\}$. If $\Pi_j : \overline{O_{X,0}} \to \mathbf{C}\{t_j\}$ denotes the canonical projection to the j-th factor then*

$$\Pi_j(O_{X,0}^s) = O_{X_j,0}^s.$$

8.5.4 Application to Local Polar Curves

Let $(X, 0) \subset (\mathbf{C}^N, 0)$ be a reduced equidimensional germ of complex analytic space. Consider linear projections $\pi \colon \mathbf{C}^N \to \mathbf{C}^2$ and the critical locus of π restricted to the smooth part X^0 of X. It is proved in [LT81], where the theory of (absolute)[3] local polar varieties was initiated, that for a Zariski dense open set U in the space $G(N, N-2)$ of linear projection, this critical locus is either empty or a reduced curve. The *closure* of this curve in X is an (absolute) **polar curve** of X and is denoted by $P_{d-1}(X, \pi)$ where d is the dimension of X. It is also denoted by $P_{d-1}(X, D)$, where $D = \ker \pi$. These curves play an important role in the local study of singularities, and especially in the study of the Lipschitz geometry of surfaces. See [LT81], [Tei82, Chap. IV] and [NP16] for more details.

Of course, if it is not empty, $P_{d-1}(X, \pi)$ varies with the projection $\pi \in U$ and a *priori* it could be that π remains constantly a non generic projection for $P_{d-1}(X, \pi)$. That seems unlikely but still we need a proof for the following:

Theorem 8.5.33 (See [Tei82, Chap. V, Lemme 1.2.2]) *Given $(X, 0) \subset (\mathbf{C}^N, 0)$ as above and assuming that $P_{d-1}(X, \pi)$ is a reduced curve for $\pi \in U \subset G(N, N-2)$, there exists a non empty Zariski open set $V \subset U$ such that for $\pi \in V$, the projection $\pi \colon \mathbf{C}^N \to \mathbf{C}^2$ is a generic projection for the curve $P_{d-1}(X, \pi) \subset \mathbf{C}^N$.*

The proof, which we only sketch, gives an example of the notion of Lipschitz equisaturation , which is found in [PT69, §6]. Fixing coordinates $z_1, \ldots, z_N$ on $\mathbf{C}^N$ and x, y on $\mathbf{C}^2$, we can parametrize by $\mathbf{C}^{2(N-2)}$ a dense open set of the space of linear projections $\mathbf{C}^N \to \mathbf{C}^2$ as follows:

$$x = z_1 + \sum_3^N a_i z_i, \quad y = z_2 + \sum_3^N b_i z_i , \quad (\underline{a}, \underline{b}) \in \mathbf{C}^{2(N-2)}.$$

To simplify notations while keeping the ideas, we assume that X is a hypersurface defined by $f(z_1, \ldots, z_N) = 0$. One can also consult [BH80, Lemme 3.7] which gives the proof for isolated singularities of surfaces in $\mathbf{C}^3$.

For any series $h(z_1, \ldots, z_N) \in \mathbf{C}\{z_1, \ldots, z_N\}$ let us denote by $h_{\underline{a}, \underline{b}}$ the series

$$h_{\underline{a}, \underline{b}}(\underline{z}, \underline{a}, \underline{b}) = h(x - \sum_3^N a_i z_i, y - \sum_3^N b_i z_i, z_3, \ldots, z_N).$$

---

[3]This precision refers to a distinction between absolute and relative polar varieties, which is not relevant here but should be mentioned to avoid confusions. See [Tei82, Chap. IV, p. 417].

The equation $f_{\underline{a},\underline{b}} = 0$ defines a germ of hypersurface $\mathcal{Z}$ in $\mathbf{C}^N \times \mathbf{C}^{2(N-2)}$ and if we consider the family of projections $\pi_{\underline{a},\underline{b}} \colon \mathbf{C}^N \times \mathbf{C}^{2(N-2)} \to \mathbf{C}^2 \times \mathbf{C}^{2(N-2)}$ defined by

$$x = z_1 + \sum_3^N a_i z_i, \quad y = z_2 + \sum_3^N b_i z_i, \quad \underline{a} \mapsto \underline{a}, \quad \underline{b} \mapsto \underline{b},$$

and the closure of its critical locus on the non singular part of $\mathcal{Z}$, we obtain a subspace which, over a Zariski open subset of $\mathbf{C}^{2(N-2)}$, contains the family of polar curves associated to the family of projections $\pi_{\underline{a},\underline{b}}$. Over a possibly smaller Zariski open subset V of $\mathbf{C}^{2(N-2)}$ this family of curves is equisingular in the sense that it has a simultaneous parametrization. The number r of irreducible components of $\mathcal{Z}$ at points of $\{0\} \times V \subset \mathbf{C}^N \times V$ is constant and after choosing as origin of $\mathbf{C}^{2(N-2)}$ a point of V we can parametrize each irreducible component in a neighborhood of $\{0\} \times \{0\}$ by:

$$z_1 = t_\ell^{n_\ell}, \quad z_2 = \upsilon(t_\ell, \underline{a}, \underline{b}), \quad z_i = \zeta_i(t_\ell, \underline{a}, \underline{b}),$$

with $\upsilon(t_\ell, \underline{a}, \underline{b}), \zeta_i(t_\ell, \underline{a}, \underline{b}) \in \mathbf{C}\{t_\ell, \underline{a}, \underline{b}\}$ for $i = 3, \ldots, N$. The normalization of $\mathcal{O}_{\mathcal{Z},0}$ being $\overline{\mathcal{O}_{\mathcal{Z},0}} = \prod_{i=1}^r \mathbf{C}\{t_\ell, \underline{a}, \underline{b}\}$.

By definition of $\mathcal{Z}$ we have for each $\ell = 1, \ldots, r$ the identity in $\mathbf{C}\{t_\ell, \underline{a}, \underline{b}\}$

$$f(t^{n_\ell} - \sum_3^N a_i \zeta_i(t_\ell, \underline{a}, \underline{b}), \upsilon(t_\ell, \underline{a}, \underline{b}) - \sum_3^N a_i \zeta_i(t_\ell, \underline{a}, \underline{b}), \zeta_3(t_\ell, \underline{a}, \underline{b}), \ldots, \zeta_N(t_\ell, \underline{a}, \underline{b})) \equiv 0.$$

Differentiating $f_{\underline{a},\underline{b}} = 0$ with respect to z_i gives the following equations on $\mathcal{Z}$:

$$-a_i \frac{\partial f_{\underline{a},\underline{b}}}{\partial z_1} - b_i \frac{\partial f_{\underline{a},\underline{b}}}{\partial z_2} + \frac{\partial f_{\underline{a},\underline{b}}}{\partial z_i} \equiv 0,$$

for $i = 3, \ldots, N$. which by definition are satisfied on the polar curve.

Differentiating the first identity with respect to b_k and taking into account the second set of identities, we obtain that the equation

$$\left(\zeta_k(t_\ell, \underline{a}, \underline{b}) - \frac{\partial \upsilon(t_\ell, \underline{a}, \underline{b})}{\partial b_k} \right) \frac{\partial f_{\underline{a},\underline{b}}}{\partial z_2} = 0$$

must be satisfied in each $\mathbf{C}\{t_\ell, \underline{a}, \underline{b}\}$. By general transversality results found in [LT81, Cor. 4.1.6] and [Tei82, Chap. IV, 5.1], $\frac{\partial f_{\underline{a},\underline{b}}}{\partial z_2}$ does not vanish because it would entail a lack of C_3-transversality (see Definition 8.4.1) of the polar curve with the kernel of the projection which defines it. So we must have on $\mathcal{Z}$ the identity $z_k = \frac{\partial \upsilon}{\partial b_k}$.

By [Tei82, Proposition 6.4.2], after perhaps shrinking V to a smaller Zariski open dense subset V_1 of $\mathbf{C}^{2(N-2)}$ we have that over V_1 the family $\mathcal{Z}_1$ of plane curves given parametrically by the parametrizations $x = t^{n_\ell}, y = \upsilon(t_\ell, \underline{a}, \underline{b})$, which consists of the plane projections of our polar curves, is equisaturated. This implies that the derivations $\frac{\partial}{\partial b_k}$ of $\mathbf{C}\{\underline{a}, \underline{b}\}$ extend to derivations D_k of $\overline{O_{\mathcal{Z}_1,(0,v)}} = \overline{O_{\mathcal{Z},(0,v)}} = \prod_{i=1}^r \mathbf{C}\{t_\ell, \underline{a}, \underline{b}\}$ into itself which preserve the relative saturated ring (see [PT69]). Since of course the functions $\upsilon(t_\ell, \underline{a}, \underline{b})$ belong to the relative saturation of $\overline{O_{\mathcal{Z}_1,(0,v)}}$, so do the $\zeta_k(t_\ell, \underline{a}, \underline{b})$ which are their images by D_k. But ζ_k belonging to this relative saturation means precisely that for $v \in V_1$, the saturations of the rings $\overline{O_{\mathcal{Z}_1(v)}}$ and $\overline{O_{\mathcal{Z}(v)}}$ of the fibers over v of $\mathcal{Z}$ and $\mathcal{Z}_1$ are equal for $v \in V_1$, which is the condition for C_5-genericity according to Proposition 8.5.11.

The fact that the plane projection of a generic polar curve by the map which defines it is generic plays an important role in the following three domains: the comparison of Zariski equisingularity and Whitney equisingularity for surfaces (see [BH80, NP16]), the comparison of Zariski equisingularity and Lipschitz equisingularity for surfaces (see [NP16, PP20]), the numerical characterization of Whitney equisingularity (see [Tei82, Chap. V]) and the valuative study of the metric geometry of surface singularities in view of their biLipschitz classification (see [BFP19]).

Acknowledgments Arturo Giles Flores was supported by CONACyT Grant 221635.

Otoniel Nogueira da Silva would like to thank CONACyT (grant 282937) for the financial support by Fordecyt 265667, and also UNAM/DGAPA for support by PAPIIT IN 113817.

The authors are grateful to the referee for his or her very careful reading and numerous useful observations.

References

[Ada12] J. Adamus, in *Topics in Complex Analytic Geometry Part II*. Lecture Notes. https://www.uwo.ca/math/faculty/adamus/adamus_publications/AGII.pdf 252

[BH80] J. Briançon, J.P.G. Henry, Equisingularité générique des familles de surfaces à singularités isolées. Bull. S.M.F. **108**(2), 259–281 (1980) 267, 269

[BGG80] J. Briançon, A. Galligo, M. Granger, Déformations équisingulières des germes de courbes gauches réduites. Mém. Soc. Math. France, 2ème serie (1) **69** (1980) 242, 243, 244, 261, 263, 264, 265, 266

[BFP19] A. Belotto da Silva, L. Fantini, A. Pichon, Inner geometry of complex surfaces: a valuative approach (2019). arXiv:1905.01677v1 [math.AG] 269

[BK86] E. Brieskorn, H. Knörrer, *Plane Algebraic Curves* (Birkhäuser/Springer Basel AG, Basel, 1986) 229

[BG80] R. Buchweitz, G.M. Greuel, The Milnor number and deformations of complex curve singularities Invent. Math. **58**, 241–281 (1980) 234

[Che78] A. Chenciner, *Courbes Algébriques Planes* (Publications Mathématiques de l'Université Paris VII, 1978) 228, 229

[Chi89] E.M. Chirka, *Complex Analytic Sets*. (Kluwer, Dordecht, 1989) 239, 240, 241

[DGPS19] W. Decker, G.M. Greuel, G. Pfister, H. Schönemann, Singular 4-1-2—A computer algebra system for polynomial computations (2019). http://www.singular.uni-kl.de 238

[deJP00] T. de Jong, G. Pfister, *Local Analytic Geometry: Basic Theory and Applications.* Advanced Lectures in Mathematics (Springer, Berlin, 2000) 223, 224, 225, 226, 227, 228, 229, 239, 251, 256

[Eis95] D. Eisenbud, *Commutative Algebra with a View Toward Algebraic Geometry.* Graduate Texts in Mathematics, n° 150 (Springer, Berlin, 1995) 230

[GT18] A.G. Flores, B. Teissier, Local polar varieties in the geometric study of singularities. Ann. Fac. Sci. Toulouse Math. (6) **4**, 679–775 (2018) 239, 251, 265

[GF02] H. Grauert, K. Fritzsche, *From Holomorphic Functions to Complex Manifolds.* Graduate Texts in Mathematics (Springer, Berlin, 2002) 225

[GLS07] G.M. Greuel, C. Lossen, E. Shustin, *Introduction to Singularities and Deformations.* Springer Monographs in Mathematics (Springer, Berlin, 2007) 227, 235, 236, 237, 238, 244

[GP07] G.M. Greuel, G. Pfister, *A Singular Introduction to Commutative Algebra* (Springer, Berlin, 2007) 252, 255

[HMP18] M.E. Hernandes, A.J. Miranda, G. Peñafort-Sanchis, A presentation matrix algorithm for $f_*O_{X,x}$. Topol. Appl. **234**, 440–451 (2018) 238

[HS06] C. Huneke, I. Swanson, *Integral Closure of Ideals, Rings and Modules.* London Mathematical Society Lecture Note Series, n° 336 (Cambridge University Press, Cambridge, 2006) 246, 247, 248, 251

[KK83] L. Kaup, B. Kaup, *Holomorphic Functions of Several Variables. An Introduction to the Fundamental Theory* De Gruyter Studies in Mathematics (1983) 227

[Le15] C.-T. Lê, Equinormalizability and topological triviality of deformations of isolated curve singularities over smooth base spaces. Kodai Math. J. (38) **3**, 642–657 (2015) 238

[LR76] D.T. Lê, C.P. Ramanujam, The invariance of Milnor's number implies the invariance of topological type. Am. J. Math. (91) **1**, 67–78 (1976) 244

[LT81] D.T. Lê, B. Teissier, Variétés polaires locales et classes de Chern des variétés singulières. Ann. Math. (2) **114**, 457–491 (1981) 267, 268

[Lej73] M. Lejeune-Jalabert, Sur l'équivalence des singularités des courbes algébriques planes. Coefficients de Newton, *ıIntroduction à la théorie des singularités*, Thesis, Paris 7, 1973. Published in *Travaux en Cours*, vol. 36 (Hermann, 1988), pp. 49–124 221

[LT08] M. Lejeune-Jalabert, B. Teissier, Clôture intégrale des idéaux et équisingularité. Ann. Fac. Sci. Toulouse Math. (6) **4**, 781–859 (2008) 246, 247, 248

[Lip82] J. Lipman, Equimultiplicity, reduction and blowing up, in *Commutative Algebra: Analytical Methods, Conf. Fairfax/Va. 1979.* Lecture Notes in Pure and Applied Mathematics, vol. 68 (1982), pp. 111–147 246, 251

[Loj91] S. Łojasiewicz, *Introduction to Complex Analytic Geometry* (Birkhäuser, Basel, 1991) 227, 230

[Mat89] H. Matsumura, *Commutative Ring Theory.* Cambridge Studies in Advanced Mathematics (Cambridge University Press, Cambridge, 1989) 250

[Mir19] A.J. Miranda. https://sites.google.com/site/aldicio/publicacoes 238

[MP89] D. Mond, R. Pellikaan, *Fitting Ideals and Multiple Points of Analytic Mappings.* Lecture Notes in Mathematics, vol. 1414 (Springer, Berlin, 1989), pp. 107–161 232, 235, 236, 238, 239, 244, 245

[NP16] W.D. Neumann, A. Pichon, Lipschitz geometry of complex surfaces, analytic invariants and equisingularity (2016). arXiv:1211.4897v3 [math.AG] 267, 269

[PT69] F. Pham, B. Teissier, Lipschitz fractions of a complex analytic algebra and Zariski saturation, in *Introduction to Lipschitz Geometry of Singularities: Lecture Notes of the International School on Singularity Theory and Lipschitz Geometry, Cuernavaca, June 2018,* ed. by W. Neumann, A. Pichon. Lecture Notes in Mathematics, vol. 2280 (Springer, Cham, 2020), pp. 309–337. https://doi.org/10.1007/978-3-030-61807-0_10 246, 257, 258, 261, 263, 264, 265, 267, 269

[PP20] A. Parusiński, L. Păunescu, Lipschitz stratification of complex hypersurfaces in codimension 2 (2020). arXiv:1909.00296v3 269

[Ree61] D. Rees, α-transforms of local rings and a theorem on multiplicities of ideals. Proc. Camb. Philos. Soc. **57**, 8–17 (1961) 251

[Rim72] D.S. Rim, Formal deformation theory, in *SGA 7 I, Groupe de Monodromie en Géométrie Algébrique*. Lecture Notes in Mathematics, vol. 288 (Springer, Berlin, 1972), pp. 32–132 230

[RG09] J.C. Rosales, P.A. García-Sánchez, *Numerical Semigroups*. Developments in Mathematics (Springer, Berlin, 2009) 262, 265

[Sno20] J. Snoussi, A quick trip into local singularities of complex curves and surfaces, in *Introduction to Lipschitz Geometry of Singularities: Lecture Notes of the International School on Singularity Theory and Lipschitz Geometry, Cuernavaca, June 2018*, ed. by W. Neumann, A. Pichon. Lecture Notes in Mathematics, vol. 2280 (Springer, Cham, 2020), pp. 45–71. https://doi.org/10.1007/978-3-030-61807-0_10 218

[Stu72a] J. Stutz, Analytic sets as branched coverings. Trans. Am. Math. Soc. **166**, 241–259 (1972) 240, 241

[Stu72b] J. Stutz, Equisingularity and equisaturation in codimension 1. Am. J. Math. **94**, 1245–1268 (1972) 240

[Stu77] J. Stutz, Equisingularity and local analytic geometry, in *Several Complex Variables (Proc. Sympos. Pure Math.)* XXX Part 1 (1977), pp. 77–84 241

[Tei73] B. Teissier, Cycles évanescents, sections planes et conditions de Whitney. *Singularités à Cargèse, Astérisque 7–8* (1973), pp. 285–362 230, 235, 246, 263

[Tei77] B. Teissier, The hunting of invariants in the geometry of discriminants, in *Real and Complex singularities (Proc. Ninth Nordic Summer School/ NAVF Sympos. Math. Oslo, 1976)* (1977), pp. 565–678 230, 234, 235, 238, 244

[Tei80] B. Teissier, Résolution simultanée, I, II, in *Séminaire sur les Singularités des Surfaces*. Lecture Notes in Mathematics n° 777 (Springer, Berlin, 1980), pp. 72–146 237, 264

[Tei82] B. Teissier, Variétés polaires. II: Multiplicités polaires, sections planes et conditions de Whitney, in *Algebraic geometry, Proc. int. Conf., La Rábida/Spain 1981*. Lecture Notes in Mathematics, vol. 961 (Springer, Berlin, 1982), pp. 314–491 248, 257, 259, 264, 267, 268, 269

[Tei04] B. Teissier, Monomial ideals, binomial ideals, polynomial ideals, in *Trends in Commutative Algebra*. Math. Sci. Res. Inst. Publ., vol. 51 (Cambridge University Press, Cambridge, 2004), pp. 211–246 248

[Tei07] B. Teissier, Complex curve singularities: a biased introduction, in *Singularities in Geometry and Topology, World Sci. Publ., Hackensack* (2007), pp. 825–887 218, 228, 262

[Tou72] J.C. Tougeron, *Idéaux de fonctions différentiables*. Ergebnisse der Math., Band 71 (Springer, Berlin, 1972) 230

[Wal78] R.J. Walker, *Algebraic Curves* (Springer, Berlin, 1978) 229

[Whi65] H. Whitney, Local properties of analytic varieties, in *Differ. and Combinat. Topology, Sympos. Marston Morse, Princeton* (1965), pp. 205–244 238, 240

[Zar39] O. Zariski, Some results in the arithmetic theory of algebraic varieties. Am. J. Math. **61**, 249–294 (1939) 223

[Zar71] O. Zariski, General Theory of saturation and of saturated local rings, II: saturated local rings of dimension1. Am. J. Math. **936**(4), 872–964 (1971) 221

[Zar06] O. Zariski, *The Moduli Problem for Plane Branches* University Lecture Series, vol. 39 (American Mathematical Society, Providence, 2006) 222, 228

[Zar68] O. Zariski, Studies in Equisingularity, III. Saturation of local rings and equisingularity. Am. J. Math. **90**, 961–1023 (1968) 221

Chapter 9
Ultrametrics and Surface Singularities

Patrick Popescu-Pampu

Abstract The present lecture notes give an introduction to works of García Barroso, González Pérez, Ruggiero and the author. The starting point of those works is a theorem of Płoski, stating that one defines an ultrametric on the set of branches drawn on a smooth surface singularity by associating to any pair of distinct branches the quotient of the product of their multiplicities by their intersection number. We show how to construct ultrametrics on certain sets of branches drawn on any normal surface singularity from their mutual intersection numbers and how to interpret the associated rooted trees in terms of the dual graphs of adapted embedded resolutions. The text begins by recalling basic properties of intersection numbers and multiplicities on smooth surface singularities and the relation between ultrametrics on finite sets and rooted trees. On arbitrary normal surface singularities one has to use Mumford's definition of intersection numbers of curve singularities drawn on them, which is also recalled.

9.1   Introduction

This paper is an expansion of my notes prepared for the course with the same title given at the *International school on singularities and Lipschitz geometry*, which took place in Cuernavaca (Mexico) from June 11th to 22nd 2018.

If S denotes a **normal surface singularity**, that is, a germ of normal complex analytic surface, a **branch** on it is an irreducible germ of analytic curve contained in S. In his 1985 paper [Pło85], Arkadiusz Płoski proved that if one associates to every pair of distinct branches on the singularity $S = (\mathbb{C}^2, 0)$ the quotient $\dfrac{A \cdot B}{m(A) \cdot m(B)}$ of their intersection number by the product of their multiplicities, then for every triple of pairwise distinct branches, two of those quotients are equal and the third one is

P. Popescu-Pampu (✉)
University of Lille, CNRS, Lille, France
e-mail: patrick.popescu-pampu@univ-lille.fr

W. Neumann, A. Pichon (eds.), *Introduction to Lipschitz Geometry of Singularities*, Lecture Notes in Mathematics 2280, https://doi.org/10.1007/978-3-030-61807-0_9

not smaller than them. An equivalent formulation is that the inverses $\dfrac{m(A) \cdot m(B)}{A \cdot B}$ of the previous quotients define an **ultrametric** on the set of branches on $(\mathbb{C}^2, 0)$.

Using the facts that the multiplicity of a branch is equal to its intersection number with a smooth branch L transversal to it, and that a given function is an ultrametric on a set if and only if it is so in restriction to all its finite subsets, one deduces that Płoski's theorem is a consequence of:

Theorem 9.1.1 *Let L be a smooth branch on the smooth surface singularity S and let $\mathcal{F}$ be a finite set of branches on S, transversal to L. Then the function $u_L : \mathcal{F} \times \mathcal{F} \to [0, \infty)$ defined by $u_L(A, B) := \dfrac{(L \cdot A) \cdot (L \cdot B)}{A \cdot B}$ if $A \neq B$ and $u_L(A, A) := 0$ is an ultrametric on $\mathcal{F}$.*

This may be seen as a property of the pair (S, L) and one may ask whether it extends to other pairs consisting of a normal surface singularity and a branch on it. It turns out that this property characterizes the so-called **arborescent singularities**, that is, the normal surface singularities such that the dual graph of every good resolution is a tree. Namely, one has the following theorem, which combines [GBGPPP18, Thm. 85] and [GBPPPR19, Thm. 1.46]:

Theorem 9.1.2 *Let L be a branch on the normal surface singularity S. Then the function u_L defined as before is an ultrametric on any finite set $\mathcal{F}$ of branches on S distinct from L if and only if S is an arborescent singularity.*

It is possible to think topologically about ultrametrics on finite sets in terms of certain types of decorated rooted trees. In particular, any such ultrametric determines a rooted tree. One may try to describe this tree directly from the pair $(S, \mathcal{F} \cup \{L\})$, when S is arborescent and the ultrametric is the function u_L associated to a branch L on it. In order to formulate such a description, we need the notion of **convex hull** of a finite set of vertices of a tree: it is the union of the paths joining those vertices pairwise.

The following result was obtained in [GBGPPP18, Thm. 87]:

Theorem 9.1.3 *Let L be a branch on the arborescent singularity S and let $\mathcal{F}$ be a finite set of branches on S distinct from L. Then the rooted tree determined by the ultrametric u_L on $\mathcal{F}$ is isomorphic to the convex hull of the strict transform of $\mathcal{F} \cup \{L\}$ in the dual graph of its preimage by an embedded resolution of it, rooted at the vertex representing the strict transform of L.*

Even when the singularity S is not arborescent, the function u_L becomes an ultrametric in restriction to suitable sets $\mathcal{F}$ of branches on S. Those sets are defined only in terms of convex hulls taken in the so-called **brick-vertex tree** of the dual graph of an embedded resolution of $\mathcal{F} \cup \{L\}$, and do not depend on any numerical parameter of the exceptional divisor of the resolution, be it a genus or a self-intersection number. The brick-vertex tree of a connected graph is obtained canonically by replacing each **brick**—a maximal inseparable subgraph which is not an edge—by a star, whose central vertex is called a **brick-vertex**. One has the following generalization of Theorem C (see [GBPPPR19, Thm. 1.42]):

Theorem 9.1.4 *Let L be a branch on the normal surface singularity S and let $\mathcal{F}$ be a finite set of branches on S distinct from L. Consider an embedded resolution of $\mathcal{F} \cup \{L\}$. Assume that the convex hull of its strict transform in the brick-vertex tree of the dual graph of its preimage does not contain brick-vertices of valency at least 4 in the convex hull. Then the function u_L is an ultrametric in restriction to $\mathcal{F}$ and the associated rooted tree is isomorphic to the previous convex hull, rooted at the vertex representing the strict transform of L.*

If S is not arborescent, there may exist other sets of branches on which u_L restricts to an ultrametric. Unlike the sets described in the previous theorem, in general they do not depend only on the topology of the dual graph of their preimage on some embedded resolution, but also on the self-intersection numbers of the components of the exceptional divisor (see [GBPPPR19, Ex. 1.44]).

The aim of the present notes is to introduce the reader to the previous results. Note that in the article [GBPPPR19, Part 2], these results were extended to the space of real-valued semivaluations of the local ring of S.

Let us describe briefly the structure of the paper. In Sect. 9.2 are recalled basic facts about **multiplicities** and **intersection numbers** of plane curve singularities. In Sect. 9.3 are stated two equivalent formulations of Płoski's theorem. In Sect. 9.4 is explained the relation between ultrametrics and rooted trees mentioned above, an intermediate concept being that of **hierarchy** on a finite set. Using this relation, Sect. 9.5 presents a proof of Theorem 9.1.1. This proof uses the so-called **Eggers-Wall tree** of a plane curve singularity relative to a smooth reference branch L, constructed using associated Newton–Puiseux series. Section 9.6 explains the notions used in the formulation of Theorem 9.1.2, that is, those of **good resolution**, **embedded resolution**, associated **dual graph** and **arborescent singularity**. In Sect. 9.7 are described the related notions of **cut-vertex** and **brick-vertex tree** of a finite connected graph. Section 9.8 explains and illustrates the statement of Theorem 9.1.4. In Sect. 9.9 is explained Mumford's intersection theory of divisors on normal surface singularities, after a proof of a fundamental property of such singularities, stating that the intersection form of any of their resolutions is negative definite. In Sect. 9.10 the ultrametric inequality concerning the restriction of u_L to a triple of branches is reexpressed in terms of the notion of **angular distance** on the dual graph of an adapted resolution. A crucial property of this distance is stated, which relates it to the cut-vertices of the dual graph. In Sect. 9.11 is sketched the proof of a theorem of pure graph theory, relating distances satisfying the previous crucial property and the brick-vertex tree of the graph. This theorem implies Theorem 9.1.4.

9.2 Multiplicity and Intersection Numbers for Plane Curve Singularities

In this section we recall the notions of multiplicity of a plane curve singularity and intersection number of two such singularities. One may find more details in [dJP00, Sect. 5.1] or [Fis01, Chap. 8].

Let (S, s) be a **smooth surface singularity**, that is, a germ of smooth complex analytic surface. Denote by $\boxed{\mathcal{O}_{S,s}}$ its local $\mathbb{C}$-algebra and by $\boxed{\mathfrak{m}_{S,s}}$ its maximal ideal, containing the germs at s of holomorphic functions vanishing at s.

A **local coordinate system** on S at s is a pair $(x, y) \in \mathfrak{m}_{S,s} \times \mathfrak{m}_{S,s}$ establishing an isomorphism between a neighborhood of s in S and a neighborhood of the origin in $\mathbb{C}^2$. Algebraically speaking, this is equivalent to the fact that the pair (x, y) generates the maximal ideal $\mathfrak{m}_{S,s}$, or that it realizes an isomorphism $\mathcal{O}_{S,s} \simeq \mathbb{C}\{x, y\}$. This isomorphism allows to see each germ $f \in \mathcal{O}_{S,s}$ as a convergent power series in the variables x and y.

A **curve singularity on** (S, s) is a germ $(C, s) \hookrightarrow (S, s)$ of not necessarily reduced curve on S, passing through s. As the germ (S, s) is isomorphic to the germ of the affine plane $\mathbb{C}^2$ at any of its points, one says also that (C, s) is a **plane curve singularity** A **defining function** of (C, s) is a function $f \in \mathfrak{m}_{S,s}$ such that $\mathcal{O}_{C,s} = \mathcal{O}_{S,s}/(f)$, where (f) denotes the principal ideal of $\mathcal{O}_{S,s}$ generated by f. We write then $C = \boxed{Z(f)}$.

The curve singularity (C, s) may also be seen as an effective principal divisor on (S, s). This allows to write $C = \sum_{i \in I} p_i C_i$, where $p_i \in \mathbb{N}^*$ for all $i \in I$ and the curve singularities C_i are pairwise distinct and irreducible. We say in this case that the C_i's are the **branches of** C. A **branch on** (S, s) is an irreducible curve singularity on (S, s).

Next definition introduces the simplest invariant of a plane curve singularity:

Definition 9.2.1 Assume that $f \in \mathcal{O}_{S,s}$. Its **multiplicity** is the vanishing order of f at s:

$$\boxed{m_s(f)} := \sup\{n \in \mathbb{N}, \ f \in \mathfrak{m}_{S,s}^n\} \in \mathbb{N} \cup \{\infty\}.$$

If (C, s) is the curve singularity defined by f, we say also that $\boxed{m_s(C)} := m_s(f)$ is its **multiplicity** at s.

It is a simple exercise to check that the multiplicity of a curve singularity is independent of the function defining it. If one chooses local coordinates (x, y) on (S, s), then $m_s(f)$ is the smallest degree of the monomials appearing in the expression of f as a convergent power series in the variables x and y. One has $m_s(f) = \infty$ if and only if $f = 0$ and $m_s(f) = 1$ if and only if f defines a *smooth* branch on (S, s).

The following definition describes a measure of the way in which two curve singularities intersect:

Definition 9.2.2 Let $C, D \hookrightarrow (S, s)$ be two plane curve singularities defined by $f, g \in \mathfrak{m}_{S,s}$. Then their **intersection number** is defined by:

$$\boxed{C \cdot D} := \dim_{\mathbb{C}} \frac{\mathcal{O}_{S,s}}{(f, g)} \in \mathbb{N} \cup \{\infty\},$$

where (f, g) denotes the ideal of $\mathcal{O}_{S,s}$ generated by f and g.

Note that $C \cdot D < +\infty$ if and only if C and D do not share common branches, which is also equivalent to the existence of $n \in \mathbb{N}^*$ such that one has the following inclusion of ideals: $(f, g) \supseteq \mathfrak{m}_{S,s}^n$. Nevertheless, unlike the multiplicity, the intersection number $C \cdot D$ is not always equal to the smallest exponent n having this property. For instance, if one takes $f := x^3$ and $g := y^2$, then $C \cdot D = 6$ but $(f, g) \supseteq (x, y)^5$. We leave the verification of the previous facts as an exercise.

The following proposition, which may be proved using Proposition 9.2.5 below, relates multiplicities and intersection numbers:

Proposition 9.2.3 *If $(C, s) \hookrightarrow (S, s)$ is a plane curve singularity, then $C \cdot L \geq m_s(C)$ for any smooth branch L through s, with equality if and only if L is transversal to C. More generally, if D is a second curve singularity on (S, s), then $C \cdot D \geq m_s(C) \cdot m_s(D)$, with equality if and only if C and D are transversal.*

Let us explain the notion of *transversality* used in the previous proposition, as it is more general than the standard notion of transversality, which applies only to smooth submanifolds of a given manifold. If C is a branch on (S, s) and one chooses a local coordinate system (x, y) on (S, s), as well as a defining function f of C, it may be shown that the lowest degree part of f is a power of a complex linear form in x and y. This linear form defines a line in the tangent plane $T_s S$ of S at s, which is by definition the **tangent line** of C at s. One may show that it is independent of the choices of local coordinates and defining function of C. If C is now an arbitrary curve singularity, then its **tangent cone** is the union of the tangent lines of its branches. Given two plane curve singularities on the same smooth surface singularity S, one says that they are **transversal** if each line of the tangent cone of one of them is transversal (in the classical sense) to each line of the tangent cone of the other one.

Let us pass now to the question of *computation* of intersection numbers. A basic method consists in breaking the symmetry between the two curve singularities, by working with a defining function of one of them and by *parametrizing* the other one. One has to be cautious and choose a *normal parametrization*, in the following sense:

Definition 9.2.4 A **normal parametrization** of the branch (C, s) is a germ of holomorphic morphism $\nu : (\mathbb{C}, 0) \to (C, s)$ which is a normalization morphism, that is, it has topological degree one.

For instance, if the branch $(C, 0)$ on $(\mathbb{C}^2, 0)$ is defined by the function $y^2 - x^3$, then $t \to (t^2, t^3)$ is a normal parametrization of C, but $u \to (u^4, u^6)$ is not. A

normal parametrization of a branch (C, s) may be also characterized by asking it to establish a homeomorphism between suitable representatives of the germs $(\mathbb{C}, 0)$ and (C, s).

Normalization morphisms may be defined more generally for reduced germs (X, x) of arbitrary dimension (see [dJP00, Sect. 4.4]), by considering the multi-germ whose multi-local ring is the integral closure of the local ring $O_{X,x}$ in its total ring of fractions. Except for curve singularities, the source of a normalization morphism is not smooth in general.

The following proposition is classical and states the announced expression of intersection numbers in terms of a parametrization of one germ and a defining function of the second one (see [BHPVdV04, Prop. II.9.1] or [dJP00, Lemma 5.1.5]):

Proposition 9.2.5 *Let C be a branch on the smooth surface singularity (S, s) and D be a second curve singularity, not necessarily reduced. Let $\nu : (\mathbb{C}, 0) \to (C, s)$ be a normal parametrization of C and let $g \in \mathfrak{m}_{S,s}$ be a defining function of D. Then:*

$$C \cdot D = \mathrm{ord}_t\, (g \circ \nu(t)),$$

where $\boxed{\mathrm{ord}_t}$ *denotes the order of a power series in the variable t.*

Proof This proof is adapted from that of [dJP00, Lemma 5.1.5].

The order of the zero power series is equal to ∞ by definition, therefore the statement is true when C is a branch of D.

Let us assume from now on that *C is not a branch of D.*

Consider a defining function $f \in \mathfrak{m}_{S,s}$ of C. By Definition 9.2.2:

$$C \cdot D = \dim_{\mathbb{C}} \frac{O_{S,s}}{(f, g)} = \dim_{\mathbb{C}} \frac{O_{S,s}/(f)}{(g_C)} = \dim_{\mathbb{C}} \frac{O_{C,s}}{(g_C)}, \tag{9.1}$$

where we have denoted by $g_C \in O_{C,s}$ the restriction of g to the branch C.

Algebraically, the normal parametrization $\nu : (\mathbb{C}, 0) \to (C, s)$ corresponds to a morphism of local $\mathbb{C}$-algebras $O_{C,s} \hookrightarrow \mathbb{C}\{t\}$, isomorphic to the inclusion morphism of $O_{C,s}$ into its integral closure taken inside its quotient field. In order to distinguish them, denote from now on by $g_C\, O_{C,s}$ the principal ideal generated by g_C inside $O_{C,s}$ and by $g_C\, \mathbb{C}\{t\}$ its analog inside $\mathbb{C}\{t\}$. One has the following equality inside the local $\mathbb{C}$-algebra $\mathbb{C}\{t\}$:

$$g \circ \nu(t) = g_C.$$

As a consequence:

$$g_C\, \mathbb{C}\{t\} = t^{\mathrm{ord}_t\,(g \circ \nu(t))}\mathbb{C}\{t\}.$$

Therefore:

$$\operatorname{ord}_t (g \circ v(t)) = \dim_{\mathbb{C}} \frac{\mathbb{C}\{t\}}{g_C \, \mathbb{C}\{t\}}. \tag{9.2}$$

By comparing Eqs. (9.1) and (9.2), we see that the desired equality is equivalent to:

$$\dim_{\mathbb{C}} \frac{O_{C,s}}{g_C \, O_{C,s}} = \dim_{\mathbb{C}} \frac{\mathbb{C}\{t\}}{g_C \, \mathbb{C}\{t\}}. \tag{9.3}$$

The two quotients appearing in (9.3) are the cokernels of the two injective multiplication maps $O_{C,s} \xrightarrow{\cdot g_C} O_{C,s}$ and $\mathbb{C}\{t\} \xrightarrow{\cdot g_C} \mathbb{C}\{t\}$. The associated short exact sequences may be completed into a commutative diagram in which the first two vertical maps are the inclusion map $O_{C,s} \hookrightarrow \mathbb{C}\{t\}$:

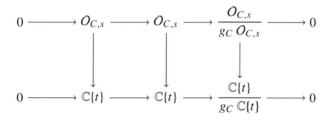

The last vertical map is not necessarily an isomorphism. We want to show that its source and its target have the same dimension. Let us complete it into an exact sequence by considering its kernel K_1 and cokernel K_2:

$$0 \longrightarrow K_1 \longrightarrow \frac{O_{C,s}}{g_C \, O_{C,s}} \longrightarrow \frac{\mathbb{C}\{t\}}{g_C \, \mathbb{C}\{t\}} \longrightarrow K_2 \longrightarrow 0.$$

For every finite exact sequence of finite-dimensional vector spaces, the alternating sum of dimensions vanishes. Therefore:

$$\dim_{\mathbb{C}} K_1 - \dim_{\mathbb{C}} \frac{O_{C,s}}{g_C \, O_{C,s}} + \dim_{\mathbb{C}} \frac{\mathbb{C}\{t\}}{g_C \, \mathbb{C}\{t\}} - \dim_{\mathbb{C}} K_2 = 0.$$

This shows that the desired equality (9.3) would result from the equality $\dim_{\mathbb{C}} K_1 = \dim_{\mathbb{C}} K_2$. This last equality is a consequence of the so-called "snake lemma" (see for instance [AM69, Prop. 2.10]), applied to the previous commutative diagram. Indeed, by this lemma, one has an exact sequence:

$$0 \longrightarrow K_1 \longrightarrow \frac{\mathbb{C}\{t\}}{O_{C,s}} \longrightarrow \frac{\mathbb{C}\{t\}}{O_{C,s}} \longrightarrow K_2 \longrightarrow 0.$$

Reapplying the previous argument about alternating sums of dimensions, one gets the needed equality $\dim_{\mathbb{C}} K_1 = \dim_{\mathbb{C}} K_2$. □

Note that the previous proof shows in fact that for any abstract branch (C, s), not necessarily planar, one has the equality:

$$\dim_{\mathbb{C}} \frac{O_{C,s}}{(g)} = \mathrm{ord}_t (g \circ \nu(t)), \tag{9.4}$$

for any $g \in O_{C,s}$ and for any normal parametrization $\nu : (\mathbb{C}, 0) \to (C, s)$ of (C, s). If the branch (C, s) is contained in an ambient germ (X, s) and H is an effective principal divisor on (X, s) which does not contain the branch, then equality (9.4) shows that the intersection number of C and H at s may be computed as the order of the series obtained by composing a defining function of (H, s) and a normal parametrization of (C, s).

Example 9.2.6 Consider the branches:

$$\begin{cases} A := Z(y^2 - x^3), \\ B := Z(y^3 - x^5), \\ C := Z(y^6 - x^5) \end{cases}$$

on the smooth surface singularity $(\mathbb{C}^2, 0)$. Denoting by m_0 the multiplicity function at the origin of $\mathbb{C}^2$, we have:

$$m_0(A) = 2, \ m_0(B) = 3, \ m_0(C) = 5,$$

as results from Definition 9.2.1. Using Proposition 9.2.5 and the fact that whenever m and n are coprime positive integers, $t \to (t^n, t^m)$ is a normal parametrization of $Z(y^n - x^m)$, one gets the following values for the intersection numbers of the branches A, B, C:

$$B \cdot C = 15, \ C \cdot A = 10, \ A \cdot B = 9.$$

Therefore:

$$\begin{cases} \dfrac{B \cdot C}{m_0(B) \cdot m_0(C)} = 1, \\[3mm] \dfrac{C \cdot A}{m_0(C) \cdot m_0(A)} = 1, \\[3mm] \dfrac{A \cdot B}{m_0(A) \cdot m_0(B)} = \dfrac{3}{2}. \end{cases}$$

One notices that two of the previous quotients are equal and the third one is greater than them. Płoski discovered that this is a general phenomenon for plane branches, as explained in the next section.

9.3   The Statement of Płoski's Theorem

In this section we state a theorem of Płoski of 1985 and a reformulation of it in terms of the notion of *ultrametric*.

Denote simply by $\boxed{S}$ the germ of *smooth* surface (S, s) and by $\boxed{m(A)}$ the multiplicity of a branch $(A, s) \hookrightarrow (S, s)$.

In his 1985 paper [Pło85], Płoski proved the following theorem:

Theorem 9.3.1 *If A, B, C are three pairwise distinct branches on a smooth surface singularity S, then one has the following relations, up to a permutation of the three fractions:*

$$\frac{A \cdot B}{m(A) \cdot m(B)} \geq \frac{B \cdot C}{m(B) \cdot m(C)} = \frac{C \cdot A}{m(C) \cdot m(A)}.$$

Denote by $\boxed{\mathcal{B}(S)}$ the infinite set of branches on S. By inverting the fractions appearing in the statement of Theorem 9.3.1, it may be reformulated in the following equivalent way:

Theorem 9.3.2 *Let S be a smooth surface singularity. Then the map $\mathcal{B}(S) \times \mathcal{B}(S) \to [0, \infty)$ defined by*

$$(A, B) \to \begin{cases} \dfrac{m(A) \cdot m(B)}{A \cdot B} & \text{if } A \neq B, \\ 0 & \text{otherwise} \end{cases}$$

is an ultrametric.

What does it mean that a function is an ultrametric? We explain this in the next section and we show how to think topologically about ultrametrics on finite sets in terms of certain kinds of decorated rooted trees. This way of thinking is used then in Sect. 9.5 in order to prove the reformulation 9.3.2 of Płoski's theorem.

9.4   Ultrametrics and Rooted Trees

In this section we define the notion of *ultrametric* and we explain how to think about an ultrametric on a finite set in topological terms, as a special kind of rooted and decorated tree. This passes through understanding that the closed balls of an

ultrametric form a *hierarchy* and that finite hierarchies are equivalent to special types
of decorated rooted trees. For more details, one may consult [GBGPPP18, Sect. 3.1].

Definition 9.4.1 Let (M, d) be a metric space. It is called **ultrametric** if one has
the following strong form of the triangle inequality:

$$d(A, B) \leq \max\{d(A, C), d(B, C)\}, \text{ for all} A, B, C \in M.$$

In this case, one says also that d is **an ultrametric** on the set M.

In any metric space (M, d), a **closed ball** is a subset of M of the form:

$$\boxed{\mathcal{B}(A, r)} := \{P \in M, \ d(P, A) \leq r\}$$

where the **center** $A \in M$ and the **radius** $r \in [0, \infty)$ are given. As we will see
shortly, given a closed ball, neither its center nor its radius are in general well-
defined, contrary to an intuition educated only by Euclidean geometry.

One has the following characterizations of ultrametrics:

Proposition 9.4.2 *Let (M, d) be a metric space. Then the following properties are
equivalent:*

1. (M, d) *is ultrametric.*
2. *The triangles are all isosceles with two equal sides not less than the third side.*
3. *All the points of a closed ball are centers of it.*
4. *Two closed balls are either disjoint, or one is included in the other.*

Proof All the equivalences are elementary but instructive to check. We leave their
proofs as exercises.   □

Example 9.4.3 Consider a set $M = \{A, B, C, D\}$ and a distance function d on
it such that: $d(B, C) = 1, d(A, B) = d(A, C) = 2, d(A, D) = d(B, D) =
d(C, D) = 5$. Note that one may embed (M, d) isometrically into a 3-dimensional
Euclidean space by choosing an isosceles triangle ABC with the given edge lengths,
and by choosing then the point D on the perpendicular to the plane of the triangle
passing through its circumcenter. Let us look for the closed balls of this finite metric
space. For radii less than 1, they are singletons. For radii in the interval $[1, 2)$, we get
the sets $\{B, C\}, \{A\}, \{D\}$. Note that both B and C are centers of the ball $\{B, C\}$, that
is, $\mathcal{B}(B, r) = \mathcal{B}(C, r) = \{B, C\}$ for every $r \in [1, 2)$. Once the radius belongs to the
interval $[2, 5)$, the balls are $\{A, B, C\}$ and $\{D\}$. Finally, for every radius $r \in [5, \infty)$,
there is only one closed ball, the whole set. Figure 9.1 depicts the set $\{A, B, C, D\}$
as well as the mutual distances and the associated set of closed balls.

Example 9.4.3 illustrates the fact that neither the center nor the radius of a closed
ball of a finite ultrametric space is well-defined, once the ball has more than one
element. Instead, every closed ball has a well-defined *diameter*:

Fig. 9.1 The balls of an ultrametric space with four points

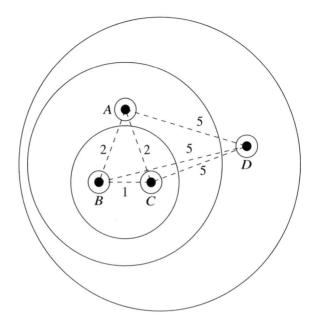

Definition 9.4.4 The **diameter** of a closed ball in a finite metric space is the maximal distance between pairs of not necessarily distinct points of it.

The last characterization of ultrametrics in Proposition 9.4.2 shows that the set $\boxed{\text{Balls}(M, d)}$ of closed balls of an ultrametric space (M, d) is a *hierarchy* on M, in the following sense:

Definition 9.4.5 A **hierarchy** on a set M is a subset $\mathcal{H}$ of its power set $\mathcal{P}(M)$, satisfying the following properties:

- $\emptyset \notin \mathcal{H}$.
- The singletons belong to $\mathcal{H}$.
- M belongs to $\mathcal{H}$.
- Two elements of $\mathcal{H}$ are either disjoint, or one is included into the other.

If $\mathcal{H}$ is a hierarchy on a set M, it may be endowed with the inclusion partial order. We will consider instead its reverse partial order $\boxed{\preceq_\mathcal{H}}$, defined by:

$$A \preceq_\mathcal{H} B \iff A \supseteq B, \quad \text{for all } A, B \in \mathcal{H}.$$

Reversing the inclusion partial order has the advantage of identifying the leaves of the corresponding rooted tree with the maximal elements of the poset $(\mathcal{H}, \preceq_\mathcal{H})$ (see Proposition 9.4.8 below).

When M is *finite*, one may represent the poset $(\mathcal{H}, \preceq_\mathcal{H})$ using its associated *Hasse diagram*:

Definition 9.4.6 Let $(X, \preceq)$ be a finite poset. Its **Hasse diagram** is the directed graph whose set of vertices is X, two vertices $a, b \in X$ being joined by an edge oriented from a to b whenever $a \prec b$ and the two points are **directly comparable**, that is, there is no other element of X lying strictly between them.

Hasse diagrams of finite posets are abstract oriented acyclic graphs. This means that they have no directed cycles, which is a consequence of the fact that a partial order is antisymmetric and transitive. Hasse diagrams are not necessarily planar, but, as all finite graphs, they may be always immersed in the plane in such a way that any pair of edges intersect transversely. When drawing a Hasse diagram in the plane as an immersion, we will use the convention to place the vertex a of the Hasse diagram *below* the vertex b whenever $a \prec b$. This is always possible because of the absence of directed cycles. This convention makes unnecessary adding arrowheads along the edges in order to indicate their orientations.

Example 9.4.7 Consider the finite set $\{1, 2, 3, 4, 6, 12\}$ of positive divisors of 12, partially ordered by divisibility: $a \preceq b$ if and only if a divides b. Its Hasse diagram is drawn in Fig. 9.2.

The Hasse diagrams of finite hierarchies are special kinds of graphs:

Proposition 9.4.8 *The Hasse diagram of a hierarchy $(\mathcal{H}, \preceq_H)$ on a finite set M is a tree in which the maximal directed paths start from M and terminate at the singletons. Moreover, for each vertex which is not a singleton, there are at least two edges starting from it.*

Proof We sketch a proof, leaving the details to the reader.

The first statement results from the fact that the singletons of M are exactly the maximal elements of the poset $(\mathcal{H}, \preceq_H)$, that M itself is the unique minimal element and that all the elements of a hierarchy which contain a given element are totally ordered by inclusion.

Fig. 9.2 The Hasse diagram of the set of positive divisors of 12

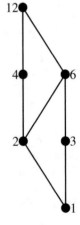

Let us prove the second statement. Consider $B_1 \in \mathcal{H}$ and assume that it is not a singleton. This means that it is not minimal for inclusion, therefore there exists $B_2 \in \mathcal{H}$ such that $B_2 \subsetneq B_1$ and B_2 is directly comparable to B_1. Let A be a point of $B_1 \setminus B_2$. Consider $B_3 \in \mathcal{H}$ which contains the point A, is included into $\mathcal{B}_1$ and is directly comparable to it. As $A \in B_3 \setminus B_2$, this shows that B_3 is not included in B_2. We want to show that the two sets B_2 and B_3 are disjoint. Otherwise, by the definition of a hierarchy, we would have $B_2 \subsetneq B_3 \subsetneq B_1$, which contradicts the assumption that B_1 and B_2 are directly comparable.   □

Example 9.4.9 Consider the ultrametric space of Example 9.4.7, represented in Fig. 9.1. We repeat it on the left of Fig. 9.3. The Hasse diagram of the hierarchy of its closed balls is drawn on the right of Fig. 9.3. Near each vertex is represented the diameter of the corresponding ball. We have added a *root vertex*, connected to the vertex representing the whole set. It may be thought as a larger ball, obtained by adding formally to $M = \{A, B, C, D\}$ a point ω, infinitely distant from each point of M. This larger ball is the set $\overline{M} := M \cup \{\omega\}$.

One may formalize in the following way the construction performed in Example 9.4.9:

Definition 9.4.10 The **tree** of a hierarchy $(\mathcal{H}, \preceq_H)$ on a finite poset M is its Hasse diagram, completed with a root representing the set $\boxed{\overline{M}} := M \cup \{\omega\}$, joined with the vertex representing M and rooted at $\overline{M}$. Here ω is a point distinct from the points of M.

The tree of a hierarchy is a **rooted tree** in the following sense:

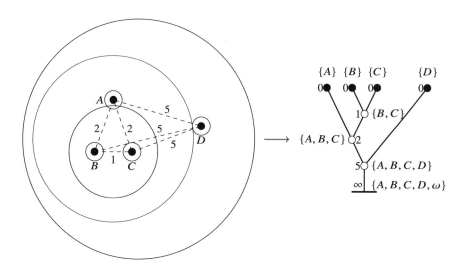

Fig. 9.3 The tree of the hierarchy of closed balls of Example 9.4.9

Definition 9.4.11 A **rooted tree** is a tree with a distinguished vertex, called its **root**. If Θ is a rooted tree with root r, then the vertex set of Θ gets partially ordered by declaring that $\boxed{a \preceq_r b}$ if and only if the unique segment $[r, a]$ joining r to a in the tree is contained in $[r, b]$.

When Θ is the rooted tree of a hierarchy $\mathcal{H}$ on a finite set M, then the partial order $\preceq_{\overline{M}}$ defined by choosing $\overline{M}$ as root restricts to the partial order $\preceq_{\mathcal{H}}$ if one identifies the set $\mathcal{H}$ with the set of vertices of Θ which are distinct from the root.

Proposition 9.4.8 may be reformulated in the following way as a list of properties of the tree of the hierarchy:

Proposition 9.4.12 *Let Θ be the tree of a hierarchy on a finite set, and let r be its root. Then r is a vertex of valency 1 and there are no vertices of valency 2.*

This proposition motivates the following definition:

Definition 9.4.13 A rooted tree whose root is of valency 1 and which does not possess vertices of valency 2 is a **hierarchical tree**. The **hierarchy** of a hierarchical tree (Θ, r) is constructed in the following way:

- Define M to be the set of **leaves** of the rooted tree (Θ, r), that is, the set of vertices of valency 1 which are distinct from the root r.
- For each vertex p of Θ different from the root, consider the subset of M consisting of the leaves a such that $p \preceq_r a$.

We leave as an exercise to prove:

Proposition 9.4.14 *The constructions of Definitions 9.4.10 and 9.4.13, which associate a hierarchical tree to a hierarchy on a finite set and a hierarchy to a hierarchical tree are inverse of each other.*

As a preliminary to the proof, one may test the truth of the proposition on the example of Fig. 9.3.

Let us return to finite ultrametric spaces (M, d). We saw that the set Balls(M, d) of its closed balls is a hierarchy on M. Proposition 9.4.14 shows that one may think about this hierarchy as a special kind of rooted tree, namely, a hierarchical tree. This hierarchical tree alone does not allow to get back the distance function d. How to encode it on the tree?

The idea is to look at the function defined on Balls(M, d), which associates to each ball its diameter (see Definition 9.4.4):

Proposition 9.4.15 *Let (M, d) be a finite ultrametric space. Then the map which sends each closed ball to its diameter is a strictly decreasing $[0, \infty)$-valued function defined on the poset $(\mathrm{Balls}(M, d), \preceq)$, taking the value 0 exactly on the singletons of M. Equivalently, it is a strictly decreasing $[0, \infty]$-valued function on the set of vertices of the tree of the hierarchy, vanishing on the set M of leaves and taking the value ∞ on the root.*

As an example, one may look again at Fig. 9.3. The value taken by the previous diameter function is written near each vertex of the hierarchical tree.

If (Θ, r) is a hierarchical tree, denote by $\boxed{V(\Theta)}$ its set of vertices and by $\boxed{a \wedge_r b}$ the infimum of a and b relative to $\preceq_r$, whenever $a, b \in V(\Theta)$. This infimum may be characterized by the property that $[r, a] \cap [r, b] = [r, a \wedge_r b]$. The following is a converse of Proposition 9.4.15:

Proposition 9.4.16 *Let (Θ, r) be a hierarchical tree and $\lambda : V(\Theta) \to [0, \infty]$ be a strictly decreasing function relative to the partial order $\preceq_r$ on Θ induced by the root. Assume that λ vanishes on the set M of leaves of Θ and takes the value ∞ at r. Then the map*

$$
d : M \times M \to [0, \infty)
$$
$$
(a, b) \to \lambda(a \wedge_r b)
$$

is an ultrametric on M.

Let us introduce a special name for the functions appearing in Proposition 9.4.16:

Definition 9.4.17 Let (Θ, r) be a hierarchical tree. A **depth function** on it is a function $\lambda : V(\Theta) \to [0, \infty]$ which satisfies the following properties:

- it is strictly decreasing relative to the partial order $\preceq_r$ on Θ induced by the root r;
- it vanishes on the set of leaves of Θ;
- it takes the value ∞ at the root r.

Note that the first two conditions of Definition 9.4.17 imply that a depth function vanishes *exactly* on the set of leaves of the underlying hierarchical tree.

One has the following analog of Proposition 9.4.14:

Proposition 9.4.18 *The constructions of Propositions 9.4.15 and 9.4.16 are inverse of each other. That is, giving an ultrametric on a finite set M is equivalent to giving a depth function on a hierarchical tree whose set of leaves is M.*

It is this proposition which allows to think about an ultrametric as a special kind of rooted and decorated tree. We leave its proof as an exercise (see [BD98]).

9.5 A Proof of Płoski's Theorem Using Eggers-Wall Trees

In this section we sketch a proof of Płoski's theorem 9.3.1 using the equivalence between ultrametrics on finite sets and certain kinds of rooted trees formulated in Proposition 9.4.18. The rooted trees used in this proof are the *Eggers-Wall trees* of a plane curve singularity relative to smooth reference branches. The precise definition of Eggers-Wall trees is not given, because the proofs of the subsequent generalizations of Płoski's theorem will be of a completely different spirit.

Instead of working both with multiplicities and intersection numbers as in Płoski's original statement, we will work only with the latest ones.

Let S be a smooth germ of surface and $L \hookrightarrow S$ be a *smooth branch*. Define the following function on the set of branches on S which are different from L:

$$\boxed{u_L} : (\mathcal{B}(S) \setminus \{L\})^2 \to \mathbb{R}_+$$
$$(A, B) \quad \to \quad \begin{cases} \dfrac{(L \cdot A) \cdot (L \cdot B)}{A \cdot B} & \text{if } A \neq B, \\ 0 & \text{otherwise.} \end{cases} \tag{9.5}$$

In the remaining part of this section we will sketch a proof of:

Theorem 9.5.1 *The function u_L is an ultrametric.*

We leave as an exercise to show using Proposition 9.2.3 that Theorem 9.5.1 implies the reformulation given in Theorem 9.3.2 of Płoski's Theorem 9.3.1.

Our proof of Theorem 9.5.1 will pass through the notion of *Eggers-Wall tree* associated to a plane curve singularity relative to a smooth branch of reference L (see the proof of Theorem 9.5.5 below). Let us illustrate it by an example.

Example 9.5.2 Consider again the branches $A = Z(y^2 - x^3)$, $B = Z(y^3 - x^5)$, $C = Z(y^6 - x^5)$ on $S = (\mathbb{C}^2, 0)$ of Example 9.2.6. Assume that the branch L is the germ at 0 of the y-axis $Z(x)$. The defining equations of the three branches A, B, C may be considered as polynomial equations in the variable y. As such, they admit the following roots which are fractional powers of x:

$$A : x^{3/2},$$
$$B : x^{5/3},$$
$$C : x^{5/6}.$$

Associate to the root $x^{3/2}$ a compact segment $\Theta_L(A)$ identified with the interval $[0, \infty]$ using an **exponent function** $e_L : \Theta_L(A) \to [0, \infty]$ and mark on it the point $e_L^{-1}(3/2)$ with exponent 3/2. Define also an **index function** $i_L : \Theta_L(A) \to \mathbb{N}^*$, constantly equal to 1 on the interval $[e_L^{-1}(0), e_L^{-1}(3/2)]$ and to 2 on the interval $(e_L^{-1}(3/2), e_L^{-1}(\infty)]$ (see the left-most segment of Fig. 9.4). Here the number 2 is to be thought as the minimal positive denominator of the exponent 3/2 of the monomial $x^{3/2}$. The segment $\Theta_L(A)$ endowed with the two functions e_L and i_L is the **Eggers-Wall tree** of the branch A relative to the branch L. It is considered as a rooted tree with root $e_L^{-1}(0)$, labeled with the branch L. Its leaf $e_L^{-1}(\infty)$ is labeled with the branch A. Consider analogously the Eggers-Wall trees $\Theta_L(B)$ and $\Theta_L(C)$, endowed with pairs of exponent and index functions and labeled roots and leaves (see the left part of Fig. 9.4).

Look now at the plane curve singularity $A + B + C$. Its Eggers-Wall tree $\Theta_L(A + B + C)$ relative to the branch L is obtained from the individual trees

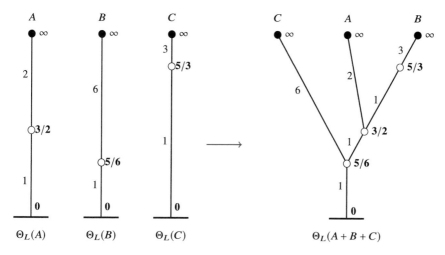

Fig. 9.4 The Eggers-Wall tree of the plane curve singularity of Example 9.5.2

$\Theta_L(A), \Theta_L(B), \Theta_L(C)$ by a gluing process, which identifies two by two initial segments of those trees.

Consider for instance the segments $\Theta_L(A), \Theta_L(B)$. Look at the order of the difference $x^{3/2} - x^{5/3}$ of the roots which generated them, seen as a series with fractional exponents. This order is the fraction $3/2$, because $3/2 < 5/3$. Identify then the points with the same exponent $\leq 3/2$ of the segments $\Theta_L(A), \Theta_L(B)$. One gets a rooted tree $\Theta_L(A+B)$ with root labeled by L and with two leaves, labeled by the branches A, B. The exponent and index functions of the trees $\Theta_L(A), \Theta_L(B)$ descend to functions with the same name e_L, i_L defined on $\Theta_L(A+B)$. Endowed with those functions, $\Theta_L(A+B)$ is the **Eggers-Wall tree** of the curve singularity $A+B$.

If one considers now the curve singularity $A+B+C$, then one glues analogously the three pairs of trees obtained from $\Theta_L(A), \Theta_L(B), \Theta_L(C)$. The resulting Eggers-Wall tree $\Theta_L(A+B+C)$ is drawn on the right side of Fig. 9.4. It is also endowed with two functions e_L, i_L, obtained by gluing the exponent and index functions of the trees $\Theta_L(A), \Theta_L(B), \Theta_L(C)$. Its marked points are its ends, its bifurcation points and the images of the discontinuity points of the index function of the Eggers-Wall tree of each branch. Near each marked point is written the corresponding value of the exponent function. The index function is constant on each segment $(a, b]$ joining two marked points a and b, where $a \prec_L b$. Here $\preceq_L$ denotes the partial order on the tree $\Theta_L(A+B+C)$ determined by the root L (see Definition 9.4.11).

One may associate analogously an **Eggers-Wall tree** $\boxed{\Theta_L(D)}$ to any plane curve singularity D, relative to a *smooth* reference branch L. It is a rooted tree endowed with an **exponent function** $\boxed{e_L}$: $\Theta_L(D) \to [0, \infty]$ and an **index function** $\boxed{i_L}$: $\Theta_L(D) \to \mathbb{N}^*$. The tree and both functions are constructed using Newton-Puiseux series expansions of the roots of a Weierstrass polynomial $f \in \mathbb{C}[[x]][y]$ defining

D in a coordinate system (x, y) such that $L = Z(x)$. The triple $(\Theta_L(D), e_L, i_L)$ is independent of the choices involved in the previous definition (see [GBGPPP18, Proposition 103]). One may find the precise definition and examples of Eggers-Wall trees in Section 4.3 of the previous reference and in [GBGPPP19, Sect. 3]. Historical remarks about this notion may be found in [GBGPPP19, Rem. 3.18] and [GBGPPP20, Sect. 6.2]. The name, introduced in author's thesis [PP01], makes reference to Eggers' 1983 paper [Egg82] and to Wall's 2003 paper [Wal03].

What allows us to prove Theorem 9.5.1 using Eggers-Wall trees is that the values $u_L(A, B)$ of the function u_L defined by relation (9.5) are determined in the following way from the Eggers-Wall tree $\Theta_L(D)$, for each pair of distinct branches (A, B) of D (recall from the paragraph preceding Proposition 9.4.16 that $A \wedge_L B$ denotes the infimum of A and B relative to the partial order $\preceq_L$ induced by the root L of $\Theta_L(D)$):

Theorem 9.5.3 *For each pair (A, B) of distinct branches of D and every smooth reference branch L different from the branches of D, one has:*

$$\frac{1}{u_L(A, B)} = \int_L^{A \wedge_L B} \frac{de_L}{i_L}.$$

Example 9.5.4 Let us verify the equality stated in Theorem 9.5.3 on the branches of Example 9.5.2. Looking at the Eggers-Wall tree $\Theta_L(A + B + C)$ on the right side of Fig. 9.4, we see that:

$$\int_L^{A \wedge_L B} \frac{de_L}{i_L} = \int_0^{3/2} \frac{de}{1} = \frac{3}{2}.$$

But $1/u_L(A, B) = (A \cdot B)/((L \cdot A)(L \cdot B)) = (A \cdot B)/(m(A) \cdot m(B)) = 3/2$, as was computed in Example 9.2.6. The equality is verified. We have used the fact that both A and B are transversal to L, which implies that $L \cdot A = m(A)$ and $L \cdot B = m(B)$.

In equivalent formulations which use so-called *characteristic exponents*, Theorem 9.5.3 goes back to Smith [Smi75, Section 8], Stolz [Sto79, Section 9] and Max Noether [Noe90]. A modern proof, based on Proposition 9.2.5, may be found in [Wal04, Thm. 4.1.6].

As a consequence of Theorem 9.5.3, we get the following strengthening of Theorem 9.5.1:

Theorem 9.5.5 *Let D be a plane curve singularity. Denote by $\mathcal{F}(D)$ the set of branches of D. Let L be a reference smooth branch which does not belong to $\mathcal{F}(D)$. Then the function u_L is an ultrametric in restriction to $\mathcal{F}(D)$ and its associated rooted tree is isomorphic as a rooted tree with labeled leaves to the Eggers-Wall tree $\Theta_L(D)$.*

Proof Consider $\Theta_L(D)$ as a topological tree with vertex set equal to its set of ends and of ramification points. Root it at L. Then it becomes a hierarchical tree in the sense of Definition 9.4.13. The function

$$P \to \left(\int_L^P \frac{de_L}{i_L} \right)^{-1}$$

is a depth function on it, in the sense of Definition 9.4.17. Using Theorem 9.5.3 and Proposition 9.4.18, we get Theorem 9.5.5.   □

For more details about the proof of Płoski's theorem presented in this section, see [GBGPPP18, Sect. 4.3].

9.6   An Ultrametric Characterization of Arborescent Singularities

In this section we state a generalization of Theorem 9.5.1 for all *arborescent singularities* and the fact that it characterizes this class of normal surface singularities. We start by recalling the needed notions of *embedded resolution* and associated *dual graph* of a finite set of branches contained in a normal surface singularity.

From now on, S denotes an arbitrary **normal surface singularity**, that is, a germ of normal complex analytic surface. Let us recall the notion of *resolution* of such a singularity:

Definition 9.6.1 Let (S, s) be a normal surface singularity. A **resolution** of it is a proper bimeromorphic morphism $\pi : S^\pi \to S$ such that S^π is smooth. Its **exceptional divisor** $\boxed{E^\pi}$ is the reduced preimage $\pi^{-1}(s)$. The resolution is **good** if its exceptional divisor has normal crossings and all its irreducible components are smooth. The **dual graph** $\boxed{\Gamma(\pi)}$ of the resolution π is the finite graph whose set of vertices $\boxed{\mathcal{P}(\pi)}$ is the set of irreducible components of E^π, two vertices being joined by an edge if and only if the corresponding components intersect.

Every normal surface singularity admits resolutions and even good ones. This result, for which partial proofs appeared already at the end of the nineteenth century, was proved first in the analytical context by Hirzebruch in his 1953 paper [Hir53]. His proof was inspired by previous works of Jung [Jun08] and Walker [Wal35], done in an algebraic context.

Assume now that $\mathcal{F}$ is a finite set of branches on S. It may be also seen as a reduced divisor on S, by thinking about their sum. The notion of **embedded resolution** of $\mathcal{F}$ is an analog of that of good resolution of S:

Definition 9.6.2 Let (S, s) be a normal surface singularity and let $\pi : S^\pi \to S$ be a resolution of S. If A is a branch on S, then its **strict transform by** π is the closure inside S^π of the preimage $\pi^{-1}(A \setminus s)$. Let $\mathcal{F}$ be a finite set of branches on S. Its **strict transform by** π is the set or, depending on the context, the divisor formed by the strict transforms of the branches of $\mathcal{F}$. The **preimage** $\boxed{\pi^{-1}\mathcal{F}}$ **of** $\mathcal{F}$ **by** π is the sum of its strict transform and of the exceptional divisor of π. The morphism π is an **embedded resolution of** $\mathcal{F}$ if it is a good resolution of S and the preimage of $\mathcal{F}$ by π is a normal crossings divisor. The **dual graph** $\boxed{\Gamma(\pi^{-1}\mathcal{F})}$ of the preimage $\pi^{-1}\mathcal{F}$ is defined similarly to the dual graph $\Gamma(\pi)$ of π, taking into account all the irreducible components of $\pi^{-1}\mathcal{F}$.

In the previous definition, the preimage $\pi^{-1}\mathcal{F}$ of $\mathcal{F}$ by π is seen as a reduced divisor. We will see in Definition 9.9.4 below that there is also a canonical way, due to Mumford, to define canonically a not necessarily reduced rational divisor supported by $\pi^{-1}\mathcal{F}$, called the **total transform** of $\mathcal{F}$ by π, and denoted by $\pi^*\mathcal{F}$.

The notion of dual graph of a resolution allows to define the following class of *arborescent singularities*, whose name was introduced in the paper [GBGPPP18], even if the class had appear before, for instance in Camacho's work [Cam88]:

Definition 9.6.3 Let S be a normal surface singularity. It is called **arborescent** if the dual graphs of its good resolutions are trees.

Remark that in the previous definition we ask nothing about the genera of the irreducible components of the exceptional divisors.

By using the fact that any two resolutions are related by a sequence of blow ups and blow downs of their total spaces (see [Har77, Thm. V.5.5]), one sees that the dual graphs of all good resolutions are trees if and only if this is true for one of them.

Consider now an *arbitrary* branch L on the normal surface singularity S. We may define the function u_L by the same formula (9.5) as in the case when both S and L were assumed smooth. Intersection numbers of branches still have a meaning, as was shown by Mumford. We will explain this in Sect. 9.9 below (see Definition 9.9.5).

The following generalization of Theorem 9.5.1 both gives a characterization of arborescent singularities and extends Theorem 9.5.5 to *all* arborescent singularities S and *all*—not necessarily smooth—reference branches L on them (recall that $\mathcal{B}(S)$ denotes the set of branches on S):

Theorem 9.6.4 *Let S be a normal surface singularity and $L \in \mathcal{B}(S)$. Then:*

1. *u_L is ultrametric on $\mathcal{B}(S) \setminus \{L\}$ if and only if S is arborescent.*
2. *In this case, for any finite set $\mathcal{F}$ of branches on S not containing L, the rooted tree of the restriction of u_L to $\mathcal{F}$ is isomorphic to the convex hull of $\mathcal{F} \cup \{L\}$ in the dual graph of the preimage of $\mathcal{F} \cup \{L\}$ by any embedded resolution of $\mathcal{F} \cup \{L\}$, rooted at L.*

We do not prove in the present notes that if u_L is an ultrametric on $\mathcal{B}(S)\setminus\{L\}$, then S is arborescent. The interested reader may find a proof of this fact in [GBPPPR19, Sect. 1.6]. The remaining implication of point (1) and point (2) of Theorem 9.6.4 are, taken together, a consequence of Theorem 9.8.1 below. For this reason, we do not give a separate proof of them, the rest of this paper being dedicated to the statement and a sketch of proof of Theorem 9.8.1. The notion of *brick-vertex tree* of a finite connected graph being crucial in this theorem, we dedicate next section to it.

By combining Theorems 9.5.5 and 9.6.4 one gets (see [GBGPPP18, Thm. 112]):

Proposition 9.6.5 *Whenever S and L are both smooth, the Eggers-Wall tree $\Theta_L(D)$ of a plane curve singularity $D \hookrightarrow S$ not containing L is isomorphic to the convex hull of the strict transform of $\mathcal{F}(D) \cup \{L\}$ in the dual graph of its preimage by any of its embedded resolutions.*

A prototype of this fact was proved differently in the author's thesis [PP01, Thm. 4.4.1], then generalized in two different ways by Wall in [Wal04, Thm. 9.4.4] (see also Wall's comments in [Wal04, Sect. 9.10]) and by Favre and Jonsson in [FJ04, Prop. D.1].

9.7   The Brick-Vertex Tree of a Connected Graph

In this section we introduce the notion of *brick-vertex tree* of a connected graph, which is crucial in order to state Theorem 9.8.1 below, the strongest known generalization of Płoski's theorem.

Definition 9.7.1 A **graph** is a compact cell complex of dimension ≤ 1. If Γ is a graph, its set of vertices is denoted $\boxed{V(\Gamma)}$ and its set of edges is denoted $\boxed{E(\Gamma)}$.

In the sequel it will be crucial to look at the vertices which disconnect a given graph:

Definition 9.7.2 Let Γ be a connected graph. A **cut-vertex** of Γ is a vertex whose removal disconnects Γ. A **bridge** of Γ is an edge such that the removal of its interior disconnects Γ. If a, b, c are three not necessarily distinct vertices of Γ, one says that b **separates** a **from** c if either $b \in \{a, c\}$ or if a and c belong to different connected components of the topological space $\Gamma \setminus \{b\}$.

Note that an end of a bridge is a cut-vertex if and only if it has valency at least 2 in Γ, that is, if and only if it is not a leaf of Γ. It will be important to distinguish the class of graphs which cannot be disconnected by the removal of one vertex, as well as the maximal graphs of this class contained in a given connected graph:

Definition 9.7.3 A connected graph is called **inseparable** if it does not contain cut-vertices. A **block** of a connected graph Γ is a maximal inseparable subgraph of it. A **brick** of Γ is a block which is not a bridge.

Note that all the bridges of a connected graph are blocks of it.

Example 9.7.4 In Fig. 9.5 is represented a connected graph. Its cut-vertices are surrounded in red. Its bridges are represented as black segments. It has three bricks, the edges of each brick being colored in the same way.

Fig. 9.5 A connected graph,
its cut-vertices, its bridges
and its bricks

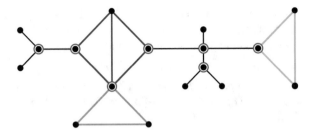

By replacing each brick of a connected graph by a star-shaped graph, one gets canonically a tree associated to the given graph:

Definition 9.7.5 The **brick-vertex tree of a graph** $\mathcal{B}\mathcal{V}(\Gamma)$ of a connected graph Γ is the tree whose set of vertices is the union of the set of vertices of Γ and of a set of new **brick-vertices** corresponding bijectively to the bricks of Γ, its edges being either the bridges of Γ or new edges connecting each brick-vertex to the vertices of the corresponding brick. Formally, this may be written as follows:

- $V(\mathcal{B}\mathcal{V}(\Gamma)) = V(\Gamma) \sqcup \{\text{bricks of } \Gamma\}$.
- $E(\mathcal{B}\mathcal{V}(\Gamma)) = \{\text{bridges of } \Gamma\} \sqcup \{[\overline{v}, \overline{b}], v \in V(\Gamma), b \text{ is a brick of } \Gamma, v \in V(b)\}$.

We denoted by $\boxed{\overline{v}}$ the vertex v of Γ when it is seen as a vertex of $\mathcal{B}\mathcal{V}(\Gamma)$ and $\boxed{\overline{b}} \in V(\mathcal{B}\mathcal{V}(\Gamma))$ the brick-vertex representing the brick b of Γ.

The notion of brick-vertex tree was introduced in [GBPPPR19, Def. 1.34]. It is strongly related to other notions introduced before either in general topology or in graph theory, as explained in [GBPPPR19, Rems. 1.35, 2.50].

Note that whenever Γ is a tree, $\mathcal{B}\mathcal{V}(\Gamma)$ is canonically isomorphic to it, as Γ has no bricks.

Example 9.7.6 On the left side of Fig. 9.6 is repeated the graph Γ of Fig. 9.5, with its cut-vertices and bricks emphasized. On its right side is represented its associated brick-vertex tree $\mathcal{B}\mathcal{V}(\Gamma)$. Each representative vertex of a brick is drawn with the same color as its corresponding brick. The edges of $\mathcal{B}\mathcal{V}(\Gamma)$ which are not bridges of Γ are represented in magenta and thicker than the other edges.

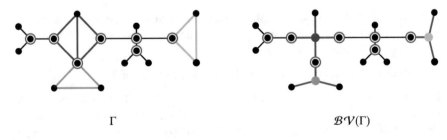

$$\Gamma \qquad\qquad\qquad \mathcal{B}\mathcal{V}(\Gamma)$$

Fig. 9.6 The connected graph of Example 9.7.6 and its brick-vertex tree

The importance of the brick-vertex tree in our context stems from the following property of it (see [GBGPPP19, Prop. 1.36]), formulated using the vocabulary introduced in Definition 9.7.2 and the notations introduced in Definition 9.7.5:

Proposition 9.7.7 *Let Γ be a finite graph and $a, b, c \in V(\Gamma)$. Then b separates a from c in Γ if and only if $\overline{b}$ separates $\overline{a}$ from $\overline{c}$ in $\mathcal{BV}(\Gamma)$.*

We are ready now to state the strongest known generalization of Płoski's theorem (see Theorem 9.8.1 below).

9.8 Our Strongest Generalization of Płoski's Theorem

In this section we formulate Theorem 9.8.1, which generalizes Theorem 9.5.5 to *all* normal surface singularities and *all* branches on them, using the notion of *brick-vertex tree* introduced in the previous section.

Recall that the notion of *brick-vertex tree* of a connected graph was introduced in Definition 9.7.5. A fundamental property of normal surface singularities is that the dual graphs of their resolutions are connected (which is a particular case of the so-called *Zariski's main theorem*, whose statement may be found in [Har77, Thm. V.5.2]). This implies that the dual graph of the preimage (see Definition 9.6.2) of any finite set of branches on such a singularity is also connected. Therefore, one may speak about its corresponding brick-vertex tree. The **convex hull** of a finite set of vertices of it is the union of the segments which join them pairwise.

Here is the announced generalization of Theorem 9.5.5, which is a slight reformulation of [GBPPPR19, Thm. 1.42]:

Theorem 9.8.1 *Let S be a normal surface singularity. Consider a finite set $\mathcal{F}$ of branches on it and an embedded resolution $\pi : S^{\pi} \to S$ of $\mathcal{F}$. Let Γ be the dual graph of the preimage $\pi^{-1}\mathcal{F}$ of $\mathcal{F}$ by π. Assume that the convex hull $\mathrm{Conv}_{\mathcal{BV}(\Gamma)}(\mathcal{F})$ of the strict transform of $\mathcal{F}$ by π in the brick-vertex tree $\mathcal{BV}(\Gamma)$ does not contain brick-vertices of valency at least 4 in $\mathrm{Conv}_{\mathcal{BV}(\Gamma)}(\mathcal{F})$. Then for all $L \in \mathcal{F}$, the restriction of u_L to $\mathcal{F} \setminus \{L\}$ is an ultrametric and the corresponding rooted tree is isomorphic to $\mathrm{Conv}_{\mathcal{BV}(\Gamma)}(\mathcal{F})$, rooted at L.*

Example 9.8.2 Assume that the dual graph Γ of $\pi^{-1}\mathcal{F}$ is as shown on the left side of Fig. 9.7. The vertices representing the strict transforms of the branches of the set $\mathcal{F}$ are drawn arrowheaded. Note that the subgraph which is the dual graph of the exceptional divisor is the same as the graph of Fig. 9.5. On the right side of Fig. 9.7 is represented using thick red segments the convex hull $\mathrm{Conv}_{\mathcal{BV}(\Gamma)}(\mathcal{F})$. We see that the hypothesis of Theorem 9.8.1 is satisfied. Indeed, the convex hull contains only two brick-vertices, which are of valency 2 and 3 in $\mathrm{Conv}_{\mathcal{BV}(\Gamma)}(\mathcal{F})$. Note that the blue one is of valency 4 in the dual graph Γ, which shows the importance of looking at the valency in the convex hull $\mathrm{Conv}_{\mathcal{BV}(\Gamma)}(\mathcal{F})$, not in Γ.

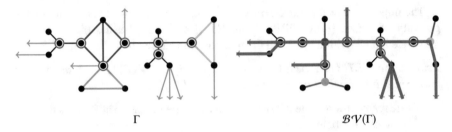

$$\Gamma \qquad\qquad\qquad \mathcal{BV}(\Gamma)$$

Fig. 9.7 An example where the hypothesis of Theorem 9.8.1 is satisfied

As shown in [GBPPPR19, Ex. 1.44], the condition about valency is not necessary in general for u_L to be an ultrametric on $\mathcal{F} \setminus \{L\}$.

Note that we have expressed Theorem 9.8.1 in a slightly different form than the equivalent Theorem 9.1.4 of the introduction. Namely, we included L in the branches of $\mathcal{F}$. This formulation emphasizes the symmetry of the situation: all the choices of reference branch inside $\mathcal{F}$ lead to the same tree, only the root being changed. In fact, we will obtain Theorem 9.8.1 as a consequence of Theorem 9.10.10, in which no branch plays any more a special role.

Before that, we will explain in the next section Mumford's definition of intersection number of two curve singularities drawn on an arbitrary normal surface singularity, which allows to define in turn the functions u_L appearing in the statement of Theorem 9.8.1.

9.9 Mumford's Intersection Theory

In this section we explain Mumford's definition of intersection number of Weil divisors on a normal surface singularity, introduced in his 1961 paper [Mum61]. It is based on Theorem 9.9.1, stating that the intersection form of any resolution of a normal surface singularity is negative definite. This theorem being fundamental for the study of surface singularities, we present a detailed proof of it.

Let $\pi : S^\pi \to S$ be a resolution of the normal surface singularity S. Denote by $(E_u)_{u \in \mathcal{P}(\pi)}$ the collection of irreducible components of the exceptional divisor E^π of π (see Definition 9.6.1).

Denote by:

$$\boxed{\mathcal{E}(\pi)_{\mathbb{R}}} := \bigoplus_{u \in \mathcal{P}(\pi)} \mathbb{R} E_u$$

the real vector space freely generated by those prime divisors, that is, the space of real divisors supported by E^π. It is endowed with a symmetric bilinear form $(D_1, D_2) \to D_1 \cdot D_2$ given by intersecting the corresponding compact cycles on S^π.

We call it **the intersection form**. Its following fundamental property was proved by Du Val [DV44] and Mumford [Mum61]:

Theorem 9.9.1 *The intersection form on $\mathcal{E}(\pi)_\mathbb{R}$ is negative definite.*

Proof The following proof is an expansion of that given by Mumford in [Mum61].

The singularity S being normal, the exceptional divisor E^π is connected (this is a particular case of *Zariski's main theorem*, see [Har77, Thm. V.5.2]). Therefore:

$$\text{The dual graph } \Gamma(\pi) \text{ is connected.} \tag{9.6}$$

Consider any germ of holomorphic function f on (S, s), vanishing at s, and look at the divisor of its lift to the surface S^π:

$$(\pi^* f) = \sum_{u \in \mathcal{P}(\pi)} a_u E_u + (\pi^* f)_{str}. \tag{9.7}$$

Here $(\pi^* f)_{str}$ denotes the strict transform of the divisor defined by f on S. Denote also:

$$\begin{cases} \boxed{e_u} := a_u E_u \in \mathcal{E}(\pi)_\mathbb{R}, \text{ for all } u \in \mathcal{P}(\pi), \\ \boxed{\sigma} := \sum_{u \in \mathcal{P}(\pi)} e_u \in \mathcal{E}(\pi)_\mathbb{R}. \end{cases} \tag{9.8}$$

As f vanishes at the point s, its lift $\pi^* f$ vanishes along each component E_u of E^π, therefore $a_u > 0$ for every $u \in \mathcal{P}(u)$. We deduce that $(e_u)_{u \in \mathcal{P}(\pi)}$ is a basis of $\mathcal{E}(\pi)_\mathbb{R}$ and that:

$$e_u \cdot e_v \geq 0, \text{ for all } u, v \in \mathcal{P}(\pi) \text{ such that } u \neq v. \tag{9.9}$$

The divisor $(\pi^* f)$ being principal, its associated line bundle is trivial. Therefore:

$$(\pi^* f) \cdot E_u = 0 \text{ for every } u \in \mathcal{P}(\pi), \tag{9.10}$$

because this intersection number is equal by definition to the degree of the restriction of this line bundle to the curve E_u. By combining the relations (9.7), (9.8) and (9.10), we deduce that:

$$\sigma \cdot e_u = -a_u (\pi^* f)_{str} \cdot E_u, \text{ for every } u \in \mathcal{P}(\pi). \tag{9.11}$$

As the germ of effective divisor $(\pi^* f)_{str}$ along E^π has no components of E^π in its support, the intersection numbers $(\pi^* f)_{str} \cdot E_u$ are all non-negative. Moreover, at least one of them is positive, because the divisor $(\pi^* f)_{str}$ is non-zero. By combining this fact with relations (9.11) and with the inequalities $a_u > 0$, we get:

$$\begin{cases} \sigma \cdot e_u \leq 0, & \text{for every } u \in \mathcal{P}(\pi), \\ \text{there exists } u_0 \in \mathcal{P}(\pi) \text{ such that } \sigma \cdot e_{u_0} < 0. \end{cases} \tag{9.12}$$

Consider now an arbitrary element $\tau \in \mathcal{E}(\pi)_{\mathbb{R}} \setminus \{0\}$. One may develop it in the basis $(e_u)_{u \in \mathcal{P}(\pi)}$:

$$\tau = \sum_{u \in \mathcal{P}(\pi)} x_u e_u. \tag{9.13}$$

We will show that $\tau^2 < 0$. As τ was chosen as an arbitrary non-zero vector, this will imply that the intersection form on $\mathcal{E}(\pi)_{\mathbb{R}}$ is indeed negative definite. The trick is to express the self-intersection τ^2 using the expansion (9.13), then to develop it by linearity and to replace the vectors e_u by $\sigma - \sum_{v \neq u} e_v$ in a precise place:

$$\begin{aligned} \tau^2 &= \left(\sum_u x_u e_u \right)^2 = \\ &= \sum_u x_u^2 e_u^2 + 2 \sum_{u < v} x_u x_v e_u \cdot e_v = \\ &= \sum_u x_u^2 \left(\sigma - \sum_{v \neq u} e_v \right) \cdot e_u + 2 \sum_{u < v} x_u x_v e_u \cdot e_v = \\ &= \sum_u x_u^2 (\sigma \cdot e_u) - \sum_{u \neq v} x_u^2 e_u \cdot e_v + 2 \sum_{u < v} x_u x_v e_u \cdot e_v = \\ &= \sum_u x_u^2 (\sigma \cdot e_u) - \sum_{u < v} (x_u - x_v)^2 e_u \cdot e_v. \end{aligned}$$

We got the equality:

$$\tau^2 = \sum_u x_u^2 (\sigma \cdot e_u) - \sum_{u < v} (x_u - x_v)^2 e_u \cdot e_v. \tag{9.14}$$

Using the inequalities (9.9) and (9.12), we deduce that its right-hand side is non-positive, therefore the intersection form is negative semi-definite.

It remains to show that $\tau^2 < 0$. Assume by contradiction that $\tau^2 = 0$. Equality (9.14) shows that the following equalities are simultaneously satisfied:

$$\sum_u x_u^2 (\sigma \cdot e_u) = 0, \tag{9.15}$$

$$(x_u - x_v)^2 e_u \cdot e_v = 0, \quad \text{for all } u < v. \tag{9.16}$$

The relations (9.16) imply that $x_u = x_v$ whenever $e_u \cdot e_v > 0$. As $e_u = a_u E_u$ with $a_u > 0$, the inequality $e_u \cdot e_v > 0$ is equivalent with $E_u \cdot E_v > 0$, that is, with the fact that $[u, v]$ is an edge of the dual graph $\Gamma(\pi)$. This dual graph being connected (see (9.6)), we see that $x_u = x_v$ for all $u, v \in \mathcal{P}(\pi)$. Consider now an index u_0 satisfying the second condition of relations (9.12). Equation (9.15) implies that $x_{u_0} = 0$. Therefore all the coefficients x_u vanish, which contradicts the hypothesis that $\tau \neq 0$. $\qquad\qquad\qquad\qquad\qquad\qquad\qquad\qquad\qquad\qquad\qquad\qquad\qquad\qquad$ $\square$

As a consequence of Theorem 9.9.1, one may define the **dual basis** $\left(\boxed{E_u^\vee} \right)_{u \in \mathcal{P}(\pi)}$ of the basis $(E_u)_{u \in \mathcal{P}(\pi)}$ by the following relations, in which δ_{uv} denotes Kronecker's delta-symbol:

$$E_u^\vee \cdot E_v = \delta_{uv}, \text{ for all } (u, v) \in \mathcal{P}(\pi)^2. \tag{9.17}$$

By associating to each prime divisor E_u the corresponding valuation of the local ring $O_{S,s}$, computing the orders of vanishing along E_u of the pull-backs $\pi^* f$ of the functions $f \in O_{S,s}$, one injects the set $\mathcal{P}(\pi)$ in the set of real-valued valuations of $O_{S,s}$. This allows to see the index u of E_u as a valuation. Such valuations are called *divisorial*. If u denotes a divisorial valuation, it has a center on any resolution, which is either a point or an irreducible component of the exceptional divisor. In the second case, one says that the valuation **appears in the resolution** . The following notion, inspired by approaches of Favre and Jonsson [FJ04, App. A] and [Jon15, Sect. 7.3.6], was introduced in [GBPPPR19, Def. 1.6]:

Definition 9.9.2 Let u, v be two divisorial valuations of S. Consider a resolution of S in which both u and v appear. Then their **bracket** is defined by:

$$\boxed{\langle u, v \rangle} := -E_u^\vee \cdot E_v^\vee.$$

The bracket $\boxed{\langle u, v \rangle}$ may be interpreted as the intersection number of two Weil divisors on S associated to the divisors E_u and E_v (see Proposition 9.9.7 below). As a consequence, it is well-defined. That is, if the divisorial valuations u, v are fixed, then their bracket does not depend on the resolution in which they appear. This fact may be also proved using the property that any two resolutions of S are related by a sequence of blow ups and blow downs (see [GBPPPR19, Prop. 1.5]).

It is a consequence of Theorem 9.9.1 that the brackets are all non-negative (see [GBPPPR19, Prop. 1.4]). Moreover, by the Cauchy–Schwarz inequality applied to the opposite of the intersection form:

Lemma 9.9.3 *For every* $a, b \in \mathcal{P}(\pi)$:

$$\langle a, b \rangle^2 \leq \langle a, a \rangle \langle b, b \rangle,$$

with equality if and only if $a = b$.

Let now D be a Weil divisor on S, that is, a formal sum of branches on S. If D is principal, that is, the divisor (f) of a meromorphic germ on S, then one may lift it to a resolution S^π as the principal divisor $(\pi^* f)$. This divisor decomposes as the sum of an exceptional part $(\pi^* D)_{ex}$ supported by E^π and the strict transform of D. The crucial property of the lift $(\pi^* f)$, already used in the proof of Theorem 9.9.1 (see relation (9.10)), is that its intersection numbers with all the components E_u of E^π vanish. In [Mum61, Sect. II (b)], Mumford imposed this property in order to define a lift $\pi^* D$ for *any* Weil divisor D on S:

Definition 9.9.4 Let D be a Weil divisor on S. Its **total transform** $\boxed{\pi^* D}$ is the unique sum $\boxed{(\pi^* D)_{ex}} + \boxed{(\pi^* D)_{str}}$ such that:

1. $(\pi^* D)_{ex} \in \mathcal{E}(\pi)_{\mathbb{Q}}$.
2. $(\pi^* D)_{str}$ is the strict transform of D by π.
3. $(\pi^* D) \cdot E_u = 0$ for all $u \in \mathcal{P}(\pi)$.

The divisor $(\pi^* D)_{ex}$ supported by the exceptional divisor of π is the **exceptional transform** of D by π.

The divisor $\pi^* D$ is well-defined, as results from Theorem 9.9.1. The point is to show that $(\pi^* D)_{ex}$ exists and is unique with the property (3). Write it as a sum $\sum_{v \in \mathcal{P}(\pi)} x_v E_v$. The last condition of Definition 9.9.4 may be written as the system:

$$\sum_{v \in \mathcal{P}(\pi)} (E_v \cdot E_u) x_v = -(\pi^* D)_{str} \cdot E_u, \quad \text{for all } u \in \mathcal{P}(\pi).$$

This is a square linear system in the unknowns x_v, whose matrix is the matrix of the intersection form in the basis $(E_u)_{u \in \mathcal{P}(\pi)}$. As the intersection form is negative definite, it is non-degenerate, therefore this system has a unique solution. Moreover, all its coefficients being integers, its solution has rational coordinates, which shows that $(\pi^* D)_{ex} \in \mathcal{E}(\pi)_{\mathbb{Q}}$.

Using Definition 9.9.4 and the standard definition of intersection numbers on smooth surfaces recalled in Sect. 9.2, Mumford defined in the following way in [Mum61, Sect. II (b)] the intersection number of two Weil divisors on S:

Definition 9.9.5 Let A, B be two Weil divisors on S without common components, and π be a resolution of S. Then the **intersection number** of A and B is defined by:

$$\boxed{A \cdot B} := \pi^* A \cdot \pi^* B.$$

Using the fact that any two resolutions of S are related by a sequence of blow ups and blow downs (see [Har77, Thm. V.5.5]), it may be shown that the previous notion is independent of the choice of resolution, similarly to that of bracket of two divisorial valuations introduced in Definition 9.9.2. In particular, if S is smooth, one may choose π to be the identity. This shows that in this case Mumford's definition gives the same intersection number as the standard Definition 9.2.2.

Example 9.9.6 Let S be the germ at the origin 0 of the quadratic cone $Z(x^2 + y^2 + z^2) \hookrightarrow \mathbb{C}^3$ (it is the so-called $\mathbf{A_1}$ surface singularity). Let A and B be the germs at 0 of two distinct generating lines of the cone. One may resolve S by blowing up 0. This morphism $\pi : S^{\pi} \to S$ separates all the generators, therefore it is an embedded resolution of $\{A, B\}$. The exceptional divisor of π is the projectivisation of the cone, that is, it is a smooth rational curve E. Its self-intersection number is the opposite of the degree of the curve seen embedded in the projectivisation of the ambient space $\mathbb{C}^3$. Therefore, $E^2 = -2$. Let us compute the total transform $\pi^*A = (\pi^*A)_{str} + xE$. The imposed constraint $\pi^*A \cdot E = 0$ becomes $1 - 2x = 0$, therefore $x = 1/2$. We have used the fact that the strict transform $(\pi^*A)_{str}$ of A by π is smooth and transversal to E, which implies that $(\pi^*A)_{str} \cdot E = 1$.

We obtained $\pi^*A = (\pi^*A)_{str} + (1/2)E$ and similarly, $\pi^*B = (\pi^*B)_{str} + (1/2)E$. Using Definition 9.9.5, we get:

$$
\begin{aligned}
A \cdot B = \pi^*A \cdot \pi^*B = \\
= ((\pi^*A)_{str} + (1/2)E) \cdot ((\pi^*B)_{str} + (1/2)E) = \\
= (\pi^*A)_{str} \cdot (\pi^*B)_{str} + (1/2)((\pi^*A)_{str} + (\pi^*B)_{str}) \cdot E + (1/2)^2 E^2 = \\
= 0 + (1/2) \cdot 2 + (1/2)^2 \cdot (-2) = \\
= 1/2.
\end{aligned}
$$

This example shows in particular that *the intersection number of two curve singularities depends on the normal surface singularity on which it is computed.* Indeed, the branches A and B are also contained in a smooth surface (any two generators of the quadratic cone are obtained as the intersection of the cone with a plane passing through its vertex). In such a surface, their intersection number is 1 instead of 1/2.

Definition 9.9.5 allows to give the following interpretation of the notion of bracket introduced in Definition 9.9.2 (see [GBPPPR19, Prop. 1.11]):

Proposition 9.9.7 *Let A, B be two distinct branches on S. Consider an embedded resolution π of their sum. Denote by E_a, E_b the components of the exceptional divisor E^{π} which are intersected by the strict transforms $(\pi^*A)_{str}$ and $(\pi^*B)_{str}$ respectively. Then:*

$$
A \cdot B = \langle a, b \rangle.
$$

Proof This proof uses directly Definition 9.9.4.

As π is an embedded resolution of $A + B$, the strict transforms $(\pi^*A)_{str}$ and $(\pi^*B)_{str}$ are disjoint. Therefore $(\pi^*A)_{str} \cdot (\pi^*B)_{str} = 0$. Using the last condition in the Definition 9.9.4 of the total transform of a divisor, we know that $(\pi^*A) \cdot$

$(\pi^*B)_{ex} = (\pi^*A)_{ex} \cdot (\pi^*B) = 0$. Combining both equalities, we deduce that:

$$
\begin{aligned}
A \cdot B &= (\pi^*A) \cdot (\pi^*B) = \\
&= (\pi^*A) \cdot ((\pi^*B)_{ex} + (\pi^*B)_{str}) = \\
&= (\pi^*A) \cdot (\pi^*B)_{str} = \\
&= ((\pi^*A)_{ex} + (\pi^*A)_{str}) \cdot (\pi^*B)_{str} = \\
&= (\pi^*A)_{ex} \cdot (\pi^*B)_{str} = \\
&= (\pi^*A)_{ex} \cdot (\pi^*B - (\pi^*B)_{ex}) = \\
&= -(\pi^*A)_{ex} \cdot (\pi^*B)_{ex} = \\
&= -(-E_a^\vee) \cdot (-E_b^\vee) = \\
&= \langle a, b \rangle.
\end{aligned}
$$

At the end of the computation we have used the equality $(\pi^*A)_{ex} = -E_a^\vee$, which results from the fact that π is an embedded resolution of A. Indeed, this implies that $((\pi^*A)_{str} + E_a^\vee) \cdot E_u = 0$ for every $u \in \mathcal{P}(\pi)$, which shows that one has indeed the stated formula for $(\pi^*A)_{ex}$. $\square$

9.10   A Reformulation of the Ultrametric Inequality

In this section we explain the notion of *angular distance* on the set of vertices of the dual graph of a good resolution of S. Theorem 9.10.2 states a crucial property of this distance, relating it to the cut-vertices of the dual graph. Then the ultrametric inequality is reexpressed in terms of the angular distance. This allows to show that Theorem 9.8.1 is a consequence of Theorem 9.10.10, which is formulated only in terms of the angular distance.

Let $\pi : S^\pi \to S$ be a good resolution of the normal surface singularity S. Recall that $\mathcal{P}(\pi)$ denotes the set of irreducible components of its exceptional divisor E^π. Using the notion of bracket from Definition 9.9.2, one may define (see [GR, Sect. 2.7] and [GBPPPR19, Sect. 1.2]):

Definition 9.10.1 The **angular distance** is the function $\rho : \mathcal{P}(\pi) \times \mathcal{P}(\pi) \to [0, \infty)$ given by:

$$
\boxed{\rho(a, b)} :=
\begin{cases}
-\log \dfrac{\langle a, b \rangle^2}{\langle a, a \rangle \langle b, b \rangle} & \text{if } a \neq b, \\
\qquad 0 & \text{if } a = b.
\end{cases}
$$

The fact that the function ρ takes values in the interval $[0, \infty)$ is a consequence of Lemma 9.9.3. The attribute "angular" was chosen by Gignac and Ruggiero because their definition in [GR, Sect. 2.7] was more general, applying to any pair of real-valued semivaluations of the local ring $O_{S,s}$, and that it depended only on those valuations up to homothety, similarly to the angle of two vectors. It is a distance by the following theorem of Gignac and Ruggiero [GR, Prop. 1.10] (recall that the notion of vertex separating two other vertices was introduced in Definition 9.7.2):

Theorem 9.10.2 *The function ρ is a distance on the set $\mathcal{P}(\pi)$. Moreover, for every $a, b, c \in \mathcal{P}(\pi)$, the following properties are equivalent:*

- *one has the equality $\rho(a, b) + \rho(b, c) = \rho(a, c)$;*
- *b separates a and c in the dual graph $\Gamma(\pi)$.*

This theorem explains the importance of cut-vertices of the dual graph $\Gamma(\pi)$ for understanding the angular distance.

Theorem 9.10.2 is a reformulation of the following theorem, which was first proved by in [GBGPPP18, Prop. 79, Rem. 81] for arborescent singularities, then in [GR, Prop. 1.10] for arbitrary normal surface singularities (see also [GBPPPR19, Prop. 1.18] for a slightly different proof):

Theorem 9.10.3 *Let $a, b, c \in \mathcal{P}(\pi)$. Then:*

$$\langle a, b\rangle\langle b, c\rangle \le \langle b, b\rangle\langle a, c\rangle,$$

with equality if and only if b separates a and c in the dual graph $\Gamma(\pi)$.

Theorem 9.10.3 may be also reformulated in terms of spherical geometry using the spherical Pythagorean theorem (see [GBPPPR19, Prop. 1.19.III]).

Using Proposition 9.9.7 and Definition 9.10.1 of the angular distance, one may reformulate in the following way the ultrametric inequality for the restriction of the function u_L to a set of three branches:

Proposition 9.10.4 *Let L, A, B, C be pairwise distinct branches on S. Consider an embedded resolution of their sum and let E_l, E_a, E_b, E_c the irreducible components of its exceptional divisor which intersect the strict transforms of L, A, B and C respectively. Then the following (in)equalities are equivalent:*

1. $u_L(A, B) \le \max\{u_L(A, C), u_L(B, C)\}$.
2. $(A \cdot B) \cdot (L \cdot C) \ge \min\{(A \cdot C)(L \cdot B), (B \cdot C)(L \cdot A)\}$.
3. $\langle a, b\rangle\langle l, c\rangle \ge \min\{\langle a, c\rangle\langle l, b\rangle, \langle b, c\rangle\langle l, a\rangle\}$.
4. $\rho(a, b) + \rho(l, c) \le \max\{\rho(a, c) + \rho(l, b), \rho(b, c) + \rho(l, a)\}$.

We leave the easy proof of this proposition to the reader. It uses the definitions of the function u_L, of the angular distance, as well as Proposition 9.9.7. Note that excepted the first one, all the inequalities are symmetric in the four branches L, A, B, C. The fourth one is a well-known condition in combinatorics, whose name was introduced by Bunemann in his 1974 paper [Bun74]:

Definition 9.10.5 Let (X, δ) be a finite metric space. One says that it satisfies the **four points condition** if whenever $a, b, c, d \in X$, one has the following inequality:

$$\delta(a, b) + \delta(c, d) \le \max\{\delta(a, c) + \delta(b, d), \ \delta(a, d) + \delta(b, c)\}.$$

In the same way in which a finite ultrametric may be thought as a special kind of decorated *rooted* tree (see Proposition 9.4.18), a finite metric space satisfying the four points condition may be thought as a special kind of decorated *unrooted* tree (see [BD98]):

Proposition 9.10.6 *The metric space (X, δ) satisfies the four points condition if and only if δ is induced by a length function on a tree containing the set X among its set of vertices. If, moreover, one constrains X to contain all the vertices of the tree of valency 1 or 2, then this tree is unique up to a unique isomorphism fixing X.*

Let us introduce supplementary vocabulary in order to deal with the special trees appearing in Proposition 9.10.6:

Definition 9.10.7 Let X be a finite set. An X**-tree** is a tree whose set of vertices contains the set X and such that each vertex of valency at most 2 belongs to X. If (X, δ) is a finite metric space which satisfies the four points condition, then the unique X-tree characterized in Proposition 9.10.6 is called the **tree hull** of (X, δ).

The basic idea of the proof of Proposition 9.10.6 is that an X-tree is characterized by the shapes of the convex hulls of the quadruples of points of X, and that those shapes are determined by the cases of equality in the 12 triangle inequalities and the 3 four points conditions associated to each quadruple. In Fig. 9.8 are represented the five possible shapes. For instance, the H-shape is the generic one, characterized by the fact that one has no equality in the previous inequalities.

Let us come back to our normal surface singularity S. One has the following property (see [GBPPPR19, Prop. 1.24]):

Proposition 9.10.8 *Let $\mathcal{F}$ be a finite set of branches on S. If u_L is an ultrametric on $\mathcal{F} \setminus \{L\}$ for one branch L in $\mathcal{F}$, then the same is true for any branch of $\mathcal{F}$.*

By Proposition 9.10.4, if u_L is an ultrametric on $\mathcal{F} \setminus \{L\}$ for one branch L in $\mathcal{F}$, then one has the symmetric relation (2) for every quadruple of branches of $\mathcal{F}$ containing L. The subtle point of the proof of Proposition 9.10.8 is to deduce from this fact that (2) is satisfied by *all* quadruples.

Given Proposition 9.10.8, it is natural to try to relate the rooted trees associated to the ultrametrics obtained by varying L among the branches of $\mathcal{F}$. By looking at

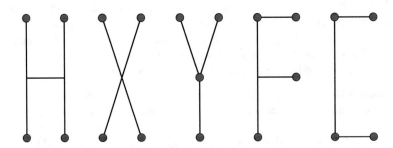

Fig. 9.8 The possible shapes of an X-tree, when X has four elements

quadruples of branches from $\mathcal{F}$, one may prove using Propositions 9.10.4 and 9.10.8 that:

Proposition 9.10.9 *Let $\mathcal{F}$ be a finite set of branches on S. Consider an embedded resolution of $\mathcal{F}$ such that the map associating to each branch A of $\mathcal{F}$ the component E_a of the exceptional divisor intersected by its strict transform is injective. Denote by $\mathcal{F}^\pi$ the set of divisorial valuations a appearing in this way. Then:*

1. *The function u_L is an ultrametric on $\mathcal{F} \setminus \{L\}$ for some branch $L \in \mathcal{F}$ if and only if the angular distance ρ satisfies the four points condition in restriction to the set $\mathcal{F}^\pi$.*
2. *Assume that the previous condition is satisfied. Then the rooted tree associated to u_L on $\mathcal{F} \setminus \{L\}$ is isomorphic to the tree hull of $(\mathcal{F}^\pi, \rho)$ by an isomorphism which sends each end marked by a branch A of $\mathcal{F}$ to the vertex a of the tree hull.*

Proposition 9.10.9 implies readily that Theorem 9.8.1 is a consequence of the following fact (see [GBPPPR19, Cor. 1.40]):

Theorem 9.10.10 *Let S be a normal surface singularity. Consider a set $\mathcal{G}$ of vertices of the dual graph Γ of a good resolution $\pi : S^\pi \to S$ of S. Assume that the convex hull $\mathrm{Conv}_{\mathcal{BV}(\Gamma)}(\mathcal{G})$ of $\mathcal{G}$ in the brick-vertex tree of Γ does not contain brick-vertices of valency at least 4 in $\mathrm{Conv}_{\mathcal{BV}(\Gamma)}(\mathcal{G})$. Then the restriction of the angular distance ρ to $\mathcal{G}$ satisfies the four points condition and the associated tree hull is isomorphic as a $\mathcal{G}$-tree to $\mathrm{Conv}_{\mathcal{BV}(\Gamma)}(\mathcal{G})$.*

In turn, Theorem 9.10.10 is a consequence of a graph-theoretic result presented in the next section (see Theorem 9.11.1).

9.11   A Theorem of Graph Theory

In this final section we state a pure graph-theoretical theorem, which implies Theorem 9.10.10 of the previous section. As we explained before, that theorem implies in turn our strongest generalization of Płoski's theorem, that is, Theorem 9.8.1.

Theorem 9.10.10 is a consequence of Theorem 9.10.2 and of the following graph-theoretic result:

Theorem 9.11.1 *Let Γ be a finite connected graph and δ be a distance on the set $V(\Gamma)$ of vertices of Γ, such that for every $a, b, c \in V(\Gamma)$, the following properties are equivalent:*

- *one has the equality $\delta(a, b) + \delta(b, c) = \delta(a, c)$;*
- *b separates a and c in Γ.*

Let X be a set of vertices of Γ such that the convex hull $\mathrm{Conv}_{\mathcal{BV}(\Gamma)}(X)$ of X in the brick-vertex tree of Γ does not contain brick-vertices of valency at least 4 in $\mathrm{Conv}_{\mathcal{BV}(\Gamma)}(X)$. Then δ satisfies the 4 points condition in restriction to X and the tree hull of (X, δ) is isomorphic to $\mathrm{Conv}_{\mathcal{BV}(\Gamma)}(X)$ as an X-tree.

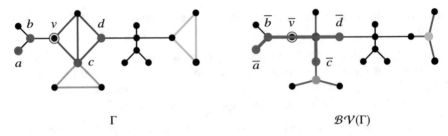

$$\Gamma \qquad\qquad\qquad\qquad \mathcal{B}\mathcal{V}(\Gamma)$$

Fig. 9.9 A convex hull of four vertices

The idea of the proof of Theorem 9.11.1 is to show that, under the given hypotheses, the equalities among the triangle inequalities and four points conditions are as described by the brick-vertex tree. It is writtend in a detailed way in [GBPPPR19, Thm. 1.38].

Example 9.11.2 Let us consider again the connected graph Γ of Example 9.7.6. Look at its vertices a, b, c, d shown on the left of Fig. 9.9. The corresponding vertices $\bar{a}, \bar{b}, \bar{c}, \bar{d}$ of the brick-vertex tree $\mathcal{B}\mathcal{V}(\Gamma)$ are shown on the right side of the figure. Denoting $X := \{a, b, c, d\}$, the convex hull $\mathrm{Conv}_{\mathcal{B}\mathcal{V}(\Gamma)}(X)$ is also drawn on the right side using thick red segments. We see that the hypothesis of Theorem 9.11.1 about the valencies of brick-vertices is satisfied, as the only brick-vertex contained in $\mathrm{Conv}_{\mathcal{B}\mathcal{V}(\Gamma)}(X)$ is of valency 3 in this convex hull.

As shown by the F-shape of $\mathrm{Conv}_{\mathcal{B}\mathcal{V}(\Gamma)}(X)$, one should have the following equalities and inequalities in the four points conditions concerning X:

$$\delta(a, d) + \delta(b, c) = \delta(a, c) + \delta(b, d) > \delta(a, b) + \delta(c, d). \qquad (9.18)$$

Let us prove that this is indeed the case. Consider the cut vertex v of Γ shown on the left side of Fig. 9.9. As it separates a from d, we have the equality $\delta(a, d) = \delta(a, v) + \delta(v, d)$. As v does not separate a from b, we have the strict inequality $\delta(a, v) + \delta(b, v) > \delta(a, b)$. Using similar equalities and inequalities, we get:

$$\begin{aligned} \delta(a, d) + \delta(b, c) &= \\ &= (\delta(a, v) + \delta(v, d)) + (\delta(b, v) + \delta(v, c)) = \\ &= (\delta(a, v) + \delta(v, c)) + (\delta(b, v) + \delta(v, d)) = \\ &= \delta(a, c) + \delta(b, d) = \\ &= (\delta(a, v) + \delta(b, v)) + (\delta(v, d) + \delta(v, c)) = \\ &> \delta(a, b) + \delta(c, d). \end{aligned}$$

The (in)equalities (9.18) are proved.

One proves similarly the triangle equalities $\delta(a, b) + \delta(b, c) = \delta(a, c)$, $\delta(a, b) + \delta(b, d) = \delta(a, d)$ and the fact that one has no equality among the triangle inequalities concerning the triple $\{a, c, d\}$, which shows that the tree hull of (X, δ) has indeed an F-shape, with the vertices a, b, c, d placed as in $\mathrm{Conv}_{\mathcal{B}\mathcal{V}(\Gamma)}(X)$.

Acknowledgments This work was partially supported by the French grants ANR-17-CE40-0023-02 LISA and Labex CEMPI (ANR-11-LABX-0007-01). I am grateful to Alexandre Fernandes, Adam Parusiński, Anne Pichon, Maria Ruas, José Seade and Bernard Teissier, who formed the scientific committee of the *International school on singularities and Lipschitz geometry*, for having invited me to give a course on the relations between ultrametrics and surface singularities. I am very grateful to my co-authors Evelia García Barroso, Pedro González Pérez and Matteo Ruggiero for the collaboration leading to our works [GBGPPP18, GBPPPR19] presented in this paper and for their remarks on a previous version of it.

References

[AM69]   M.F. Atiyah, I.G. Macdonald, *Introduction to Commutative Algebra* (Addison-Wesley, Reading, 1969) 279

[BHPVdV04] W.P. Barth, K. Hulek, C.A.M. Peters, A. Van de Ven, *Compact Complex Surfaces*. Ergebnisse der Mathematik und ihrer Grenzgebiete. 3. Folge. A. Series of Modern Surveys in Mathematics [Results in Mathematics and Related Areas. 3rd Series. A Series of Modern Surveys in Mathematics], vol. 4, 2nd edn. (Springer, Berlin, 2004) 278

[BD98]   S. Böcker, A.W.M. Dress, Recovering symbolically dated, rooted trees from symbolic ultrametrics. Adv. Math. **138**(1), 105–125 (1998) 287, 304

[Bun74]   P. Buneman, A note on the metric properties of trees. J. Combin. Theory Ser. B **17**, 48–50 (1974) 303

[Cam88]   C. Camacho, Quadratic forms and holomorphic foliations on singular surfaces. Math. Ann. **282**(2), 177–184 (1988) 292

[dJP00]   T. de Jong, G. Pfister, *Local Analytic Geometry: Basic Theory and Applications*. Advanced Lectures in Mathematics (Friedr. Vieweg & Sohn, Braunschweig, 2000) 276, 278

[DV44]   P. Du Val, On absolute and non absolute singularities of algebraic surfaces. Rev. Fac. Sci. Univ. Istanbul **11**, 159–215 (1944) 297

[Egg82]   H. Eggers, *Polarinvarianten und die Topologie von Kurvensingularitäten*. Bonner Mathematische Schriften [Bonn Mathematical Publications], vol. 147 (Universität Bonn, Mathematisches Institut, Bonn, 1982). Dissertation, Rheinische Friedrich-Wilhelms-Universität, Bonn, 1982 290

[FJ04]   C. Favre, M. Jonsson, *The Valuative Tree*. Lecture Notes in Mathematics, vol. 1853 (Springer, Berlin, 2004) 293, 299

[Fis01]   G. Fischer, *Plane Algebraic Curves*. Student Mathematical Library, vol. 15 (American Mathematical Society, Providence, 2001). Translated from the 1994 German original by Leslie Kay 276

[GBGPPP18] E.R. García Barroso, P.D. González Pérez, P. Popescu-Pampu, Ultrametric spaces of branches on arborescent singularities, in *Singularities, Algebraic Geometry, Commutative Algebra and Related Topics. Festschrift for Antonio Campillo on the Occasion of his 65th Birthday* (Springer, Berlin, 2018), pp. 55–106 274, 282, 290, 291, 292, 293, 303, 304

[GBGPPP19] E.R. García Barroso, P.D. González Pérez, P. Popescu-Pampu, The valuative tree is the projective limit of Eggers-Wall trees. Rev. R. Acad. Cienc. Exactas Fís. Nat. Ser. A Mat. RACSAM **113**(4), 4051–4105 (2019) 290, 295

[GBPPPR19] E.R. García Barroso, P.D. González Pérez, P. Popescu-Pampu, M. Ruggiero, Ultrametric properties for valuation spaces of normal surface singularities. Trans. Am. Math. Soc. **372**(12), 8423–8475 (2019) 274, 275, 292, 294, 295, 296, 299, 301, 302, 303, 304, 305, 306, 307

[GBGPPP20] E.R. García Barroso, P.D. González Pérez, P. Popescu-Pampu, The combinatorics of plane curve singularities. How Newton polygons blossom into lotuses, in *Handbook of Geometry and Topology of Singularities 1* (Springer, Berlin, 2020). https://arxiv.org/abs/1909.06974 290

[GR] W. Gignac, M. Ruggiero, *Local Dynamics of Non-invertible Maps Near Normal Surface Singularities.* Memoirs of the AMS (to appear). http://arxiv.org/abs/1704.04726 302, 303

[Har77] R. Hartshorne, *Algebraic Geometry.* Graduate Texts in Mathematics, vol. 52 (Springer, New York, 1977) 292, 295, 297, 300

[Hir53] F. Hirzebruch, Über vierdimensionale Riemannsche Flächen mehrdeutiger analytischer Funktionen von zwei komplexen Veränderlichen. Math. Ann. **126**, 1–22 (1953) 291

[Jon15] M. Jonsson, Dynamics on Berkovich spaces in low dimensions, in *Berkovich Spaces and Applications.* Lecture Notes in Mathematics, vol. 2119 (Springer, Cham, 2015), pp. 205–366 299

[Jun08] H.W.E. Jung, Darstellung der Funktionen eines algebraischen Körpers zweier unabhängigen Veränderlichen x, y in der Umgebung einer Stelle $x = a$, $y = b$. J. Reine Angew. Math. **133**, 289–314 (1908) 291

[Mum61] D. Mumford, The topology of normal singularities of an algebraic surface and a criterion for simplicity. Inst. Hautes Études Sci. Publ. Math. **9**, 5–22 (1961) 296, 297, 300

[Noe90] M. Noether, Les combinaisons caractéristiques dans la transformation d'un point singulier. Rend. Circ. Mat. Palermo **4**, 89–108, 300–301 (1890) 290

[Pło85] A. Płoski, Remarque sur la multiplicité d'intersection des branches planes. Bull. Polish Acad. Sci. Math. **33**(11–12), 601–605 (1985) 273, 281

[PP01] P. Popescu-Pampu, Arbres de contact des singularités quasi-ordinaires et graphes d'adjacence pour les 3-variétés réelles. Thèse, Univ. Paris 7 (2001) 290, 293

[Smi75] H.J.S. Smith, On the higher singularities of plane curves. Proc. Lond. Math. Soc. **6**, 153–182 (1875) 290

[Sto79] O. Stolz, Die multiplicität der schnittpunkte zweier algebraischer curven. Math. Ann. **15**, 122–160 (1879) 290

[Wal35] R.J. Walker, Reduction of the singularities of an algebraic surface. Ann. Math. **36**(2), 336–365 (1935) 291

[Wal03] C.T.C. Wall, Chains on the Eggers tree and polar curves, in *Proceedings of the International Conference on Algebraic Geometry and Singularities (Spanish) (Sevilla, 2001)*, vol. 19, no. 2 (2003), pp. 745–754 290

[Wal04] C.T.C. Wall, *Singular Points of Plane Curves.* London Mathematical Society Student Texts, vol. 63 (Cambridge University Press, Cambridge, 2004) 290, 293

Chapter 10
Lipschitz Fractions of a Complex Analytic Algebra and Zariski Saturation

Frédéric Pham and Bernard Teissier

Abstract This text is about the algebra of germs of Lipschitz meromorphic functions on a germ of reduced complex analytic space $(X, 0)$. It is shown to be an analytic algebra, the Lipschitz saturation of the algebra of $(X, 0)$, which in some important cases coincides with Zariski's algebraic saturation. In the case of reduced germs of plane curves, the results in Sect. 10.6 imply that two such germs are topologically equivalent if and only if their Lipschitz saturations are analytically isomorphic. Applications to bi-Lipschitz equisingularity are given.

10.1 Introduction

While seeking to define a good notion of equisingularity (see [Zar65a, Zar65b]), Zariski was led to define in [Zar68] what he calls the **saturation** of a local ring: the saturated ring $\widetilde{A}$ of a ring A contains A and is contained in its normalization $\overline{A}$, and for a complete integral ring of dimension 1, the datum of the saturated ring is equivalent to the datum of the set of Puiseux characteristic exponents of the corresponding algebroid curve.

In the case of complex analytic algebras, it is well known that the normalization $\overline{A}$ coincides with the set of germs of meromorphic functions with bounded module; among the intermediate algebras between A and $\overline{A}$, there is one which can be introduced quite naturally: it is the algebra of the germs of Lipschitz meromorphic functions. We propose to study this algebra, first formally (Sect. 10.3), then geometrically (Sect. 10.4), and to prove (Sects. 10.5 and 10.7) that at least in the case of hypersurfaces, it coincides with the Zariski saturation. In Sect. 10.6, in the case of a reduced but not necessarily irreducible curve, we show how the constructions of

F. Pham
Université de Nice Sophia Antipolis, CNRS, LJAD, Nice, France
e-mail: bernard.teissier@imj-prg.fr

B. Teissier (✉)
Institut mathématique de Jussieu-Paris Rive Gauche, UP7D, Paris, France

© The Author(s), under exclusive license to Springer Nature Switzerland AG 2020
W. Neumann, A. Pichon (eds.), *Introduction to Lipschitz Geometry of Singularities*,
Lecture Notes in Mathematics 2280, https://doi.org/10.1007/978-3-030-61807-0_10

Sects. 10.3 and 10.4 provide a sequence of rational exponents (defined intrinsically, without reference to any coordinates system), which generalizes the sequence of characteristic Puiseux exponents of an irreducible curve. Finally, in Sect. 10.8, we recover in a very simple way the result of Zariski which states that the equisaturation of a family of hypersurfaces implies their topological equisingularity (we even obtain the Lipschitz equisingularity, realized by a Lipschitz deformation of the ambient space).

All the arguments are based on the techniques of normalized blowups (recalled in the Preliminary Section; see also [Hir64b]), and we thank Professor H. Hironaka who taught it to us.[1]

10.2 Preliminaries

(Reminders on the techniques of normalized blowups and majorations of analytic functions)

10.2.1 Conventions

In what follows the rings are commutative, unitary and noetherian. A ring A is said to be **normal** if it is integrally closed in its total ring of fractions $\mathrm{tot}(A)$. An analytic space (X, O_X) is said to be **normal** if at every point $x \in X$, $O_{X,x}$ is normal. We will denote by $\overline{A}$ the integral closure of a ring A in $\mathrm{tot}(A)$.

10.2.2 Universal Property of the Normalisation

Let $n : \overline{X} \to X$ be the **normalisation** of an analytic space $\overline{X}$, i.e., $\overline{X} = \mathrm{specan}_X \overline{O_X}$, where $\overline{O_X}$ is a finite O_X-algebra satisfying $(\overline{O_X})_x = \overline{O_{X,x}}$.

Definition 10.2.1 For every normal analytic space $Y \xrightarrow{f} X$ above X such that the f-image of any irreducible component of Y is not contained in $N = \mathrm{supp}\ \overline{O_X}/O_X$ (the analytic subspace of points in X where $O_{X,x}$ is not normal), there exists a unique

---

[1]We are also very grateful to Mr. Naoufal Bouchareb who brilliantly and expertly translated our 1969 manuscript from French to English and from typewriting to LaTeX.

factorization:

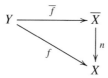

Proof

(a) Algebraic version:

Let $\varphi : A \to B$ be a homomorphism of rings. Let $(\mathfrak{p}_i)_{i=1,\dots,k}$ be the prime ideals of 0 in B.

We suppose that:

  (i) B is normal;

  (ii) for every $i = 1, \dots, k$, $C_{\overline{A}}(A)$ is not included in $\varphi^{-1}(\mathfrak{p}_i)$, where $C_{\overline{A}}(A)$ denotes the conductor of A in $\overline{A}$:

$$C_{\overline{A}}(A) = \{g \in A / g\overline{A} \subset A\}.$$

Then, there is a unique factorization:

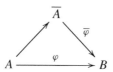

Indeed, by the Prime Avoidance Lemma (see [Bou61, §1]), there exists $g \in C_{\overline{A}}(A)$ such that $g \notin \varphi^{-1}(\mathfrak{p}_i)$ for all $i = 1, \dots, k$. This implies that $\varphi(g)$ is not a divisor of 0 in B. For every $h \in \overline{A}$, set:

$$\overline{\varphi}(h) = \frac{\varphi(g.h)}{\varphi(g)} \in \text{tot}(B)$$

Since h is integral on A, $\overline{\varphi}(h)$ is integral on $\varphi(A)$, and thus also on B. Hence, $\overline{\varphi}(h) \in B$ and $\overline{\varphi}$ is the desired factorization. The uniqueness is obvious.

(b) Geometric version:

Let $Y \xrightarrow{f} X$ satisfy the conditions of the statement. The conditions of the statement remain true locally at $y \in Y$ since if $\varphi^{-1}(N)$ contains locally an irreducible component of Y, it contains it globally. We deduce from this that the local homomorphism:

$$O_{X,f(y)} \longrightarrow O_{Y,y}$$

satisfies the conditions of the algebraic version. We then have the unique factorization:

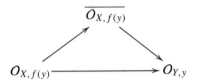

and by the coherence of $\overline{O}_X$, the existence and uniqueness of the searched morphism.   □

10.2.3   Universal Property of the Blowing-up (see [Hir64a])

Proposition 10.2.2 *Let $Y \hookrightarrow X$ be two analytic spaces and let $\mathcal{I}$ be the ideal of Y in X. There exists a unique analytic space $Z \xrightarrow{\pi} X$ over X such that:*

(i) $\pi^{-1}(Y)$ is a divisor of Z, i.e., $\mathcal{I}.O_Z$ is invertible.

(ii) for every morphism $T \xrightarrow{\varphi} X$ such that $\mathcal{I}.O_T$ is invertible, there is a unique factorization:

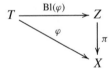

*The morphism $Z \xrightarrow{\pi} X$ is called the **blowup** of X along Y. Recall that π is bimeromorphic, proper and surjective and that $\pi|Z \setminus \pi^{-1}(Y)$ is an isomorphism on $X \setminus Y$.*

10.2.4   Universal Property of the Normalized Blowup

Proposition 10.2.3 *Let $Y \hookrightarrow X$ such that X is normal outside of Y. Then, for every morphism $T \xrightarrow{\varphi} X$ such that:*

(i) T is normal;

(ii) $\mathcal{I}.O_T$ is invertible,

there exists a unique factorization.

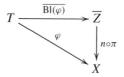

Proof It is sufficient to check that the factorization $T \xrightarrow{\mathrm{Bl}(\varphi)} Z$ satisfies the conditions of Sect. 10.2.3. Since $\pi | Z \setminus \pi^{-1}(Y)$ is an isomorphism, $Z \setminus \pi^{-1}(Y)$ is normal and it is sufficient to verify that the image of each irreducible component of T meets $Z \setminus \pi^{-1}(Y)$. But the inverse image of $\pi^{-1}(Y)$ by $\mathrm{Bl}(\varphi)$ is a divisor by assumption. Since T is normal, this divisor cannot contain any irreducible component. $\qquad \square$

10.2.5 Normalized Blowup and Integral Closure of an Ideal

(See also [Lip69, Chap. II].)

Let A be the analytic algebra of an analytic space germ $(X, 0)$, let I be an ideal of A and let $Y \hookrightarrow X$ be the corresponding sub-germ. It is known that the blowup of the germ Y in the germ X^2 is the projective object $Z = \mathrm{Proj}_A E$ over X associated with the graded algebra $E = \underset{n \geqslant 0}{\oplus} I^n$. The normalization of Z can be written $\overline{Z} = \mathrm{Proj}_A \overline{E}$, with $\overline{E} = \underset{n \geqslant 0}{\oplus} \overline{I^n}$ (where, for an ideal J of A, we define:

$$\overline{J} = \left\{ h \in \mathrm{tot}(A) \mid \exists j_1 \in J, j_2 \in J^2, \ldots, j_k \in J^k : h^k + j_1 h^{k-1} + \cdots + j_k = 0 \right\},$$

which is the ideal of $\overline{A}$ called the **integral closure of the ideal J in $\overline{A}$**).

As an object over $\overline{X}$, the space $\overline{Z}$ equals $\mathrm{Proj}_{\overline{A}} \overline{E}$. But since $\overline{E}$ is a graded $\overline{A}$-algebra of finite type, there exists a positive integer s such that the graded algebra

$$\overline{E}^{(s)} = \underset{n \geqslant 0}{\oplus} \overline{I^{n.s}}$$

is generated by its degree 1 elements: $\overline{E}_1^{(s)} = \overline{I^s}$. But then, $\overline{E}_n^{(s)} = (\overline{I^s})^n$, and as we know that there is a canonical isomorphism $\overline{Z} = \mathrm{Proj}_{\overline{A}} \overline{E}^{(s)}$, we see that the normalized blowup $\overline{Z}$ of I in A, with its canonical morphism to $\overline{X}$, coincides with the blowup of $\overline{I^s}$ in $\overline{A}$.

---

[2]Here, as in other places, we abuse language to identify the germ $(X, 0)$ with one of its representatives.

Proposition 10.2.4 *I and $\overline{I}$ generate the same ideal of $O_{\overline{Z}}$, i.e., $IO_{\overline{Z}} = \overline{I}O_{\overline{Z}}$.*

Proof $\overline{E}$ is a finite type E-module, so for N big enough, $I.\overline{I^N} = \overline{I^{N+1}}$. But $\overline{I}.\overline{I^N} \subset \overline{I^{N+1}}$, therefore:

$$\overline{I}O_{\overline{Z}} . \overline{I^N}O_{\overline{Z}} \subset IO_{\overline{Z}} . \overline{I^N}O_{\overline{Z}}. \tag{10.1}$$

But if $N = k.s$, then $\overline{I^N} . O_{\overline{Z}} = (\overline{I^s})^k . O_{\overline{Z}}$. The latter ideal being invertible, we can simplify by $\overline{I^N}O_{\overline{Z}}$ in the inclusion (10.1). Then $\overline{I}.O_{\overline{Z}} \subset I.O_{\overline{Z}}$. The reverse inclusion is obvious. $\qquad\qquad\square$

Proposition 10.2.5 *$\overline{I}$ coincides with the set of elements of $\overline{A}$ which define a section of $I.O_{\overline{Z}}$.*

Proof If $f \in \overline{I}$, then f obviously defines a section of $\overline{I}O_{\overline{Z}}$. But $\overline{I}O_{\overline{Z}} = IO_{\overline{Z}}$ according to Proposition 10.2.4. Conversely, suppose that $f \in \overline{A}$ defines a section of $IO_{\overline{Z}}$; by writing what this means in some affine open sets $\overline{Z}_{(g_k)} \subset \overline{Z}$, where $g_k \in \overline{I^s}$, one finds that there must exist some integers μ_k such that $f.g_k^{\mu_k} \in I.(\overline{I^s})^{\mu_k}$.

Let (g_k) be a finite family of generators of $\overline{I^s}$. For N large enough, every monomial of degree N in the g_k's will contain one of the $g_k^{\mu_k}$ as a factor, so:

$$f. (\overline{I^s})^N \subset I. (\overline{I^s})^N,$$

i. e., by choosing a base (e_i) of $(\overline{I^s})^N$,

$$f. e_i = \sum_j a_{ij} e_j, \quad a_{ij} \in I.$$

Since $\overline{A}$ can be supposed to be integral, we deduce from this that

$$\det(f. 1 - \|a_{ij}\|) = 0,$$

which is an equation of integral dependence for f on I. $\qquad\qquad\square$

10.2.6 Majoration Theorems

Theorem 10.2.6 (Well Known, see for Example [Abh64]) *Let A be a reduced complex analytic algebra and let $(X,0)$ be the associated germ. For every $h \in \mathrm{tot}(A)$, the following properties are equivalent:*

(i) $h \in \overline{A}$
(ii) h defines on X^{red} a function germ with bounded module.

Theorem 10.2.7 *Let A be a complex analytic algebra, let $(X, 0)$ be the associated germ, let $I = (x_1, \ldots, x_p)$ be an ideal of A and let $\overline{Z}$ be the normalized blowup of I in X. For every $h \in \mathrm{tot}(A)$, the following properties are equivalent:*

(i) $h \in I.O_{\overline{Z}}$
(ii) h *defines on* X^{red} *a germ of function with module bounded by* $\sup|x_i|$ *(up to multiplication by a constant).*

__Proof__ Let A be a noetherian local ring and let $I = (x_1, \ldots, x_p)$ be a principal ideal of A. Then I is generated by one of the x_i's (easy consequence of Nakayama's lemma). Thus $\overline{Z}$ is covered by a finite number of open sets such that in each of them, one of the x_i's generates $I.O_{\overline{Z}}$.

To show that $\frac{|h|}{\sup|x_i|}$ is bounded on X, we just have to prove that it is bounded on each of these open-sets, since $\overline{Z} \to X$ is proper and surjective. In the open set where x_i generates $I.O_{\overline{Z}}$, $\frac{|h|}{\sup|x_i|}$ is bounded if and only if $\frac{|h|}{|x_i|}$ is bounded and we are back to Theorem 10.2.6. $\square$

Corollary 10.2.8 (From Preliminary 10.2.5) *For every $h \in \overline{A}$, the following properties are equivalent:*

(i) $h \in \overline{I}$
(ii) h *defines on* X^{red} *a germ of function with module bounded by* $\sup|x_i|$ *(up to multiplication by a constant).*

10.3   Algebraic Characterization of Lipschitz Fractions

Let A be a reduced complex analytic algebra and let $\overline{A}$ be its normalization ($\overline{A}$ is a direct sum of normal analytic algebras, each being therefore an integral domain, one per irreducible component of the germ associated to A). Consider the ideal:

$$I_A = \ker(\overline{A} \widehat{\underset{C}{\otimes}} \overline{A} \to \overline{A} \underset{A}{\otimes} \overline{A}),$$

where $\widehat{\otimes}$ means the operation on the algebras that corresponds to the cartesian product of the analytic spaces.

Definition 10.3.1 We will call **Lipschitz saturation** of A the algebra:

$$\widetilde{A} = \{f \in \overline{A} \mid f\widehat{\otimes}1 - 1\widehat{\otimes}f \in \overline{I_A}\}$$

where $\overline{I_A}$ denotes the integral closure of the ideal I_A (in the sense of Sect. 10.2.5).

Theorem 10.3.2 $\widetilde{A}$ *is the set of fractions of A that define Lipschitz function germs on the analytic space X, a small enough representative of the germ $(X, 0)$ associated to A.*

Proof Firstly, let us remark that that every Lipschitz function is locally bounded and that the set of bounded fractions of A constitutes the normalization $\overline{A}$ (Theorem 10.2.6). However, denoting by $\overline{X}$ the disjoint sum of germs of normal analytic spaces associated to the algebra $\overline{A}$, the Lipschitz condition $|f(x) - f(x')| \leqslant C \sup |z_i - z_i'|$ for an element $f \in \overline{A}$ is equivalent to say that on $\overline{X} \times \overline{X}$, the function $f \widehat{\otimes} 1 - 1 \widehat{\otimes} f$ has its module bounded by the supremum of the modules of the $z_i \widehat{\otimes} 1 - 1 \widehat{\otimes} z_i$, where $z_1, \ldots, z_r$ denotes a system of generators of the maximal ideal of A. But the ideal generated by $z_i \widehat{\otimes} 1 - 1 \widehat{\otimes} z_i, i = 1 \ldots, r$ is nothing but the ideal I_A defined above. Theorem 10.3.2 is therefore a simple application of Corollary 10.2.8.   □

Corollary 10.3.3 $\widetilde{A}$ *is a local algebra (and thus an analytic algebra).*

Proof Since the algebra $\widetilde{A}$ is intermediate between A and $\overline{A}$, it is a direct sum of analytic algebras. If this sum had more than one term, the element $1 \oplus 0 \oplus \cdots \oplus 0$ of $\widetilde{A}$ would define on X a germ of function equal to 1 on at least one of the irreducible components of X, and to 0 on another of these components. But such a function could not be continuous on X and a fortiori not Lipschitz.   □

The following geometric construction, which comes from Sect. 10.2.5, will play a fundamental role in the sequel. We will associate the following commutative diagram to the analytic space germ X:

$$
\begin{array}{ccc}
D_X & \hookrightarrow & E_X \\
\downarrow & & \downarrow \\
\overline{X} \underset{X}{\times} \overline{X} & \hookrightarrow & \overline{X} \times \overline{X}
\end{array}
$$

where E denotes the projective object over $\overline{X} \times \overline{X}$ obtained by the blowup with center $\overline{X} \underset{X}{\times} \overline{X}$ followed by the normalization (i.e., E_X is the normalized blowup of the ideal I_A which defines $\overline{X} \underset{X}{\times} \overline{X}$ in $\overline{X} \times \overline{X}$); the space D_X is the **exceptional divisor**, inverse image of $\overline{X} \underset{X}{\times} \overline{X}$ in E_X. According to Sect. 10.2.5, the condition:

$$
f \widehat{\otimes} 1 - 1 \widehat{\otimes} f \in \overline{I_A},
$$

which defines $\widetilde{A}$, is equivalent to:

$$
(f \widehat{\otimes} 1 - 1 \widehat{\otimes} f) | D_X = 0.
$$

In other words, the germ $\widetilde{X}$ associated with the analytic algebra $\widetilde{A}$ is nothing but the coequalizer[3] of the canonical double arrow

$$D_X \rightrightarrows \overline{X}$$

obtained by composing the natural map $D_X \to \overline{X} \times \overline{X}$ with the two projections to $\overline{X}$. This germ of analytic space $\widetilde{X}$ will be called the **the Lipschitz saturation** of the germ X.

It is easy to see that the above local construction can be globalized: it is well known for the objects E_X and D_X, which come from blowups and normalizations. Likewise for $\widetilde{X}$: it is easy to define, on an analytic space $X = (|X|, O_X)$, the sheaf $\widetilde{O}_X$ of germs of Lipschitz fractions, and to verify that it is a coherent sheaf of O_X-modules (as a subsheaf of the coherent sheaf $\overline{O}_X$); we thus define an analytic space $\widetilde{X} = (|X|, \widetilde{O}_X)$ called **the Lipschitz saturation** of $X = (|X|, O_X)$, whose underlying topological space $|X|$ coincides with that of X (in fact, the canonical morphism is bimeromorphic and with Lipschitz inverse, so it is a homeomorphism).

Question 10.1 The inclusion $\widetilde{A} \subset \overline{A}$ was obvious in the transcendental interpretation: "every Lipschitz fraction is bounded".

But if one is interested in objects other than analytic algebras, for example in algebras of formal series, there is no longer any reason for $\overline{A}$ to play a particular role in the definition of $\widetilde{A}$. For example, we can define, for any extension B of A in its total fractions ring, the **Lipschitz saturation of A in B**:

$$\widetilde{A}(B) = \{f \in B \,|\, f \otimes 1 - 1 \otimes f \in \overline{I_{A(B)}}\}$$

with

$$I_{A(B)} = \ker(B \underset{C}{\otimes} B \to B \underset{A}{\otimes} B).$$

The question then arises whether we still have the inclusion $\widetilde{A}(B) \subset \overline{A}$.

10.4 Geometric Interpretation of the Exceptional Divisor D_X: Pairs of Infinitely Near Points on X

Each point of D_X^{red} (the reduced space of the exceptional divisor D_X) will be interpreted as a pair of infinitely near points on X. The different irreducible components ${}^\tau D_X^{\mathrm{red}}$ of D_X^{red}, labelled by the index τ, will correspond to different *types* of infinitely near points. The image of ${}^\tau D_X^{\mathrm{red}}$ in X (by the canonical map

---

[3]So we have a canonical morphism of analytic spaces $D_X \to \widetilde{X}$.

$^{\tau}D_X^{\mathrm{red}} \hookrightarrow D_X \to \widetilde{X} \to X$) is an irreducible analytic subset germ $^{\tau}X \subset X$, which we can call **confluence locus of the infinitely near points of type** τ. Among the types of infinitely near points, it is necessary to distinguish the **trivial types** whose confluence points are the irreducible components of X: the **generic** point of a *trivial* $^{\tau}D_X^{\mathrm{red}}$ will be a pair obtained by making two points of X tend towards the same smooth point of X. All the other (*non-trivial*) types have their confluence locus consisting of singular points of X: for example, we will see later that every hypersurface has as non-trivial confluence locus the components of codimension 1 of its singular locus.

What do the Lipschitz fractions become in this context? We have seen in Sect. 10.3 that a Lipschitz fraction is an element $f \in \overline{A}$ such that $(f \otimes 1 - 1 \otimes f)|D_X = 0$. But, since D_X is a divisor of the normal space E_X, this condition will be satisfied everywhere if it is only satisfied in a neighbourhood of a point of each irreducible component of this divisor; or, in intuitive language: "to verify the Lipschitz condition it is enough to verify it for a pair of infinitely near points of each type". Notice that we do not need to worry about trivial types, for which the condition is trivially satisfied for all $f \in \overline{A}$ (note also that the trivial $^{\tau}D_X$ are reduced).

We deduce from this the following result.

Theorem 10.4.1 *A meromorphic function which is locally bounded on the complex analytic space X is locally Lipschitz at every point if only it is locally Lipschitz at one point in each confluence locus $^{\tau}X$.*

To give a first (very rough) idea of the shape of the $^{\tau}D_X^{\mathrm{red}}$, let us look at their images in the space $\widehat{E}_X$ defined by blowing-up the ideal I_A in $\overline{X} \times \overline{X}$. The space E_X that we are interested in is the normalization of $\widehat{E}_X$. But $\widehat{E}_X$ has a simpler geometric interpretation: it is the closure in $\overline{X} \times \overline{X} \times \mathbf{P}^{N-1}$ of the graph Γ of the map

$$(\overline{X} \times \overline{X} - \overline{X} \underset{X}{\times} \overline{X}) \longrightarrow \mathbf{P}^{N-1}$$

which maps each pair $(\overline{x}, \overline{x}')$ outside of the diagonal to the line defined, in homogeneous coordinates, by:

$$(z_1 - z_1' : z_2 - z_2' : \ldots : z_N - z_N') \,,$$

where $(z_1, z_2, \ldots, z_N)$ denotes a system of generators of the maximal ideal of $O_{X,x}$.

We will denote by $\hat{z}: \widehat{E}_X \to \mathbf{P}^{N-1}$ the underlying morphism and by $\hat{z}': \widehat{D}_X \to \mathbf{P}^{N-1}$ the restriction of $\hat{z}$ over $\overline{X} \underset{X}{\times} \overline{X}$ (these morphisms depend on the choice of the generators $(z_1, z_2, \ldots, z_N)$). The fiber $\widehat{D}_X(x)$ of the exceptional divisor $\widehat{D}_X$ over a point $x \in X$ is the disjoint sum of a finite number of algebraic varieties (as many as $\overline{X} \underset{X}{\times} \overline{X}$ has points over x) that are embedded in $\mathbf{P}^{N-1}$ by the map $\hat{z}'|\widehat{D}_X(x)$. In particular, if x is a smooth point, $\widehat{D}_X(x)$ is nothing but the projective space $\mathbf{P}^{n-1}$ associated with the tangent space to X at x.

By composition with the finite morphisms $E_{\tilde{X}} \to \widehat{E}_X$ (normalization) and $D_X \to \widehat{D}_X$, we deduce from $\hat{z}$ and $\hat{z}'$ two morphisms

$$\tilde{z} \colon E_X \to \mathbf{P}^{N-1}$$

$$\tilde{z}' = \tilde{z}|\, D_X \colon D_X \to \mathbf{P}^{N-1}$$

where $\tilde{z}'$ has the following property: its restriction to the fiber $D_X(x)$ of D_X over $x \in X$ is a finite morphism.

Corollary 10.4.2 *If $X \subset \mathbf{C}^N$ is of pure dimension n, the confluence loci $^\tau X$ are of dimension at least equal to $2n - N$.*

Proof According to the finiteness of the above morphism, $\dim D_X(x) \leqslant N - 1$, so each irreducible component $^\tau D_X^{\mathrm{red}}$ of D_X will have an image $^\tau X$ in X of dimension:

$$\dim {}^\tau X \geq \dim {}^\tau D_X^{\mathrm{red}} - (N - 1) = (2n - 1) - (N - 1) = 2n - N.$$

The Special Case of Hypersurfaces In this case, $N = n + 1$, so the confluence loci are of dimension at least equal to $n - 1$. The only non-trivial confluence loci are the codimension 1 components of the singular locus of X. Furthermore, the fibres $^\tau D_X(x)$ of the non-trivial $^\tau D_X$ are sent onto $\mathbf{P}^{N-1}$ by finite morphisms (which are surjective by a dimension argument). In the special case of hypersurfaces, Theorem 10.4.1 is thus formulated as follows:

Theorem 10.4.3 *A meromorphic function on a complex analytic hypersurface X is locally Lipschitz at every point if only it is locally Lipschitz at one point of each irreducible component (of codimension 1) of its polar locus.*

Definition 10.4.4 At a generic point of the divisor $^\tau D_X^{\mathrm{red}}$, this divisor is a smooth divisor of the smooth space E_X. Let s be its irreducible local equation. The ideal of the non reduced divisor $^\tau D_X$ is then locally of the form $(s^{\mu(\tau)})$, where $\mu(\tau)$ is a positive integer, the **multiplicity of the divisor $^\tau D_X$**".

10.5 Lipschitz Fractions Relative to a Parametrization

Let $R \subset A$ be an analytic subalgebra of A and let S be the associated analytic space germ. By considering X as a relative analytic space over S, we are going to proceed to a construction analogous to that of Sect. 10.3, where the product $\overline{X} \times \overline{X}$ is replaced by the fiber product on S. This gives a diagram:

$$
\begin{array}{ccc}
D_{X/S} & \hookrightarrow & E_{X/S} \\
\downarrow & & \downarrow \\
\underset{X}{\overline{X} \times \overline{X}} & \hookrightarrow & \underset{S}{\overline{X} \times \overline{X}}
\end{array}
$$

which enables one to define the **algebra of Lipschitz fractions relative to S**:

$$\widetilde{A}^R = \left\{ f \in \overline{A} \mid (f\widehat{\otimes}1 - 1\widehat{\otimes}f) \mid D_{X/S} = 0 \right\},$$

whose geometric interpretation is given by the "relative" analog to Theorem 10.3.2:

Theorem 10.5.1 (Relative Theorem 10.3.2) $\widetilde{A}^R$ *is the set of fractions of A that satisfy a Lipschitz condition:*

$$\left| f(x) - f\left(x'\right) \right| \le C \sup_i \left| z_i - z_i' \right|$$

for every pair of points (x, x') taken in the same fiber of X/S (with the same constant C for all fibers).

Notice the inclusion $\widetilde{A} \subset \widetilde{A}^R$, which is evident in the geometric interpretation. Formally, this inclusion can also be deduced from the existence of a "morphism" from the above relative diagram to the absolute diagram of Sect. 10.3:

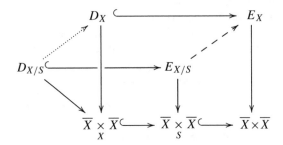

where the dotted arrow $\;\;- - - - - - \rightarrow\;\;$ is defined by the universal property of the normalized blowup (see Sect. 10.2.4, noting that $\overline{X} \times \overline{X}$ is normal).

We will now assume that X is of pure dimension n and we will be interested in the case where R is a **parametrization** of A, i.e., the regular algebra $\mathbf{C}\{z_1, z_2, \ldots, z_n\}$ generated by a system of parameters of A (an n-uple of elements of A such that the ideal generated in A contains a power of the maximal ideal). In other words, $X \to S$ is a finite morphism from X to a Euclidean space of dimension equal to that of X. Let $z = (z_1, z_2, \ldots, z_N)$ be a system of generators of the maximal ideal of A, and let us consider n linear combinations of them:

$$(az)_1 = a_{11}z_1 + a_{12}z_2 + \cdots + a_{1N}z_N$$
$$(az)_2 = a_{21}z_1 + a_{22}z_2 + \cdots + a_{2N}z_N$$
$$(az)_n = a_{n1}z_1 + a_{n2}z_2 + \cdots + a_{nN}z_N$$

$$\left(a_{i,j} \in \mathbf{C} \right)$$

The set of the $a = (a_{ij})$ for which $\mathbf{C}\{(az)_1, (az)_2, \ldots, (az)_n\}$ is a parametrization of A forms, obviously, a dense open set of the space $M_{N \times n}(\mathbf{C})$ of all the $N \times n$ matrices. We will say more generally that a family $\mathcal{P}$ of parametrizations is **generic** if for every system $z = (z_1, z_2, \ldots, z_n)$ of generators of the maximal ideal of A, the set of matrices a for which $\mathbf{C}\{(az)_1, (az)_2, \ldots, (az)_n\} \in \mathcal{P}$ contains a dense open set of $M_{N \times n}(\mathbf{C})$.

We propose to prove the:

Theorem 10.5.2 *For any generic family $\mathcal{P}$ of parametrizations,*

$$\widetilde{A} = \bigcap_{R \in \mathcal{P}} \widetilde{A}^R$$

It follows from this theorem that the following two questions admit identical answers:

Question 10.2 Is the equality $\widetilde{A} = \widetilde{A}^R$ generically true (i.e., for a generic family R of parametrizations)?

Question 10.2' Is $\widetilde{A}^R$ generically independent of R?

We will see that at least in the case of hypersurfaces the answer to these two questions is yes.

Proof *(of Theorem 10.5.2)* We have already seen that $\widetilde{A} \subset \widetilde{A}^R$ for every R. Conversely, consider a function $f \in \bigcap_{R \in \mathcal{P}} \widetilde{A}^R$; does it belong to $\widetilde{A}$?

Let us consider the family of irreducible divisors in E_X consisting of the $^\tau D_X^{\mathrm{red}}$ and of the irreducible components of $\{f \widehat{\otimes} 1 - 1 \widehat{\otimes} f = 0\}$. Let us denote by $^\tau \Delta_f$ the set of points of $^\tau D_X^{\mathrm{red}}$ which:

1. do not belong to any other irreducible divisor of the family;
2. are smooth points of $^\tau D_X^{\mathrm{red}}$ and of E_X.

Since E_X is normal, hence non-singular in codimension 1, $^\tau \Delta_f$ is a Zariski dense open set of $^\tau D_X^{\mathrm{red}}$. At every point $w \in {}^\tau D_X^{\mathrm{red}}$, the local ideal of $^\tau D_X$ in E_X is of the form $(s^{\mu(\tau)})$, where s is a coordinate function of a local chart of E_X, and $\mu(\tau)$ an integer ≥ 1 (the multiplicity of the divisor $^\tau D_X$). Moreover, the function $f \widehat{\otimes} 1 - 1 \widehat{\otimes} f$ is of the form $u s^{\nu(\tau)}$, where u is a unit of the local ring of E_X at the point w and $\nu(\tau)$ is an integer ≥ 0.

Then, it remains to prove that $\nu(\tau) \geq \mu(\tau)$ for every τ (see Sect. 10.4).

Let S be the germ associated with a parametrization $R \in \mathcal{P}$ and let us denote by $E_{X/S}^*$ (resp. $D_{X/S}^*$) the image of $E_{X/S}$ (resp. $D_{X/S}$) in E_X by the canonical map $E_{X/S} \dashrightarrow E_X$ defined at the beginning of the section. By definition, $D_{X/S}^* = E_{X/S}^* \cap D_X$, so that if $E_{X/S}^*$ contains a point $w \in {}^\tau \Delta_f$, the divisor $D_{X/S}^*$ will be given in $E_{X/S}^*$, in a neighbourhood of this point, by the ideal $(s^{\mu(\tau)})$. If this ideal is not zero, i.e., if $E_{X/S}^*$ is not included in D_X, the relative Lipschitz condition:

$$(f \widehat{\otimes} 1 - 1 \widehat{\otimes} f)| D_{X/S} = 0$$

implies that the function $(f \widehat{\otimes} 1 - 1 \widehat{\otimes} f)| E^*_{X/S}$ is divisible by $s^{\mu(\tau)}$ in a neighbourhood of w.

By writing $f \widehat{\otimes} 1 - 1 \widehat{\otimes} f = u s^{\nu(\tau)}$ and by remarking that u, which is a unit of E_X, remains a unit after restriction to $E^*_{X/S}$, we deduce from this that $\nu(\tau) \geq \mu(\tau)$.

On the way, we had to admit that there exists an $R \in \mathcal{P}$ such that, for every non-trivial type τ, $E^*_{X/S}$ meets ${}^\tau \Delta_f$ and is not locally included in ${}^\tau \Delta_f$. To make sure of this, and thus to complete the proof of Theorem 10.5.2, it suffices to prove:

Lemma 10.5.3 *For every Zariski dense open set ${}^\tau \Delta \subset {}^\tau D_X^{\text{red}}$ (τ non-trivial) consisting of smooth points of ${}^\tau D_X^{\text{red}}$ which are also smooth points of E_X, there exists a generic family of parametrizations R for which the map $E_{X/S} \rightarrow E_X$ intersects ${}^\tau \Delta$ in at least one point w and is an embedding transversal to ${}^\tau \Delta$ at this point.* $\qquad\square$

(The condition of "transversal embedding" is obviously stronger than what we asked, but will be more manageable).

Let $z = (z_1, z_2, \ldots, z_n)$ be a system of generators of the maximal ideal of $O_{X,0}$, and denote by ${}^\tau \tilde{z} : {}^\tau D_X^{\text{red}} \rightarrow \mathbf{P}^{N-1}$ the restriction of the morphism $\tilde{z} : E_X \rightarrow \mathbf{P}^{N-1}$ of Sect. 10.4. To every parametrization $R(a) = \mathbf{C}\{(az)_1, (az)_2, \ldots, (az)_n\}$ defined by a matrix $a \in M_{N \times n}(\mathbf{C})$, let us associate the $(N - n - 1)$-plane $\mathbf{P}^{N-n-1}(a) \subset \mathbf{P}^{N-1}$ defined as the projective subspace associated to the kernel of the matrix a.

Lemma 10.5.4 *If the map ${}^\tau \tilde{z} : {}^\tau \Delta \rightarrow \mathbf{P}^{N-1}$ is effectively transversal[4] to $\mathbf{P}^{N-1}(a)$ at the point $w \in {}^\tau \Delta$, then the map $E_{X/S(a)} \rightarrow E_X$ is an embedding effectively transversal to ${}^\tau \Delta$ at this point.* $\qquad\square$

Proof *(of Lemma 10.5.4)* Since ${}^\tau \tilde{z}$ is the restriction of $\tilde{z} : E_X \rightarrow \mathbf{P}^{N-1}$, the transversality of ${}^\tau \tilde{z}$ implies the transversality of $\tilde{z}$.

Hence, $\tilde{z}^{-1}(\mathbf{P}^{N-n-1}(a))$ is a smooth subvariety of E_X of dimension n, which intersects ${}^\tau \Delta$ transversely along the smooth subvariety ${}^\tau \tilde{z}^{-1}(\mathbf{P}^{N-n-1}(a))$ of dimension $n - 1$. In particular, $\tilde{z}^{-1}(\mathbf{P}^{N-n-1}(a))$ is the closure of the complement of ${}^\tau \tilde{z}^{-1}(\mathbf{P}^{N-n-1}(a))$, i.e., the closure of its part located outside of the exceptional divisor. But outside of the exceptional divisor, the right vertical arrow of the following commutative diagram is an isomorphism (since $\bar{X} \times \bar{X}$ is normal), while the left vertical arrow is surjective.

---

[4]The sentence: *the map is transversal at w* expresses one of the following two possible cases:

1. w is sent outside of the subvariety into consideration;
2. w is sent into the subvariety into consideration, and the image of the tangent map to the point w is a vector subspace transversal to the tangent space of this subvariety.

In the second case, we will say that the map is **effectively transversal in** w.

$$
\begin{array}{ccc}
E_{X/S(a)} & \longrightarrow & E_X \\
\downarrow & & \downarrow \\
\overline{X} \underset{S(a)}{\times} \overline{X} & \hookrightarrow & \overline{X} \times \overline{X}
\end{array}
$$

Therefore, the image $E^*_{X/S(a)}$ of the upper arrow is identified with $\overline{X} \underset{S(a)}{\times} \overline{X}$, i.e., with $\tilde{z}^{-1}(\mathbf{P}^{N-n-1}(a))$. Since the equality

$$
E^*_{X/S(a)} = \tilde{z}^{-1}(\mathbf{P}^{N-n-1}(a))
$$

is true outside of the exceptional divisor, it is true everywhere, by taking the closure.

It remains to prove that $E_{X/S(a)} \to E^*_{X/S(a)}$ is an isomorphism (in a neighbourhood of w), but it is obvious. Indeed, it is the germ of a morphism between two smooth varieties of the same dimension which sends a smooth divisor of one onto a smooth divisor of the other and which is an isomorphism outside of these divisors.

$\square$

By Lemma 10.5.4, we can consider Lemma 10.5.3 as a simple consequence of:

Lemma 10.5.5 *There exists a dense open set of matrices $a \in M_{N \times n}(\mathbf{C})$ for which the map $^\tau \tilde{z} : {}^\tau \Delta \to \mathbf{P}^{N-1}$ is effectively transversal to $\mathbf{P}^{N-n-1}(a)$ in at least one point $w \in {}^\tau \Delta$.*

Proof Let us construct a **stratification** of $^\tau D_X^{\mathrm{red}}$ such that each of the following analytic sets is a union of strata:

1. the reduced fiber $^\tau D_X^{\mathrm{red}}(0)$ of $^\tau D_X^{\mathrm{red}}$ over the origin $0 \in X$;
2. the complement of the Zariski open set $^\tau \Delta$.

Let us denote by W be the maximal stratum of this stratification (obviously $W \subset {}^\tau \Delta$) and by W_0 the maximal stratum of one (arbitrarily chosen) of the irreducible components of $^\tau D_X^{\mathrm{red}}$. We will assume that the stratification has been chosen sufficiently fine so that every pair of strata (W_0, V) satisfies Whitney (a)-Condition [Whi65], where V belongs to the **star** of W_0 (see the appendix in the present paper). In these conditions, it follows from the appendix that if the map $^\tau \tilde{z} : {}^\tau D_X^{\mathrm{red}} \to \mathbf{P}^{N-1}$ has its restriction to W_0 effectively transversal to $\mathbf{P}^{N-n-1}(a)$ at a point $w_0 \in W_0$, then its restriction to W will be effectively transversal to $\mathbf{P}^{N-n-1}(a)$ in at least one point $w \in W$ close to w_0.

$\square$

But, we will now prove the:

Lemma 10.5.6 *There exists a dense open set of matrices $a \in M_{N \times n}(\mathbf{C})$ for which the map $^\tau \tilde{z}_{|W_0} : W_0 \to \mathbf{P}^{N-1}$ is effectively transversal to $\mathbf{P}^{N-n-1}(a)$ in at least one point $w_0 \in W_0$.*

$\square$

Proof Since τ is a non-trivial type, the image of the projection $^\tau D_X^{\mathrm{red}} \to X$ is of dimension $\leqslant n - 1$, so that the dimension of the fiber $^\tau D_X^{\mathrm{red}}(0)$ must be at least n (as $\dim{}^\tau D_X^{\mathrm{red}} = 2n - 1$). Now, we know (Sect. 10.4) that the map $^\tau \tilde{z}$ restricted to $^\tau D_X^{\mathrm{red}}(0)$ is a finite morphism. By considering the algebraic variety of dimension $\geqslant n$ in $\mathbf{P}^{N-1}$ defined as the image of a component of $^\tau D_X^{\mathrm{red}}(0)$, and the Zariski dense open set of this variety defined as the image of the set of points of W_0 where the morphism is a local isomorphism, we see that Lemma 10.5.6 is reduced to:

Lemma 10.5.7 *Consider an algebraic variety of dimension $\geqslant n$ in the projective space $\mathbf{P}^{N-1}$ and a Zariski dense open set in this variety. The set of $(N - n - 1)$-planes of $\mathbf{P}^{N-1}$ which intersect transversely this open set in at least one smooth point contains a dense open set of the Grassmann manifold.* ☐

The proof of this lemma is left to the reader. This completes the proof of Lemma 10.5.6. ☐

To summarize:

$$\text{Lemma } 10.5.7 \Longrightarrow \text{Lemma } 10.5.6 \Longrightarrow \left.\begin{array}{l} \text{Lemma } 10.5.5 \\ \text{Lemma } 10.5.4 \end{array}\right\} \Longrightarrow \text{Lemma } 10.5.3 \Longrightarrow \text{Theorem } 10.5.2$$

This completes the proof of Theorem 10.5.2. ☐

Remark 10.5.8 The arguments of Sect. 10.4 generalize without difficulty to the relative case. Thus, for every analytic subalgebra $R \subset A$, we have the notion of *confluence locus relative to R* and the relative analog of Theorem 10.4.1. If R is a parametrization of A, we can see, by an argument similar to that of Sect. 10.4, that the dimension of the relative confluence locus admits the same lower bound $2n - N$ as in the absolute case; in particular, the confluence locus of a hypersurface X relative to a parametrization are the codimension 1 components of the **relative singular locus** of X, i.e., the set of points of X where the finite morphism $X \to S$ is not a submersion of smooth varieties.

We deduce from this:

Theorem 10.5.9 (Relative Version of Theorem 10.4.3) *Let $X \to S$ be a finite morphism of a complex analytic hypersurface to a smooth variety of the same dimension. Then, a meromorphic function on X is locally Lipschitz relatively to S at every point of X if and only if it is locally Lipschitz relatively to S at one point of each irreducible component (of codimension 1) of its polar locus.*

10.6   The Particular Case of Plane Curves

Let $X \overset{(x,y)}{\hookrightarrow} \mathbf{C}^2$ be a germ of reduced analytic plane curve and let $\tilde{z} : E_X \to \mathbf{P}^1$ be the morphism corresponding to the germ of embedding (x, y) (Sect. 10.4).

Let U be the dense open set of $\mathbf{P}^1$ defined as the complement of the tangent directions of X.

Let $u \in U$. By performing a linear change of coordinates if necessary, we can assume that u corresponds to the direction of $\{x = 0\}$. In a neighbourhood of u, we take as local coordinate v in $\mathbf{P}^1$ the inverse of the slope in these coordinates.

Proposition 10.6.1 *In a neighbourhood of every point* $w \in D_X \cap \tilde{z}^{-1}(u)$, $\tilde{z}|D_X^{\mathrm{red}}$ *is an isomorphism,* E_X *is smooth, and* $E_X \cong E_{X/S(u)} \times D_X^{\mathrm{red}}$.

Proof Firstly, let us remark that for every $|v|$ and $|t|$ (and obviously every $|x|$ and $|y|$) small enough, the line $x - vy = t$ remains non-tangent to X and therefore, intersects X transversally at simple points if t is non-zero.

Let $\Gamma \subset (\overline{X} \times \overline{X} - \overline{X} \underset{X}{\times} \overline{X}) \times \mathbf{P}^1$ be the graph of the map defined in Sect. 10.4. We consider the map $\Psi_0 \colon \Gamma \to \mathbf{C} \times \mathbf{P}^1$ defined by $(P, P', v) \mapsto (x(P) - vy(P), v)$ (by noticing that, by definition, $x(P) - vy(P) = x(P') - vy(P')$). The map Ψ_0 extends to a meromorphic map $\widehat{E}_X \overset{\Psi_1}{\longrightarrow} \mathbf{C} \times \mathbf{P}^1$ which is obviously bounded, and so extends locally to a unique morphism $E_X \overset{\Psi}{\longrightarrow} \mathbf{C} \times \mathbf{P}^1$ (all this is done in a neighbourhood of a point w of $\tilde{z}^{-1}(u)$ on E_X).

It is easy to check, and moreover it is geometrically obvious, that Ψ has finite fibers. In addition, by the remark of the beginning of the proof, it is clear that Ψ is unramified outside of $\{0\} \times \mathbf{P}^1$. Therefore, the ramification locus is $\{0\} \times \mathbf{P}^1$ (unless it is empty).

Hence, the vector field $\frac{\partial}{\partial v}$ of $\mathbf{C} \times \mathbf{P}^1$ is tangent to the ramification locus of Ψ. Therefore, it lifts by Ψ to a holomorphic vector field on the normal space E_X (see [Zar65a, Theorem 2]).[5] At every point $w \in D_X \cap \tilde{z}^{-1}(u)$, the integration of this vector field in a neighbourhood of w endows locally E_X with a product structure $E_X \simeq \tilde{z}^{-1}(u) \times \Psi^{-1}(\{0\} \times \mathbf{P}^1)$.

But, on the one hand, we can now apply Lemma 10.5.4 to prove that $\tilde{z}^{-1}(u) \simeq E_{X/S(u)}$ in a neighbourhood of w, and on other hand, again by the above remark, $\tilde{z}^{-1}(u)$ does not meet any $^{\tau}D_X$ with trivial type τ.

We conclude by noticing that since the origin, which is the only possible singularity of the germ X, is the support of all non trivial confluence loci $^{\tau}X$, we have:

$$\Psi^{-1}(\{0\} \times \mathbf{P}^1) = \bigcup_{\tau \text{ non trivial}} {}^{\tau}D_X.$$

Corollary 10.6.2 (See Sect. 10.4) *In this situation, the equation of* $D_{X/S(u)}$ *in* $E_{X/S(u)}$ *is the equation of* D_X *in* E_X.

---

[5]We can also see this by an argument similar to that of Lemma 10.8.6 below.

We will now study the relative situation:

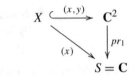

by assuming that $\{x = 0\}$ is not tangent to X at 0.

We will denote by X_α the irreducible components of X and by n_α their multiplicities.

For a local ring of dimension 1, the normalized blowup of an ideal is a regular ring which is nothing but the normalized ring. Hence, $E_{X/S} = \overline{X} \underset{S}{\times} \overline{X}$. We can easily determine the irreducible components of $E_{X/S}$ and the morphism $E_{X/S} \to S$ by using the following lemmas, after having noticed that an irreducible component of $E_{X/S}$ projects onto a pair of irreducible components of $\overline{X}$.

Lemma 10.6.3 *Set $m_{\alpha,\alpha'} = \mathrm{lcm}(n_\alpha, n_{\alpha'})$ and let $\varphi : \mathbf{C}\{x\} \to \mathbf{C}\{s\}$ be given by $\varphi(x) = s^{m_{\alpha,\alpha'}}$. The set B of $\mathbf{C}\{x\}$-homomorphisms*

$$\mathbf{C}\{x^{1/n_\alpha}\} \underset{\mathbf{C}\{x\}}{\otimes} \mathbf{C}\{x^{1/n_{\alpha'}}\} \longrightarrow \mathbf{C}\{x\}$$

can be identified with the set of pairs $\{(\beta, \beta') \in \mathbf{C}^2 : (\beta^{n_\alpha}, \beta'^{n_{\alpha'}}) = (1, 1)\}$ by the correspondance:

$$\begin{cases} x^{1/n_\alpha} \otimes 1 \longmapsto \beta s^{m_{\alpha\alpha'}/n_\alpha} \\ 1 \otimes x^{1/n_{\alpha'}} \longmapsto \beta' x^{m_{\alpha\alpha'}/n_{\alpha'}} \end{cases}$$

(the pairs (β, β') correspond to the pairs of determinations of $(x^{1/n_\alpha}, x^{1/n_{\alpha'}})$).

If we endow B with the equivalence relation: $b_1 \sim b_2$ if $b_1 - b_2$ is a $\mathbf{C}\{x\}$-automorphism of $\mathbf{C}\{s\}$ (it is the equivalence of pairs of determinations "modulo the monodromy"), then, the set $B/\sim$ has $(n_\alpha, n_{\alpha'})$ elements.

Lemma 10.6.4

$$\overline{\mathbf{C}\{x^{1/n_\alpha}\} \underset{\mathbf{C}\{x\}}{\otimes} \mathbf{C}\{x^{1/n_{\alpha'}}\}} = \underset{B/\sim}{\oplus} \mathbf{C}\{x^{1/m_{\alpha\alpha'}}\}$$

with the obvious arrows.

Lemma 10.6.4 can be proved by using Lemma 10.6.3 and the universal property of the normalization. The proof of Lemma 10.6.3 is left to the reader.

We can now determine the equation of $D_{X/S}$ in $E_{X/S}$. At a point of an irreducible component of $E_{X/S}$, the ideal of $^\tau D_{X/S}$ is generated by $y \otimes 1 - 1 \otimes y = a_\tau s^{\mu(\tau)}$ (where a_τ is a unit of $\mathbf{C}\{s\}$), which can be interpreted as the difference of $y_{\alpha\beta}(x) -$

$y_{\alpha'\beta'}(x)$ of the Puiseux expansions of y_α and y'_α computed for the "determinations" (β, β') of $(x^{1/n_\alpha}, x^{1/n_{\alpha'}})$ corresponding to the chosen irreducible component:

$$y_{\alpha\beta}(x) - y_{\alpha'\beta'}(x) = a_{\beta\beta'} x^{\mu(\beta,\beta')/m_{\alpha\alpha'}}$$

where $a_{\beta\beta'}$ is a unit of $\mathbf{C}\{x^{1/m_{\alpha\alpha'}}\}$, $a_{\beta\beta'} = a_\tau$ and $\mu(\beta, \beta') = \mu(\tau)$.

In the particular case where X is irreducible of multiplicity n at the origin, we deduce from this that the sequence of the distinct $\mu(\tau)$ (for τ non trivial), indexed in increasing order, coincides with the sequence:

$$\left\{ \frac{m_1}{n_1} n, \frac{m_2}{n_1 n_2} n, \dots, \frac{m_g}{n_1 \dots n_g} n \right\}$$

where the $\frac{m_i}{n_1 \dots n_i} n$ are the characteristic Puiseux exponents.

Now, we return back to E_X and D_X. If $^\tau D_{X/S}$ is an irreducible component of D_X, we know from Sect. 10.4 that $\tilde{z}|^\tau D_X$ is a finite morphism, and it follows from Proposition 10.6.1 that its ramification locus is contained in the set of directions of tangent lines to the irreducible components X_α and $X_{\alpha'}$ corresponding to $^\tau D_X$.

Proposition 10.6.5

(i) *If X_α and $X_{\alpha'}$ have the same tangent line, then $\deg \tilde{z}|^\tau D_X = 1$, so the number of types τ corresponding to the pair (α, α') equals $(n_\alpha, n_{\alpha'})$.*

(ii) *If X_α and $X_{\alpha'}$ have distinct tangent lines, then $\deg \tilde{z}|^\tau D_X = (n_\alpha, n_{\alpha'})$ and there is a unique type τ.*

Proof In Case (i), let $r \in \mathbf{P}^1$ be the direction of the common tangent line. Since $\mathbf{P}^1 \setminus \{r\}$ is contractible, $^\tau D_X \setminus \tilde{z}^{-1}(r)$ is a trivial fiber bundle on $\mathbf{P} \setminus \{r\}$. This fiber bundle is connected since $^\tau D_X$ is irreducible, therefore, it is a covering space of degree 1.

Case (ii) is more delicate. Let r_1 and r_2 be the two tangent directions and let $u \in \mathbf{P}^1 \setminus \{r_1, r_2\}$. We have to prove that we can join any two points of $E_{X/S(u)}$ by a path contained in $^\tau D_X$ and avoiding $\tilde{z}^{-1}(r_1) \cup \tilde{z}^{-1}(r_2)$. We can do this by looking at two pairs of points $(P_\alpha, P_{\alpha'})$ and $(Q_\alpha, Q_{\alpha'})$, where $P_\alpha, Q_\alpha \in X_\alpha \setminus \{0\}$ and $P_{\alpha'}, Q_{\alpha'} \in X_{\alpha'} \setminus \{0\}$ are close to the origin and located on the same line with slope u. It is possible to pass continuously from the pair $(P_\alpha, P_{\alpha'})$ to the pair $(Q_\alpha, Q_{\alpha'})$ in such a way that the slopes of the lines joining the intermediate pairs stay at bounded distance from r_1 and r_2. We then conclude by taking the limit.

10.7 Lipschitz Saturation and Zariski Saturation

Let $R \subset A$ be a parametrization of a complex analytic algebra A, and let $X \to S$ be the associated germ of morphism of analytic spaces. Zariski defines a **domination** relation between fractions of A which, translated into transcendental terms, can be formulated as follows:

Definition 10.7.1 f **dominates** g over R ($f \underset{R}{>} g$) if and only if, for every pair $g_\beta(x)$, $g_{\beta'}(x)$ of distinct determinations of g, considered as a multivalued function of $x \in S$, the quotient

$$\frac{f_\beta(x) - f_{\beta'}(x)}{g_\beta(x) - g_{\beta'}(x)}$$

has bounded module, where $f_\beta(x)$ and $f_{\beta'}(x)$ denote the corresponding determinations of f.

An extension B of A in its total ring of fractions is said **saturated over** R if every fraction which dominates an element of B belongs to B.

The **saturated algebra of** A (with respect to R) is defined as the smallest saturated algebra containing A.

Question 10.3 Is there a relation between the saturated algebra in the sense of Zariski and the algebra $\widetilde{A}^R$ defined in Sect. 10.5?

In the particular case of hypersurfaces, $A = R[y]$, we can easily see that the Zariski saturation coincides with the set of fractions which dominate y, i.e., in this case, with the algebra $\widetilde{A}^R$ of Lipschitz fractions relative to the parametrization R.

In the general case of an arbitrary codimension, $A = R[y_1, \ldots, y_k]$, the Zariski saturation and the Lipschitz saturation are both more complicated to define, and answering Question 10.3 does not seem easy to us.

In some cases, including the case of hypersurfaces, Zariski can prove that his saturation is independant of the chosen parametrization as long as the latter is generic. Therefore, we obtain, in the case of hypersurfaces, a positive answer to Questions 10.2 and 10.2' of Sect. 10.5. More precisely, we have:

Theorem 10.7.2[6] *Let A be the complex analytic algebra of a hypersurface germ X, and consider the (generic) family $\mathcal{P}$ of the parametrizations defined by a direction of projection transversal to X (i.e., not belonging to the tangent cone) at a generic point of each irreducible component of codimension 1 of the singular locus. Then, for every $R \in \mathcal{P}$, $\widetilde{A} = \widetilde{A}^R$, which equals the Zariski saturation.*

Indeed, this family of parametrizations $\mathcal{P}$ is the one for which Zariski proves the invariance of his saturation [Zar68, Theorem 8.2].

---

[6](Added in 2020) For a more algebraic approach, see [Lip75a, Lip75b]. For a more general result without the hypersurface assumption, see [Bog74, Bog75].

10.8   Equisaturation and Lipschitz Equisingularity

The notion of saturation used in this section is the Lipschitz saturation which, as we have just seen, coincides with the Zariski saturation in the case of hypersurfaces.

Let $r : X \to T$ be an analytic retraction of a reduced complex analytic space germ X on a germ of smooth subvariety $T \hookrightarrow X$. Denote by $X_0 = r^{-1}(0)$ the fiber of this retraction over the origin $0 \in T$.

Definition 10.8.1 We say that (X, r) is **equisaturated along** T if the saturated germ $\widetilde{X}$ admits a product structure:

$$\widetilde{X} = \widetilde{X}_0 \times T$$

compatible with the retraction r (i.e., such that the second projection is $\widetilde{X} \to X \overset{r}{\to} T$).

Theorem 10.8.2 *If (X, r) is equisaturated along T, then (X, r) is topologically (and even Lipschitz) trivial along T.*

By **topological triviality**, we mean the following property: for every embedding

of the retraction r in a retraction r^N of a euclidean space, the pair $(\mathbf{C}^N, X)$ is homeomorphic to the product $(\mathbf{C}^{N-p} \times T, X_0 \times T)$, in a compatible way with the retraction r^N.

By **Lipschitz triviality**, we mean that the above homeomorphism is Lipschitz as well as its inverse.

Proof Let $(t_1, \ldots, t_p)$ be a local coordinates system on T. By using the product structure $\widetilde{X} = \widetilde{X}_0 \times T$, let us denote by ∇_i the vector field on $\widetilde{X}$ whose first projection is zero and whose second one equals $\frac{\partial}{\partial t_i}$. Let A be the algebra of X; ∇_i is a derivation from $\widetilde{A}$ to $\widetilde{A}$. Then, by restriction, it defines a derivation from A to $\widetilde{A}$. Let us consider an embedding $X \hookrightarrow \mathbf{C}^N$, i.e., a system of N generators of the maximal ideal of A:

$$(z_1, z_2, \ldots, z_{N-p}, t_1 \circ r, t_2 \circ r, \ldots, t_p \circ r)$$

(by a change of coordinates, all the systems can be reduced to this form). The functions $\nabla_i z_1, \nabla_i z_2 \ldots, \nabla_i z_{N-p}$ are Lipschitz functions on X. Then, they can extend to Lipschitz functions $g_{i,1}, g_{i,2}, \ldots, g_{i,N-p}$ on all $\mathbf{C}^N$. Hence, for all $i =$

$1, 2, \ldots, p$, we have a Lipschitz vector field on $\mathbf{C}^N$:

$$g_{i,1}\frac{\partial}{\partial z_1} + g_{i,2}\frac{\partial}{\partial z_2} + \cdots + g_{i,N-p}\frac{\partial}{\partial z_{N-p}} + \frac{\partial}{\partial t_i}$$

which is tangent to X and projects onto the vector field $\frac{\partial}{\partial t_i}$ of T. Since they are Lipschitz, these vector fields are locally integrable and their integration realizes the topological triviality of X.   □

Relative Equisaturation

We will now define a relative notion of equisaturation. Let X/S be a germ of analytic space *relative to a parametrization*, consisting of the data of a reduced analytic germ X of pure dimension n and of the germ of a finite morphism $X \to S$ on a germ of smooth variety of dimension n. Let

$$r : X/S \to T$$

be an *analytic retraction of the relative analytic space X/S on a smooth subvariety T*. By this, we mean the datum of a commutative diagram:

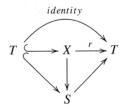

Denote by X_0/S_0 the relative analytic space defined as the inverse image of the point $0 \in T$ by this retraction.

Definition 10.8.3 (Relative Definition 10.8.1) We say that $(X/S, r)$ is **equisaturated along** T, if the germ of relative saturated space $\widetilde{X}^S/S$ admits a product structure:

$$
\begin{array}{ccccc}
\widetilde{X}^S & = & \widetilde{X_0}^{S_0} & \times & T \\
\downarrow & & \downarrow & & \downarrow{\scriptstyle id} \\
S & = & S_0 & \times & T
\end{array}
$$

which is compatible with the retraction r.

In the case where X is a hypersurface, it results immediately from Theorem 10.7.2 that if S is a generic parametrization, the equisaturation of X/S (**relative equisaturation**) implies the equisaturation of X (**absolute equisaturation**).

Question 10.4 Conversely, does the equisaturation of X imply the existence of a generic parametrization S such that X/S is equisaturated?

It would be interesting to know the answer to this question because the work of Zariski gives a lot of informations on the relative notion of equisaturation.

We assume in the sequel that X is a hypersurface. Let $R = \mathbf{C}\{z_1, \ldots, z_n\}$ be a parametrization of A. We can write:

$$A = R[y] = R[Y]/(f),$$

where f is a reduced monic polynomial in Y with coefficients in R and where $y = Y + (f)$ is the residue class of Y modulo f. The reduced discriminant of this polynomial generates an ideal in R which depends only of A and R; we will call it the **ramification ideal** of the parametrization R. We will denote by $\Sigma \subset S$ the subspace defined by this ideal; this subspace will be called the **ramification locus** of X/S.

Definition 10.8.4 We say that $(X/S, r)$ has **trivial ramification locus** along T if the pair (S, Σ) admits a product structure:

$$
\begin{array}{ccccc}
\Sigma & = & \Sigma_0 & \times & T \\
\big\uparrow & & \big\uparrow & & \big\downarrow {\scriptstyle id} \\
S & = & S_0 & \times & T
\end{array}
$$

which is compatible with the retraction r.

Theorem 10.8.5 (Zariski [Zar68]) *Let X be a hypersurface. The following two properties are equivalent:*

(i) $(X/S, r)$ is equisaturated along T;
(ii) $(X/S, r)$ has trivial ramification locus.

Moreover, these two properties imply the topological triviality along T of the hypersurface X.

Notice that in the case of a generic parametrization, where the relative equisaturation implies the absolute equisaturation, the last part of Theorem 10.8.5 is a simple Corollary of our Theorem 10.7.2. But Zariski proves Theorem 10.8.5 for any parametrization.

We will limit ourselves to the proof of the implication $(ii) \Rightarrow (i)$ and we refer to [Zar68] for the rest.

Lemma 10.8.6 *Every derivation of the ring R in itself which leaves stable the ideal of ramification extends canonically into a derivation of the relative saturation $\tilde{A}^R$ in itself.*

Proof Since A is finite over R, every derivation $\nabla: R \to R$ admits a canonical extension to the ring of fractions of A. Explicitly, we have:

$$\nabla y = -\left(\sum_{i=1}^{n} \frac{\partial f}{\partial z_i} \nabla z_i\right)\Big/ \frac{\partial f}{\partial y}.$$

We have to prove that under the hypothesis of the lemma, $\nabla g \in \tilde{A}^R$ for every $g \in \tilde{A}^R$. But the polar locus of every $g \in \overline{A}$ is obviously included in the singular locus of X, so, in the zero locus of $\frac{\partial f}{\partial y}$. By writing

$$\nabla g = \frac{\partial g}{\partial y} \nabla y + \sum_{i=1}^{n} \frac{\partial g}{\partial z_i} \nabla z_i,$$

we deduce from this that the polar locus of ∇g is included in the zero locus of $\frac{\partial f}{\partial y}$.

In order to check that $g \in \tilde{A}^R$, it is then sufficient (by Theorem 10.5.9) to check it at a generic point of each irreducible component (of codimension 1, of course) of the zero locus of $\frac{\partial f}{\partial y}$.

Let $^S X$ be such an irreducible component, restricted to a small neighbourhood of one of its points. For a generic choice of the point, we can assume that:

(1) $^S X$ is smooth and the restriction to $^S X$ of the morphism: $X \to S$ is an embedding;
(2) $^S X = (X|\Sigma)^{\text{red}}$, where $\Sigma \subset S$ denotes the image of $^S X$, i.e., the ramification locus of the morphism $X \to S$;
(3) the finite cover $D^{\text{red}}_{X/S} \to {}^S X$ is étale, i.e., $D^{\text{red}}_{X/S}$ is a disjoint union of[7] components $^{\tau} D^{\text{red}}_{X/S}$ isomorphic to $^S X$.

By (1), Σ is a smooth divisor of S and we can choose local coordinates $(x, t_1, t_2, \ldots, t_{n-1})$ in S so that $x = 0$ is a local equation of this divisor. (2) means that x does not vanish outside of $^S X$.

Locally in S, the submodule of the derivations which leave stable the ramification ideal (x) is generated by $x \frac{\partial}{\partial x}$ and the $\frac{\partial}{\partial t_i}$'s.

Therefore, it suffices to prove that the functions $x \frac{\partial g}{\partial x}$ and $\frac{\partial g}{\partial t_i}$ are Lipschitz relatively to S.

Consider the space $E_{X/S}$, that we can assume to be smooth, in a neighbourhood of one of the components $^{\tau} D^{\text{red}}_{X/S}$ of the étale cover of (3). The function x is well

---

[7](Added in 2020) ... open subsets of components $^{\tau} D^{\text{red}}_{X/S}$, isomorphic to their image in ...

defined on $E_{X/S}$ (by composition with the canonical morphism $E_{X/S} \to \overline{X} \underset{S}{\times} \overline{X} \to$ S) and by (2), it does not vanish outside of $D_{X/S}^{\text{red}}$. Therefore it is of the form $x = s^{m(\tau)}$, where $s = 0$ is an irreducible equation of the smooth divisor $^\tau D_{X/S}^{\text{red}}$. On the other hand, the ideal of the non-reduced divisor $^\tau D_{X/S}$ is generated by:

$$^\tau \Delta y = y \otimes 1 - 1 \otimes y = a_\tau(t) s^{\mu(\tau)} + \cdots ,$$

where a_τ must be a unit of the ring $\mathbf{C}\{t_1, t_2, \ldots, t_{n-1}\}$ since $^\tau \Delta y$ vanishes only on $^\tau D_X^{\text{red}}$.

Thus, for every τ, we have a series expansion whose terms are increasing powers of $x^{1/m(\tau)}$ (compare to Sect. 10.6):

$$^\tau \Delta y = a_\tau(t) x^{\frac{\mu(\tau)}{m(\tau)}} + \cdots ,$$

and a function g will be Lipschitz relatively to S if and only if for every τ, the series expansion of $^\tau \Delta g$ into rational powers of x has no terms with exponents less than $\mu(\tau)/m(\tau)$. Let g be such a function:

$$^\tau \Delta g = b_\tau(t) x^{\frac{\mu(\tau)}{m(\tau)}} + \cdots$$

We have:

$$^\tau \Delta\left(x \frac{\partial g}{\partial x}\right) = x \frac{\partial}{\partial x}(^\tau \Delta g) = x \left(\frac{\mu(\tau)}{m(\tau)} b_\tau(t) x^{\frac{\mu(\tau)}{m(\tau)} - 1} + \cdots \right)$$

and:

$$^\tau \Delta\left(\frac{\partial g}{\partial t_i}\right) = \frac{\partial}{\partial t_i}(^\tau \Delta g) = \frac{\partial b_\tau(t)}{\partial t_i} x^{\frac{\mu(\tau)}{m(\tau)}} + \cdots ,$$

so that $x\frac{\partial g}{\partial x}$ and $\frac{\partial g}{\partial t_i}$ are still functions of the same type, i.e., Lipschitz functions relative to S. This completes the proof of Lemma 10.8.6   □

Proof (of (ii) $\Rightarrow$ (i) of Theorem 10.8.5) Let us choose local coordinates $(x_1, x_2, \ldots, x_{n-p}, t_1, t_2, \ldots, t_p)$ in S compatible with the product structure $S_0 \times T$. The vector field $\frac{\partial}{\partial t_i}$ is tangent to the ramification locus $\Sigma = \Sigma_0 \times T$. Therefore, by Lemma 10.8.6, it lifts to a holomorphic vector field ∇_i on $\widetilde{X}^S$. The integration of these p vector fields $\nabla_1, \ldots, \nabla_p$ realizes the desired product structure on $\widetilde{X}^S$.   □

Speculation on Equisingularity

We would like to find a "good" definition of equisingularity of X along T, satisfying if possible the two following properties:

(TT) the equisingularity implies the topological triviality;
(OZ) the set of points of T where X is equisingular forms a dense Zariski open set.

Equisaturation satisfies (TT) (Theorem 10.7.2 above), but satisfies (OZ) only in the case where $\mathrm{codim}_X T = 1$ (equisaturation of a family of curves coincides with equisingularity). In the general case, one can find some $X \to T$ such that X is not equisaturated at any point of T.[8]

Zariski proposed a definition to the equisingularity of hypersurfaces that generalizes the idea of trivialization of the ramification locus [Zar37, Zar64]: the hypersurface X is **equisingular** along T if, for a generic parametrization, the ramification locus Σ is equisingular along (the projection of) T. Since the codimension of T in Σ is smaller than its codimension in X minus one, we therefore obtain a definition of the equisingularity by induction on the codimension.[9]

This definition satisfies (OZ), but we do not know how to prove (TT).[10]

In the case where T coincides with the singular locus of X (family of analytic spaces with isolated singularities), Hironaka found a criterion of equisingularity which satisfies (TT) and (OZ) at the same time. This criterion is defined [Hir64] in terms of the normalized blowup of an ideal (i.e., the product of the ideal of T by the Jacobian ideal of X). The topological triviality is proved by integrating a vector field,[11] but:

1. Instead of being holomorphic on $\overline{X}$, this vector field is differentiable (i.e., C^∞) on the blown-up space $\widehat{X}$ of X (the normalized blowup of the ideal mentioned above).

2. Instead of being Lipschitz on X, i.e., satisfying a Lipschitz inequality for every ordered pair of points in $X \times X$, this vector field satisfies a Lipschitz inequality only for the ordered pair of points in $T \times X$.

---

[8]Here we are thinking about the relative equisaturation characterized (Theorem 10.8.5) by the triviality of the ramification locus. But likely, the notion of absolute equisaturation leads to about the same thing—don't we want to answer yes to Question 10.4?

Added in 2020: The approaches of E. Böger in [Bog75] and J. Lipman in [Lip75b] would probably lead to a positive answer.

[9](Added in 2020) This theory was described by Zariski in [Zar79, Zar80]. The reason why equisaturation does not satisfy (OZ) in general is that it corresponds to a condition of analytical triviality of the discriminant, which of course does not satisfy (OZ) in general. See also [LT79].

[10](Added in 2020) There are now several results where Zariski equisingularity implies topological triviality sometimes via the Whitney conditions. See [Var73, Spe75].

[11]Cf. H. Hironaka (not published but see [Hir64b]).

The general solution to the problem of equisingularity will maybe use some rings of this type of functions (C^∞ in a blown-up space and "weakly Lipschitz" below).[12]

Appendix: Stratification, Whitney's (a)-property and Transversality

A **stratification**[13] of an analytic (reduced) space X is a locally finite partition of X in smooth varieties called **strata**, such that:

1. the closure $\overline{W}$ of every stratum W is an (irreducible) analytic space;
2. the boundary $\partial W = \overline{W} \setminus W$ of every stratum W is a union of strata.

We call **star** of a stratum W the set of strata which have W in their boundary.

Let (W_0, W) be an ordered pair of strata, with $W_0 \subset \partial W$. We say that this ordered pair satisfies the **property (a) of Whitney** at a point $x_0 \in W_0$ if for every sequence of points $x_i \in W$ tending to x_0 in such a way that the tangent space $T_{x_i}(W)$ admits a limit, this limit contains the tangent space $T_{x_0}(W)$ (we suppose that X is locally embedded in a Euclidean space, in such a way that the tangent spaces are realized as subspaces of the same vector space; the property (a) of Whitney is independent of the chosen embedding). For every ordered pair of strata (W_0, W) of a stratification, there exists a Zariski dense open set of points of W_0 where the property (a) of Whitney is satisfied [Whi65]. We can then refine every stratification into a stratification such that the property (a) of Whitney is satisfied at every point for every ordered pair of strata.

Proposition 10.8.7 *Let (X, x_0) be a stratified germ of complex analytic space such that the ordered pairs of strata (W_0, W) satisfy the property (a) of Whitney, where W_0 denotes the stratum which contains x_0 and where W is any stratum of the star of W_0. Let $\varphi : X \to \mathbf{C}^m$ be a morphism germ such that $\varphi|W_0$ is effectively transversal to the value 0 at the point x_0. Then, for every stratum W, $\varphi|W$ is effectively transversal to the value 0 at (at least) one point of W arbitrarily close to x_0.*

Proof The transversality of $\varphi|W$ at every point close to x_0 is an obvious consequence of the property (a) of Whitney for the ordered pair (W_0, W). It remains to prove the effective transversality i.e., to prove that $(\varphi|W)^{-1}(0)$ is not empty. But $(\varphi|\overline{W})^{-1}(0)$ is a closed analytic subset of $\overline{W}$, non empty (because it contains the point x_0) and defined by m equations. Therefore its codimension is at most m. If

---

[12](Added in 2020) The idea of considering vector fields which are differentiable on some blown-up space was used by Pham in [Pha71a] and, in real analytic geometry by Kuo who introduced *blow-analytic* equivalence of singularities; see [Kuo85].

[13]See also David Trotman's article "Stratifications, Equisingularity and Triangulation" in this volume.

$(\varphi|W)^{-1}(0)$ were empty, then ∂W would contain at least one stratum W' such that $(\varphi|W')^{-1}(0)$ is non empty and of dimension $\geqslant \dim W - m$. But on the other hand, the transversality of $\varphi|W'$ implies that $(\varphi|W')^{-1}(0)$ is a smooth variety of dimension $< \dim W' - m$, and then of dimension $< \dim W - m$. We then get a contradiction.

$\square$

Remark 10.8.8 Of course, in the statement of Proposition 10.8.7, we could replace the transversality relative to the value of 0 by the transversality relative to a smooth variety of $\mathbf{C}^m$.

References

[Abh64] S.S. Abhyankar, *Local Analytic Geometry*, in *Pure and Applied Mathematics*, vol. XIV (Academic Press, New York-London, 1964) 314

[Bou61] N. Bourbaki, *Algèbre Commutative, Chapitre II* (Hermann, Paris 1961) 311

[Hir64a] H. Hironaka, Resolution of singularities of an algebraic variety over a field of characteristic zero I, II. Ann. Math. (2) **79**, 109–203 (1964); ibid. (2), **79**, 205–326 (1964) 312

[Hir64b] H. Hironaka, Equivalences and deformations of isolated singularities, in *Woods Hole A.M.S. Summer Institute on Algebraic Geometry* (1964). https://www.jmilne.org/math/Documents/ 310, 334

[Lip69] J. Lipman, Rational singularities, with applications to algebraic surfaces and unique factorization. Inst. Hautes Études Sci. Publ. Math. **36**, 195–279 (1969) 313

[Whi65] H. Whitney, Tangents to an analytic variety. Ann. Math. (2) **81**, 496–549 (1965) 323, 335

[Zar37] O. Zariski, A theorem on the Poincaré group of an algebraic hypersurface. Ann. Math. (2) **38**(1), 131–141 (1937) 334

[Zar64] O. Zariski, Equisingularity and related questions of classification of singularities, in *Woods Hole A.M.S. Summer Institute on Algebraic Geometry* (1964). https://www.jmilne.org/math/Documents/ 334

[Zar65a] O. Zariski, Studies in equisingularity, I, Equivalent singularities of plane algebroid curves. Am. J. Math. **87**, 507–536 (1965) 309, 325

[Zar65b] O. Zariski, Studies in equisingularity, II, Equisingularity in codimension 1 (and characteristic zero). Am. J. Math. **87**, 972–1006 (1965) 309

[Zar68] O. Zariski, Studies in equisingularity, III, Saturation of local rings and equisingularity. Am. J. Math. **90**, 961–1023 (1968) 309, 328, 331

A Sample of More Recent Bibliography[14]

[Bog74] E. Böger, Zur Theorie der Saturation bei analytischen Algebren. Math. Ann. **211**, 119–143 (1974) 328

[Bog75] E. Böger, Über die Gleicheit von absoluter und relativer Lipschitz-Saturation bei analytischen Algebren. Manuscripta Math **16**, 229–249 (1975) 328, 334

[Fer03] A. Fernandes, Topological equivalence of complex curves and Bi-Lipschitz Homeomorphisms. Michigan Math. J. **51**, 593–606 (2003) Not cited.

---

[14]The references [Fer03, Gaf10, Lip76, NP14, Pha71b, Zar71a, Zar71b, Zar73, Zar75] in the sample of more recent bibliography below are not cited in the footnotes to the text, but they all add quite significantly to our understanding of saturation and equisingularity.

[Gaf10] T. Gaffney, Bi-Lipschitz equivalence, integral closure and invariants, in *Real and Complex singularities*, ed. by M. Manoel, M.C. Romero Fuster, C.T.C. Wall. London Mathematical Society Lecture Notes Series, vol. 380 (Cambridge University, Cambridge, 2010), pp. 125–137 Not cited.

[Kuo85] T.C. Kuo, On classification of real singularities. Invent. Math. **82**(2), 257–262 (1985) 335

[Lip75a] J. Lipman, Absolute saturation of one-dimensional local rings. Am. J. Math. **97**(3), 771–790 (1975) 328

[Lip75b] J. Lipman, Relative Lipschitz saturation. Am. J. Math. **97**, 791–813 (1975) 328, 334

[Lip76] J. Lipman, Errata: relative Lipschitz saturation. Am. J. Math. **98**(2), 571 (1976) Not cited.

[LT79] J. Lipman, B. Teissier, Introduction to Volume IV, in *Equisingularity on Algebraic Varieties*, ed. Zariski's Collected Papers (MIT Press, Boston, 1979) 334

[NP14] W.D. Neumann, A. Pichon, Lipschitz geometry of complex curves. J. Singul. **3**, 225–234 (2014) Not cited.

[Pha71a] F. Pham, Déformations équisingulières des idéaux Jacobiens de courbes planes, in *Proceedings of Liverpool Singularities Symposium, II (1969/70)*. Lecture Notes in Mathematical, vol. 209 (Springer, Berlin, 1971), pp. 218–233 335

[Pha71b] F. Pham, Fractions lipschitziennes et saturation de Zariski des algèbres analytiques complexes (exposé d'un travail fait avec Bernard Teissier), in *Actes du Congrès International des Mathématiciens, Nice 1970*, vol. II (Gauthier-Villars, Paris, 1971), pp. 649–654 Not cited.

[Spe75] J.-P. Speder, Equisingularité et conditions de Whitney. Am. J. Math. **97**(3), 571–588 (1975) 334

[Var73] A. Varchenko, The connection between the topological and the algebraic-geometric equisingularity in the sense of Zariski. Funkcional Anal. i Priloz. **7**(2), 1–5 (1973) 334

[Zar71a] General theory of saturation and of saturated local rings, I. Am. J. Math. **93**(3), 573–648 (1971) Not cited.

[Zar71b] General theory of saturation and of saturated local rings, II. Am. J. Math. **93**(4), 872–964 (1971) Not cited.

[Zar73] Quatre exposés sur la saturation (Notes de J-J. Risler), in *Singularités à Cargèse (Rencontre Singularités en Géométrie Analytique, Institut d'Etudes Scientifiques de Cargèse, 1972)*. Asterisque, n° 7 et 8 (Society of Mathematical France, Paris, 1973), pp. 21–39 Not cited.

[Zar75] General theory of saturation and of saturated local rings, III. Am. J. Math. **97**(2), 415–502 (1975) Not cited.

[Zar79] O. Zariski, Foundations of a general theory of equisingularity on r-dimensional algebroid and algebraic varieties, of embedding dimension r+1. Am. J. Math. **101**(2), 453–514 (1979) 334

[Zar80] O. Zariski, Addendum to my paper: "Foundations of a general theory of equisingularity on r-dimensional algebroid and algebraic varieties, of embedding dimension r+1". Am. J. Math. **102**(3), 453–514 (1980). [Am. J. Math. **101**(2), 453–514 (1979)] 334

List of Participants to the International School on Singularity Theory and Lipschitz Geometry

Jonathas Phillipe Almeida
Fuensanta Aroca
Aubin Arroyo
Paul Vladimir Barajas Guzmán
Gonzalo Barranco Mendoza
Pedro Benedini Riul
Edmundo Bernardo de Castro Martins
Lev Birbrair
Lito Bocanegra
Francisco Nicolás Cardona
Jesús Adrián Cerda Rodriguez
Ana Lucília Chaves de Toledo
Enrique Chávez Martínez
José Luis Cisneros Molina
Octave Curmi
Juan Diaz
Daniel Duarte
Miguel Fernández Duque
Erick García Ramírez
Arturo Enrique Giles Flores
Mirna Gomez
Oziel Gómez Martínez
Lilí Guadarrama Bustos

Felipe Espreafico Guelerman Ramos
Benoit Guerville-Ballé
Carlos Rodrigo Guzmán Durán
Sonja Heinze
Aaron Aparicio Hernandez
Lê Dũng Tráng
Lucía López de Medrano
Edison Marcavillaca Niño de Guzmán
Julio César Magaña Cáceres
Aurélio Menegon Neto
Maria Michalska
Helge Møller Pedersen
Edgar Mosqueda Camacho
Walter Neumann
Xuan Viet Nhan Nguyen
Otoniel Nogueira da Silva
Mutsuo Oka
Leandro Oliveira
Anne Pichon
Patrick Popescu-Pampu
Pierre Py
Ángel David Ríos Ortiz
Faustino Agustín Romano Velázquez

Cuernavaca (Mexico), June 11–22, 2018.

W. Neumann, A. Pichon (eds.), *Introduction to Lipschitz Geometry of Singularities*, Lecture Notes in Mathematics 2280, https://doi.org/10.1007/978-3-030-61807-0

Maria Aparecida Soares Ruas
Gonzalo Ruiz Stolowicz
Hellen Santana
Mateus Schmidt
José Antonio Seade
Manuel Sedrano
Jawad Snoussi

Roger Sousa
Renato Targino
Jonatán Torres Orozco Román
Bernard Teissier
Saurabh Trivedi
David Trotman
Juan Viu-Sos

Index

LECTURE NOTES IN MATHEMATICS

Editors in Chief: J.-M. Morel, B. Teissier;

Editorial Policy

1. Lecture Notes aim to report new developments in all areas of mathematics and their applications – quickly, informally and at a high level. Mathematical texts analysing new developments in modelling and numerical simulation are welcome.

   Manuscripts should be reasonably self-contained and rounded off. Thus they may, and often will, present not only results of the author but also related work by other people. They may be based on specialised lecture courses. Furthermore, the manuscripts should provide sufficient motivation, examples and applications. This clearly distinguishes Lecture Notes from journal articles or technical reports which normally are very concise. Articles intended for a journal but too long to be accepted by most journals, usually do not have this "lecture notes" character. For similar reasons it is unusual for doctoral theses to be accepted for the Lecture Notes series, though habilitation theses may be appropriate.

2. Besides monographs, multi-author manuscripts resulting from SUMMER SCHOOLS or similar INTENSIVE COURSES are welcome, provided their objective was held to present an active mathematical topic to an audience at the beginning or intermediate graduate level (a list of participants should be provided).

   The resulting manuscript should not be just a collection of course notes, but should require advance planning and coordination among the main lecturers. The subject matter should dictate the structure of the book. This structure should be motivated and explained in a scientific introduction, and the notation, references, index and formulation of results should be, if possible, unified by the editors. Each contribution should have an abstract and an introduction referring to the other contributions. In other words, more preparatory work must go into a multi-authored volume than simply assembling a disparate collection of papers, communicated at the event.

3. Manuscripts should be submitted either online at www.editorialmanager.com/lnm to Springer's mathematics editorial in Heidelberg, or electronically to one of the series editors. Authors should be aware that incomplete or insufficiently close-to-final manuscripts almost always result in longer refereeing times and nevertheless unclear referees' recommendations, making further refereeing of a final draft necessary. The strict minimum amount of material that will be considered should include a detailed outline describing the planned contents of each chapter, a bibliography and several sample chapters. Parallel submission of a manuscript to another publisher while under consideration for LNM is not acceptable and can lead to rejection.

4. In general, **monographs** will be sent out to at least 2 external referees for evaluation.

   A final decision to publish can be made only on the basis of the complete manuscript, however a refereeing process leading to a preliminary decision can be based on a pre-final or incomplete manuscript.

   Volume Editors of **multi-author works** are expected to arrange for the refereeing, to the usual scientific standards, of the individual contributions. If the resulting reports can be forwarded to the LNM Editorial Board, this is very helpful. If no reports are forwarded or if other questions remain unclear in respect of homogeneity etc, the series editors may wish to consult external referees for an overall evaluation of the volume.

5. Manuscripts should in general be submitted in English. Final manuscripts should contain at least 100 pages of mathematical text and should always include

   – a table of contents;
   – an informative introduction, with adequate motivation and perhaps some historical remarks: it should be accessible to a reader not intimately familiar with the topic treated;
   – a subject index: as a rule this is genuinely helpful for the reader.
   – For evaluation purposes, manuscripts should be submitted as pdf files.

6. Careful preparation of the manuscripts will help keep production time short besides ensuring satisfactory appearance of the finished book in print and online. After acceptance of the manuscript authors will be asked to prepare the final LaTeX source files (see LaTeX templates online: https://www.springer.com/gb/authors-editors/book-authors-editors/manuscriptpreparation/5636) plus the corresponding pdf- or zipped ps-file. The LaTeX source files are essential for producing the full-text online version of the book, see http://link.springer.com/bookseries/304 for the existing online volumes of LNM). The technical production of a Lecture Notes volume takes approximately 12 weeks. Additional instructions, if necessary, are available on request from lnm@springer.com.

7. Authors receive a total of 30 free copies of their volume and free access to their book on SpringerLink, but no royalties. They are entitled to a discount of 33.3 % on the price of Springer books purchased for their personal use, if ordering directly from Springer.

8. Commitment to publish is made by a *Publishing Agreement*; contributing authors of multiauthor books are requested to sign a *Consent to Publish form*. Springer-Verlag registers the copyright for each volume. Authors are free to reuse material contained in their LNM volumes in later publications: a brief written (or e-mail) request for formal permission is sufficient.

Addresses:
Professor Jean-Michel Morel, CMLA, École Normale Supérieure de Cachan, France
E-mail: moreljeanmichel@gmail.com

Professor Bernard Teissier, Equipe Géométrie et Dynamique,
Institut de Mathématiques de Jussieu – Paris Rive Gauche, Paris, France
E-mail: bernard.teissier@imj-prg.fr

Springer: Ute McCrory, Mathematics, Heidelberg, Germany,
E-mail: lnm@springer.com